Maintenance and Philosophy of Technology

What can we learn about the nature of technology by studying practices of maintenance and repair? This volume addresses this question by bringing together scholarship from philosophers of technology working at the forefront of this emerging and exciting topic.

The chapters in this volume explore how attending to maintenance and repair can challenge and complement existing ways of thinking about technology focused on use and design and introduce new philosophical perspectives on the relationship between technology, time and human practice. They examine the significance of maintenance and repair practices at different scales in relation to a diverse range of philosophical traditions and a wide variety of technologies, from urban infrastructure such as bridges and buildings to data technologies such as servers and software systems. Together, the contributions highlight common themes in the philosophical study of maintenance, including the role of skill, the significance of social values and the potential of these practices to transform the technologies to which they are applied. By reflecting on the different ways in which we keep technologies going, from the devices we use in our homes to the large technical systems which surround us, this volume reveals the philosophical significance of practices of maintenance, not only as a source of new insights but also as a resource for enriching our understanding of a variety of existing topics in philosophy.

Maintenance and Philosophy of Technology will appeal to scholars and advanced students working in philosophy of technology, philosophy of engineering and science & technology studies.

Mark Thomas Young is a Marie Curie Postdoctoral Fellow in the Philosophy Department at the University of Vienna. His research covers two fields: the Philosophy of Technology, where he focuses on practices of maintenance and the use of automating technologies, and the History and Philosophy of Science, where he explores instruments, craft practices and tacit knowledge in the early modern period.

Mark Coeckelbergh is a full Professor of Philosophy of Media and Technology at the University of Vienna and ERA Chair at the Institute of Philosophy of the Czech Academy of Sciences in Prague. He was President of the Society for Philosophy and Technology and is the author of numerous publications including *Environmental Skill*, *AI Ethics*, *Self-Improvement* and *The Political Philosophy of AI*.

Routledge Studies in Contemporary Philosophy

Philosophy of Mental Disorder
An Ability-Based Approach
Sanja Dembić

Historical Explanation
An Anti-Causalist Approach
Gunnar Schumann

Exploring Extended Realities
Metaphysical, Psychological, and Ethical Challenges
Edited by Andrew Kissel and Erick José Ramirez

Expected Experiences
The Predictive Mind in an Uncertain World
Edited by Tony Cheng, Ryoji Sato, and Jakob Hohwy

Hybrid Societies
Living with Social Robots
Piercosma Bisconti

Philosophy, Expertise, and the Myth of Neutrality
Edited by Mirko Farina and Andrea Lavazza

Maintenance and Philosophy of Technology
Keeping Things Going
Edited by Mark Thomas Young and Mark Coeckelbergh

For more information about this series, please visit: www.routledge.com/Routledge-Studies-in-Contemporary-Philosophy/book-series/SE0720

Maintenance and Philosophy of Technology

Keeping Things Going

Edited by Mark Thomas Young and Mark Coeckelbergh

Routledge
Taylor & Francis Group
NEW YORK AND LONDON

First published 2024
by Routledge
605 Third Avenue, New York, NY 10158

and by Routledge
4 Park Square, Milton Park, Abingdon, Oxon, OX14 4RN

Routledge is an imprint of the Taylor & Francis Group, an informa business

Funded by TU Delft.

British Library Cataloguing-in-Publication Data
A catalogue record for this book is available from the British Library

ISBN: 978-1-032-32686-3 (hbk)
ISBN: 978-1-032-32687-0 (pbk)
ISBN: 978-1-003-31621-3 (ebk)

DOI: 10.4324/9781003316213

Typeset in Sabon
by SPi Technologies India Pvt Ltd (Straive)

Contents

Contributors

Joost Alleblas is a PhD Candidate at the Faculty of Technology, Policy, and Management at Delft Technical University. Since 2019, he has been a part of the ERC Project on Value Change in the Department of Values, Technology, and Innovation. This project tries to understand how we can design large technical systems so that they can accommodate changing values in societies. His research interests lie with energy transition, nuclear energy and the role of visions in the promotion of both. Amongst others, he studies the moralisation of socio-technical visions, especially those visions that incorporate the realisation of moral ideals.

Keith Begley studied Mental and Moral Science at Trinity College Dublin, Ireland, and holds an MA and a PhD in Philosophy from the University of Dublin. He also studied Computer Science at the University College Dublin and holds an HDip and an MSc in Computer Science from the National University of Ireland. He has held positions as an Assistant Lecturer in the Department of Philosophy, Maynooth University, a Demonstrator in the School of Computer Science, UCD, and an Adjunct Assistant Professor in the Department of Philosophy, TCD. He has published on philosophy of language and linguistics, philosophy of artificial intelligence, medical epistemology, computational philology, Heraclitus, Jerrold J. Katz and Wittgenstein.

Cristina Bernabéu teaches Philosophy at Pedro Salinas high school in Spain. She has written a dissertation about technical artefacts and gender normativity and published on gender, feminist philosophy of technology and social epistemology. She is especially interested in material and feminist ways of activism.

Tiago Mesquita Carvalho is a Researcher and Lecturer at the Philosophy Institute (IF) in the Humanities Faculty in Porto working on a project concerning technology, responsibility and catastrophes. He has degrees in both Environmental Sciences and Environmental Philosophy and

Aesthetics. In 2019, he gained his PhD in FCUL with a thesis on how the good life demands the deliberate use of artifacts, crossing both virtue ethics, philosophy of technology and science and technology studies. Before joining IF, he was a research assistant in the CoLABOR team, aiming to build a technological impact assessment methodology in work and employment. For 10 years, he has also been a volunteer at a DIY bicycle workshop based in Lisbon.

Mark Coeckelbergh is a full Professor of Philosophy of Media and Technology at the University of Vienna and ERA Chair at the Institute of Philosophy of the Czech Academy of Sciences in Prague. He was a president of the Society for Philosophy and Technology and is the author of numerous publications including *environmental skill, AI ethics, self-improvement,* and *the political philosophy of AI.*

Benjamin Hofbauer is a PhD Candidate at the Faculty of Technology, Policy and Management at Delft Technical University and a Research Fellow at the Research Institute for Sustainability in Potsdam. In his work, he focuses on the moral implications of introducing and researching new and emerging technologies as well as the ethical assessment of the research process itself. While his thesis deals with solar climate engineering research governance, his interests expand throughout the intersection of climate change and technology, ranging from political philosophy and environmental ethics to the ethics and philosophy of technology.

Sanna Lehtinen (PhD 2015, University of Helsinki) works as a Research Fellow at the Aalto University School of Arts, Design and Architecture. She is also a Docent in Aesthetics at the University of Helsinki. Sanna's research focuses on the intersection of urban aesthetics, philosophy of the city and philosophy of technology. Sanna's publications include journal articles published in *Open Philosophy*, *Essays in Philosophy*, *Philosophical Inquiries*, *Sustainability Science* and *Behaviour & Information Technology*, chapters in edited volumes by Routledge, Bloomsbury, Springer and Oxford University Press as well as editing special issues for *Environmental Values*, *Contemporary Aesthetics*, *ESPES*, *East Asian Journal of Philosophy* and *Open Philosophy*. Sanna is a Codirector on the Board of the Philosophy of the City Research Group and an affiliate member of the 4TU Centre for Ethics and Technology. Sanna is an assistant editor of *Contemporary Aesthetics* and founder and Co-editor-in-chief of the new *Philosophy of the City Journal*. https://www.sannalehtinen.com

Simon Penny is an artist, teacher and theorist with a long-standing focus on emerging technologies, embodied and situated aspects of artistic practices and critical analysis of computer culture. He published

Making Sense: Cognition, Computing, Art and Embodiment in 2017 and directed *A Body of Knowledge: Embodied Cognition and the Arts* Conference UCI 2016. As a Professor of Art and Robotics at Carnegie Mellon (1993–2000), he developed VR and robotics projects. He is co-director and co-originator of the *Industrial Crafts Research Network*. He is a lifelong maker and environmental activist with a focus on sustainability, green technologies and regenerative agriculture. Penny is a professor of Electronic Art and Design (Department of Art) at the University of California, Irvine (with appointments in the depts of Music and Informatics) where he teaches Art and Sustainability, A Cultural History of the Anthropocene and How to be Clever with Stuff, among other classes.

Johannes F.M. Schick has been Scientific Coordinator of the SFB 1187 Media of Cooperation since January 2022. He has been the head of the research project "Action, Operation, Gesture: Technology as Interdisciplinary Anthropology" at the a.r.t.e.s. Graduate School for the Humanities (University of Cologne) from October 2017 to February 2021. His research focuses on interdisciplinary (techno-)anthropology (from Bergson, Espinas and Mauss to Simondon), Bergson's philosophy of life and the relation of anthropology to philosophy (and vice versa) as well as the "Category Project of the Durkheim School". He is a member of the Editorial Board of the *Durkheimian Studies* (new series). Furthermore, he is interested in phenomenological psychiatry, the phenomenon of creativity, systems theory and practice theory. In his PhD thesis, Schick focused on the relation of intuition and emotion in the philosophy of Henri Bergson (published as *Erlebte Wirklichkeit. Zum Verhältnis von Intuition zu Emotion bei Henri Bergson*, 2012).

Steffen Steinert is an Assistant Professor in the Ethics and Philosophy of Technology section at Delft University of Technology. He has studied philosophy, history and sociology at the Technical University of Dresden and the New School for Social Research in New York. He completed his PhD at the University of Munich before relocating to the Netherlands to pursue an academic career in philosophy of technology. Steffen's research and teaching focus on philosophy and ethics of technology. He has worked on brain–computer interfaces, social media and robotics. Steffen's main interest is in exploring the impact of technology on society, specifically how it influences our values.

Taylor Stone is a Senior Researcher at the Institute for Science and Ethics, University of Bonn, and a member of the Bonn Sustainable AI Lab. He also is an associate editor of the *Philosophy of the City* journal. Taylor received his PhD in Ethics of Technology from Delft University of Technology and has held postdoctoral and lecturer positions in industrial design, ethics of technology and engineering ethics education. His research focuses on the ethics of urban technologies and the incorporation of environmental values into design processes and built artefacts.

Aimee van Wynsberghe is the Alexander von Humboldt Professor for Applied Ethics of Artificial Intelligence at the University of Bonn in Germany. Aimee is the director of the Institute for Science and Ethics and the Bonn Sustainable AI lab. She is co-director of the Foundation for Responsible Robotics and a member of the European Commission's High-Level Expert Group on AI. She is a founding editor for the international peer-reviewed journal *AI & Ethics* and a member of the World Economic Forum's Global Futures Council on Artificial Intelligence and Humanity. In each of her roles, Aimee works to uncover the ethical risks associated with emerging robotics and AI. Aimee's current research, funded by the Alexander von Humboldt Foundation, brings attention to the sustainability of AI by studying the hidden environmental costs of developing and using AI.

Jesús Vega-Encabo is a Professor of Logic and Philosophy of Science at the Autonomous University of Madrid (Spain). He has published extensively on epistemology, philosophy of mind and philosophy of science and technology. His most recent research is focused on intellectual autonomy, epistemic dependence, epistemic normativity and the nature of artefacts.

Steven Vogel is a Professor Emeritus of Philosophy at Denison University in Granville, Ohio USA. He is the author of *Against Nature: The Concept of Nature in Critical Theory* (1996) and *Thinking Like a Mall: Environmental Philosophy After the End of Nature* (2015).

Mark Thomas Young is a Marie Curie Postdoctoral Fellow in the Philosophy Department at the University of Vienna. His research covers two fields: the Philosophy of Technology, where he focuses on practices of maintenance and the use of automating technologies, and the History and Philosophy of Science, where he explores instruments, craft practices and tacit knowledge in the early modern period.

Jochem Zwier is a Researcher in Philosophy of Technology at Wageningen University & Research and the Managing Director of the 4TU Centre for Ethics and Technology. He studied Philosophy of Technology at the University of Twente (MSc) and Radboud University Nijmegen (PhD). His work focuses on the intersection of philosophy of technology, ecology and environmental philosophy, particularly in the context of the Anthropocene. His current work further revolves around temporality and sustainability, the meaning of innovation and the concept of anthropocentrism.

Acknowledgements

We would like to thank the Society for the Philosophy of Technology (SPT) for their support of a special interest group on *Maintenance and Philosophy of Technology* which was established in April 2022, alongside the many participants of the group who provided feedback on many of the contributions that appear in this volume. In addition, we would like to thank the following people for their helpful comments on earlier drafts of this manuscript; Tobias Stadler, Nico Formanek, Mark Theunissen, Dan Weiskopf, Maxine Morphis-Riesbeck, Jacqueline Bellon, Marcel Siegler, Frodo Podscwadek, Tommy Lindgren, Luke Collison, Luca Possati, Scott Luan, Elena Popa, Robin K. Hill, Dirk van Leemput, Marilena Pateraki, Andrea Gammon, Madison Snider, Guru Madhaven, Bogdan Chernyakevich, Zachary Pirtle, Sveinung Sivertsen, Katheleen Wallace, Kjetil Akø, Patrick Zapata, Michel Bourban, Alexandria Poole, Jaana Parviainen, Céline Kermish, Mateusz Salwa, David Jenkins and Sandra Zákutná. We'd also like to thank the editorial team at Routledge for their help, support and patience throughout the development of this volume. Finally, we would like to thank the contributors of the volume itself for their enthusiasm and support for this project, alongside their work responding to editorial comments and reviewer feedback.

1 Keeping Things Going

Maintenance and Philosophy of Technology

Mark Thomas Young and Mark Coeckelbergh

There is currently a growing recognition among philosophers of technology, that despite a consistent effort to understand the nature of the relationship that people share with the technologies that surround them, existing work within the field has nonetheless ignored certain forms of technical practice that are very important. The practices we refer to here are known collectively as maintenance and together comprise the different activities that people must perform in order to 'keep things going'. Until now, philosophical reflections on maintenance have been few and far between. The most developed treatments of the topic within anglophone philosophy, for example, can be found in Robert Pirsig's *Zen and the Art of Motorcycle Maintenance* (Pirsig 1974) and Elizabeth Spelman's *Repair: The Impulse to Restore in a Fragile World* (Spelman 2002), both of which explored practices of maintaining technologies as a segue into the examination of wider issues relating to intellectual cultural division and social justice, respectively. However, among the topics and themes addressed by contemporary philosophy of technology, maintenance is almost nowhere to be found. With the exception of a few recent articles, maintenance is a topic which is often overlooked entirely.[1]

That maintenance has until now received so little attention from philosophers of technology should strike us as surprising for a variety of reasons. In the first place, the technological societies that we currently inhabit literally depend on the constant performance of maintenance. Consider, for example, the buildings which rise from our cities or the bridges, roads and tunnels which enable the movement of people to and from work every day, not to mention the amenities of modern urban living such as running water, electricity and information and communications technology, all of which depend on constant maintenance – not only of the material components of such systems but also the institutions and technical cultures within which they are located.[2] Furthermore, the crucial importance of maintenance in sustaining the function of technologies applies no less to technologies which purportedly operate without human involvement, such as

DOI: 10.4324/9781003316213-1

artificial intelligence and automation. The capacity of technologies such as ChatGPT to generate humanlike conversational dialogues, for example, depends on the ongoing maintenance of websites and code, telephone lines and fibre optic cables which snake their way across continents and underneath oceans, data centres and servers, alongside the maintenance of electricity networks on which these rely. The sheer importance of maintenance work is reflected in the relative proportions of tasks performed by those, such as engineers and software developers, whose professions aim to provide us with these technologies. Regardless of whether we consider boats or bridges or cloud servers or code, the amount of money and time devoted to the initial development of technologies is often vastly outweighed by what is required to keep them going.[3] As a result, the majority of those working in professions commonly thought to involve the design and production of new things instead spend their time maintaining existing things (Young 2021a). Whether we recognise it or not, maintenance and repair are 'in many ways, the engine room of modern economies and societies' (Graham & Thrift 2007). Yet this simple fact appears to have gone unnoticed in contemporary philosophy of technology which has approached the analysis of technical expertise by focusing exclusively on the knowledge required to design and build new things. As a result, what is by far the largest and most widespread form of technological expertise (Edgerton 2006, p. 80), has remained unanalysed.

Yet the neglect of maintenance by philosophers is surprising not only because of the ubiquity of this mode of human engagement with technology but also due to its unparalleled intimacy. Compared to the lived relationships we establish with technologies through the activities of design and use, maintenance very often requires forms of understanding which combine aspects of both while also going beyond them in significant ways. For example, while the designers of a technology may possess a privileged understanding of how a particular technical design contributes to the achievement of certain functional states, philosophers have long been cognizant of the inherent limitations in their capacity to predict how a particular technology will affect and be affected by the natural and human environments in which it comes to be employed.[4] Designers of a technology are very often aware that a deeper understanding of the nature of their innovations can only be gained after they have become widely used.[5] While some aspects of this deeper understanding may stem from the users of a technology, who in order to use it in the way they desire, discover things the designers could not have predicted, this deeper understanding only seems to be properly granted through the particular form of use established through the practice of maintaining a technology over time.

When we seek to maintain, we must often establish a different relationship with a technology than what is required when our goal is simply to use it to achieve the purpose it was designed for. A common way of conceptualising the nature of this relationship is through analogy to the concept of care which emerged in feminist ethics in the late twentieth century.[6] Whereas care ethics was originally conceived as applying to human relationships, recent work in science and technology studies (STS) has increasingly explored its relevance as an ethical framework for analysing the relationship between humans and the technological environments they inhabit.[7] The seeds for such approaches were sown already in early studies of care in the feminist tradition.[8] In Joan Tronto and Bernice Fisher's well-known definition of care, for example, the connection to maintenance is already clearly visible:

> On the most general level, we suggest that caring be viewed as a species activity that includes everything we do to maintain, continue and repair 'our world' so that we can live in it as well as possible. That world includes our bodies, our selves, and our environment, all of which we seek to interweave in a complex, life sustaining web.
>
> (Fisher & Tronto 1991)

According to feminist theorists, one of the basic requirements for establishing a relationship of care is forming an understanding of the needs of the cared-for. Care theorists consider this to require a different moral attitude to the introspective, disinterested stance typical to ethical analysis in the tradition of liberal individualism, whereby we attempt to imagine what we would need if we inhabited the situation of those for whom we care. In contrast, feminist theorists argue that we must adopt a stance of attentiveness in which we look beyond ourselves and cultivate a receptiveness to recognising the needs of those we intend to care for.[9] This involves resisting generalisations and paying adequate attention to the particular characteristics of individuals and the circumstances they occupy. Applied to technological rather than social relations, the requirement of attentiveness manifests itself in a receptiveness to what particular technologies require in order for them to remain in a state we desire. In many cases, this requires us to have a working understanding of how they are made and how different aspects of their physical design contribute to their different functional capacities. A poignant example of this kind of knowledge is found in *Working Knowledge: Skill and Community in a Small Shop* (1987), in which the sociologist Douglas Harper recounts his experiences documenting the practical expertise of Willie, a local repairman in North Country New York. As Harper recalls:

> Willie is alert to the machines he uses and the environment they service. This might include something as simple as the proper use of a chainsaw

> on a cold day, which I learned after I ruined a chain by not letting the lubricating oil warm up enough before I began cutting. I later noticed that after Willie started his saw he pointed it downward. Only when oil began to spurt from the end of the blade, making a small black track in the snow, did he begin cutting. He explained that when the temperature is below zero, the engine must run long enough to heat the blade oil, which is in a separate reservoir. It is a very simple idea, but it signals an understanding of how the machine responds to the environment and how its use must be modified when conditions change.
>
> (Harper 1987, p. 161)

When our goal is merely to *use* a technology, we may not care whether particular patterns of use lead to the premature degradation of certain components. Maybe we intend to replace the technology when it can no longer be used or maybe our intention was simply to use something for as long as it lasted, much as one does with disposable paper cups. Attempting to maintain a technology, on the other hand, signals a temporal commitment which influences and shapes the relationships we maintain with technologies through the act of using them. This commitment often manifests itself in specific forms of practice, such as keeping things clean or lubricated or ensuring that certain functions do not exceed particular thresholds.

As Harper's example illustrates, attempting to maintain a technology very often requires us to deepen our understanding of how it works. In addition to encouraging particular forms of use, maintaining a technology also often requires us to *open* the black box, by popping the hood and looking inside, unscrewing the casing in order to remove or clean components or raising things up to see underneath. At the same time, the attentiveness which many scholars understand to be central to the maintenance attitude also requires us to go beyond the knowledge of designers in various ways. Extended experience maintaining particular technologies may grant insights unknown to those who designed them, such as how certain components deteriorate earlier than expected under certain environmental conditions or the effects of different patterns of use on the durability of the technology as a whole.

As existing scholarship on maintenance demonstrates, while some forms of maintenance practice may proceed according to predetermined plans, such as when car manufacturers develop detailed maintenance plans for the vehicles they produce, other forms are irreducibly situated insofar as they involve regular inspection to monitor the effects of particular patterns of use or the unique environment a technology inhabits as it develops over time.[10] Such forms of attentiveness also encourage us to develop an awareness of the unique characteristics possessed by individual tokens of particular technologies as they are manifested within specific contexts of use.[11] A good example occurs in *Zen and the Art of Motorcycle Maintenance*, in which Pirsig's unnamed protagonist reflects on how the act of maintaining his motorcycle required him to develop an understanding of the individual personality of his machine:

> a friend who owns a cycle of the same make, model and even same year brought it over for repair, and when I test rode it afterward it was hard to believe it had come from the same factory years ago. You could see that long ago it had settled into its own kind of feel and ride and sound, completely different from mine. No worse, but different. I suppose you could call that a personality. Each machine *has* its own, unique personality which probably could be defined as the intuitive sum total of everything you know and feel about it. This personality constantly changes, usually for the worse, but sometimes surprisingly for the better, and it is this personality that is the real object of motorcycle maintenance.
>
> (Pirsig 1974, p. 39)

Yet whereas Pirsig considered such attentiveness to the unique characteristics and needs of different artefacts to form the basis of a desirable relationship to technology, he also recognised the extent to which this attentiveness stood in conflict with many aspects of the culture and society he inhabited. For *Zen and the Art of the Motorcycle Maintenance* addressed a concern that worried Pirsig's contemporaries as much as it worries many today, that the rise of consumerism, in which we purchase rather than make many of the things we need, has prevented many from establishing such relationships and as a result has caused people to become alienated from the material objects which surround them. When things break or stop working, many of us throw them away, buy new ones or return them to the manufacturer for replacement or repair. Furthermore, the incentives against developing the kind of intimate, engaged and active relationships with the technologies around us that Pirsig described are increasing. Artefacts are increasingly designed deliberately in ways that prevent or discourage us from repairing them when they break (Verbeek 2005, p. 226) or are constructed from materials which are difficult, if not impossible to repair.[12] Consider, for instance, contemporary mobile phones or laptops which are often designed in ways which discourage users from opening them to engage in repair activities. Moreover, undertaking repair tasks ourselves often requires specialised equipment, skills or replacement parts which can be more expensive than simply replacing the item in question. Legal contracts such as warranty agreements disincentivise repair activities insofar as performing them may release the manufacturer from the obligation to repair or replace defective products. Finally, manufacturers often attempt to ensure the obsolescence of their products long before they reach a state which we might consider unrepairable by encouraging us to view them as outdated through the marketing of products with new features, for example, or by discontinuing services such as updates that are geared towards ensuring their products' compatibility with the wider technological systems in which they are employed.

The romanticised vision of maintenance that Pirsig outlined stands in stark contrast to the passive relationship consumers are often encouraged to maintain towards their possessions in the context of industrial capitalism. Many modern commentators on the practice of maintenance have followed his lead in understanding the activity of maintenance and repair to present the possibility of establishing a more fulfilling and authentic relationship with material objects. Steven Jackson, for example, identifies repair as the activity by which 'human value is preserved and extended' (Jackson 2014). Matthew Crawford, on the other hand, understands practices of maintenance and repair to be intrinsically valuable and argues that manual engagement with technology is necessary for human flourishing (Crawford 2009, p. 64). Such claims reflect that despite its neglect in philosophical literature, there is nonetheless an enduring conviction regarding the desirability of engaging with artefacts through the practices of maintenance and repair. We, the editors of this volume, for example, were both fascinated by Pirsig's account of technological maintenance when we read *Zen and the Art of Motorcycle Maintenance* during our undergraduate study. In fact, we both recall this being one of the inspirations that attracted us specifically to the philosophy of technology and we know from talking to other philosophers that we are not alone. Many of us who were interested in learning more about the different kinds of active and engaged relationships humans establish with technology drifted inevitably towards early phenomenological accounts of technology as found in the work of Heidegger and Merleau-Ponty, alongside the extension of these ideas by later American philosophers such as Don Ihde and Hubert Dreyfus. But the practice of maintenance and repair was not a topic that any of us encountered again. So why is it that despite the ubiquity and importance of this form of technical activity, it has until now received only scant attention from philosophers of technology?

The reasons for this neglect most likely stem from a combination of causes which complicates the task of providing a complete and definitive answer to this question. However, in order to help facilitate this volume's goal of stimulating philosophical reflection on the practice of maintenance, it is nonetheless worth drawing attention to some of the main obstacles that have hindered the further development of philosophical reflection on maintenance practices so that they may be more easily overcome. In the first place, we might consider how recent trends and interests within the philosophy of technology have helped to divert attention away from the practice of maintenance. The later stages of the twentieth century witnessed a transition in the philosophy of technology, away from transcendentalist critiques of technology and towards more focused investigations of specific technologies and technical practices. This shift was promoted by its proponents under the banner of the 'empirical turn' in the philosophy of technology.[13] As Verbeek notes, the empirical turn remains clearly visible today in the two main orientations within contemporary philosophy of technology.

The *first* aims to clarify the ontological and epistemological aspects of engineering practice and employs a conceptual vocabulary and methodology which owes much to analytic metaphysics and philosophy of science. A common conviction among those writing within this orientation is to consider the practice of designing as the essence of engineering itself. Not surprisingly then, despite the ubiquity and importance of maintenance within engineering practice, this orientation has nonetheless tended to conflate engineering with innovation: the practice of building something *new*. This is reflected not only in the common tendency among philosophers to analyse the practice and methods of engineering by focusing exclusively on the activity of design but also in the dominant view that the production of technical artefacts occurs exclusively through the material realisations of their makers' intentions (Young 2021a).

The *second* orientation within the empirical turn focuses on the *social implications* of technologies and draws inspiration from phenomenology and more recently, STS. This research line is more attuned to the use of technologies and explores the relationships between the design of a technology and the way in which it enables, encourages or hinders different forms of practice alongside the effects of the widespread use of particular technologies within society. Yet, while work within this orientation recognises that new forms of use or values may emerge around existing technologies, the fact that artefacts themselves may require constant work or must often undergo customisation and modification in order to remain functional is almost always overlooked. Instead, it is often implicitly assumed that the nature and capacities of a technology are determined and fixed once and for all through the initial processes of design and production. It should come as no surprise then to find that the ethics of technology which has emerged in the wake of the empirical turn has tended to focus exclusively on practices of design and the ethical issues which arise around the activity of making a new artefact or technology. Stemming from the widespread conviction that the moral or political status of a technology depends on the way it was designed, philosophers are increasingly looking towards the practice of design as both the source and solution of ethical problems surrounding technology. This is reflected in the recent emergence of a wide range of design for values approaches[14] alongside the common tendencies to focus exclusively on design in ethical discourses surrounding specific technologies such as artificial intelligence. Dominant trends and interests in contemporary philosophy of technology have therefore left few opportunities for the investigation of the different kinds of ongoing work that is required *after* technologies are produced. Despite all the work on human–technology relations, for example, one of the most basic and pervasive ways in which humans relate to technologies, through the activities of maintenance and repair, has been overlooked entirely. Much existing work in the philosophy of technology seems to assume that everything that's worth exploring about technologies occurs in relation to how they are *made* and consequently *used*.

This volume seeks to challenge this assumption by exploring a variety of ontological, epistemological and ethical questions which emerge from attending to the practice of maintenance. As such, it represents the first systematic investigation of the topic of maintenance within the philosophy of technology. The chapters collected here consider maintenance in relation to a wide variety of different technologies – from artificial intelligence to socks (with many in between). Together, they not only demonstrate the relevance and potential of maintenance for existing topics and discussions in the philosophy of technology but also identify new areas for philosophical investigation which emerge when we shift our focus to the ongoing work technologies require after production. As will become clear, unlike many other topics in the philosophy of technology, maintenance does not immediately align with a particular subfield, philosophical approach or school of thought. The broad relevance of this emerging topic is reflected in the diverse range of theoretical approaches and styles represented by each of the contributions in this volume. By elaborating on how maintenance matters for a variety of different topics and orientations in the contemporary philosophy of technology, we aim to clear a space for what we hope will be an ongoing discussion on the topic.

There are reasons to think that the time is ripe for the emergence of maintenance as a central topic in the philosophy of technology. A growing awareness of our impact on the natural environment is leading to an increasing interest in questioning and scrutinising the relationships we maintain with the technologies that we create. The rising interest among philosophers in topics such as the Anthropocene and sustainability reflects how such concerns are increasingly perceived to be within the purview of philosophical reflection itself.[15] Luckily, however, the task of establishing a philosophical discourse on maintenance doesn't need to occur in a vacuum. Outside of philosophy, there is already a well-established body of literature on the topic of maintenance and repair in disciplines such as STS, sociology and geography. Over the past two decades, this interdisciplinary field of maintenance and repair studies (MRS) has both generated a large body of empirical accounts of maintenance practices and developed theoretical frameworks for their analysis. Yet, while it is not uncommon to find MRS theorists drawing upon philosophical ideas or theories in their accounts of maintenance and repair,[16] up until now philosophers have remained absent from such discussions. By collecting a range of contributions which illustrate the relevance of maintenance to a variety of existing themes in the philosophy of technology, this volume is intended as an invitation for philosophers to join the table and contribute to developing our understanding of this important topic.

Our goals in the remainder of this introduction are as follows. First, we will provide an overview of the development of MRS in Section 1.1 while also highlighting dominant themes and theoretical approaches which characterise the literature. Then we'll turn our attention towards setting a preliminary agenda and orientation for the development of a discourse surrounding maintenance in the philosophy of technology by identifying specific questions and areas of analysis surrounding maintenance that we consider to be particularly suited to philosophical analysis. Finally, we'll provide a brief overview of the different chapters which are collected together in this volume.

1.1 Maintenance and Repair Studies

Aside from a few studies in the history of technology and anthropology,[17] a focused body of literature on maintenance and repair didn't really begin developing until the 1990s. During this time, a selection of ethnographic accounts appeared that focused on the maintenance of technologies in professional contexts.[18] These studies emerged alongside calls for a reorientation of scholarship in the history and sociology of technology away from the previously dominant focus on processes of invention and towards the active roles played by users in the development of technologies over time.[19] One of the most tangible results of these calls to explore the active engagements of users with technologies was the development of the field of the social construction of technology (SCOT). However, despite the rapid increase in popularity of such approaches, shaking the emphasis on innovation turned out to be easier said than done. Critics pointed out how while SCOT may indeed have acknowledged certain ways in which users play an active role in the development of technologies over time, such approaches still failed to look beyond periods of innovation (Mackay & Gillespie 1992; Edgerton 2010; Jackson 2014).

Shortly after the turn of the twenty-first century, the growth of the field gained momentum from renewed calls to begin correcting for the disproportionate scholarly focus on processes of invention, innovation and production by examining practices of maintenance and repair specifically (Edgerton 2006; Graham and Thrift 2007). Those promoting a turn towards maintenance have often argued that the common tendency to neglect consideration of the ongoing forms of activity required to sustain technologies over time increasingly has produced a body of scholarship that ultimately fails to explain how the objects, practices and institutions which together comprise technology come into being. The reasons for this neglect were sometimes identified in underlying assumptions and attitudes surrounding the nature of technology itself – normal modes of picturing and theorising technology that rendered maintenance invisible (Jackson 2014), for example, or cultural tendencies to glorify innovation which devalued maintenance work or cast it as uninteresting (Vinsel & Russell 2020). In other cases, scholars have examined how specific theoretical concepts and approaches in individual disciplines, such as the traditional emphasis on processes of closure and stability in STS, have worked to divert attention from the role maintenance plays in sustaining technologies over time (Denis 2020; Denis, Mongili & Pontille 2015). This most recent phase saw the popularity of the topic expand beyond the history and sociology of technology to an eclectic range of disciplines and fields such as geography, media studies, STS, art and craft studies, architecture and design theory[20] and coincided with the establishment by historians Lee Vinsel and Andy Russel in 2015 of an interdisciplinary academic community devoted to the topic called *The Maintainers*.[21] There is now a growing body of literature on the topic of maintenance and repair which is expansive enough to secure the status of maintenance and repair as an emerging field of interdisciplinary study. The following provides a short overview of key themes and common approaches within this literature.

1.1.1 Maintenance Beyond Materiality

The dominant strain of work in the field consists of ethnographic accounts of maintenance and repair in different contexts which explore how order is restored in cases of breakdown.[22] Studies of this kind have long drawn inspiration from a theoretical orientation in sociology known as ethnomethodology, an approach to the study of everyday practices which understands social order to arise as the result of constant, ubiquitous acts of repair.[23] It is long been common for studies in ethnomethodology to take the form of conversation analysis (Sharrock 1989).[24] Stemming from the premise that normal conversation is rife with moments of breakdown in which participants fail to understand each other, these studies explore how successful communication is nonetheless achieved through numerous and constant acts of repair, such as the addition of short clarifying dialogues or hand motions, which are performed by participants during a conversation.

Through the analysis of conversations, proponents of ethnomethodology produced what is perhaps the most extensive body of work on the phenomena of repair within sociology (Henke & Sims 2020, p. 18).

During the 90s, these approaches came to be applied to the activity of repair in technological contexts. Answering the growing calls to pay attention to the use of technology, sociologists began applying the methods and insights of ethnomethodology to the analysis of work, focusing in particular on human interactions with technology. These studies prepared the ground for a number of classic ethnographic accounts of maintenance and repair within professional contexts which emerged shortly after.[25] Today, many proponents of MRS understand the field in terms of an extension of ethnomethodological ideas and approaches to the analysis of technology, exploring how socio-material order emerges not as a result of the way a technology is designed, but rather of ubiquitous acts of repair performed during use (Graham & Thrift 2007; Denis, Mongili & Pontille 2015).

The influence of ethnomethodology on current work in the field of MRS is visible in various ways. For example, the emphasis in ethnomethodology on the restoration of the social order of work as the goal of repair, alongside the traditional methodological focus on the analysis of conversation, has helped encourage a much wider conception of the activity of repair in technological contexts than material engagement alone. Both Orr and Henke's seminal ethnographic accounts of repair work acknowledge the importance of forms of repair in which a technology is 'fixed' by altering the manner in which users perceive or engage with it. As Orr noted, 'the problems encountered by technicians are most fundamentally breakdowns of the interaction between customers and their machines, which may or may not include a malfunction or failure of some machine component' (Orr 1996, p. 3). Drawing on examples from which users become convinced that a technology has malfunctioned, these accounts illustrated how technicians often seek to resolve the problem by engaging with the user rather than the technology itself. This might involve offering explanations for why a technology doesn't operate as expected in certain contexts, using measuring devices to provide demonstrations that it is actually working properly or even providing instructions and training geared towards enabling users to address minor problems themselves. Scholarship on maintenance which takes its cues from ethnomethodology tends to approach the analysis of repair in ways which undermine traditional conceptions of breakdown as an objective state of an artefact. Instead, by emphasising the way artefacts may appear as broken for some users but not others, studies of this kind encourage an understanding of repair as a *relational* activity in which the status of artefacts as broken or fixed depends on the perspectives of the actors involved and is often subject to negotiation (Henke 2019).

1.1.2 The Invisibility of Maintenance

Another way in which the influence of ethnomethodology is visible in current MRS literature concerns the tendency for the activity of repair to recede from view. As is well known, in developing ethnomethodology, Harold Garfinkel drew inspiration from phenomenologists such as Martin Heidegger and Alfred Schutz, alongside the later work of Ludwig Wittgenstein in characterising central aspects of practice to be invisible to those in the midst of performance (McHoul 1998; vom Len 2016 p. 86). As is well known, Heidegger claimed that when we employ technologies that are working correctly, we don't experience them as visible isolated objects, but rather as part of a background of undifferentiated equipment upon which our activity depends (*zuhanden*). One of the ways in which technologies become visible is when they prevent us from achieving our goals because they are faulty or broken, in which case they appear to us as individual objects (*vorhanden*).[26] Wittgenstein, on the other hand, aimed to undermine conceptions of skilled practice as rule-based by examining how the creative activity necessary for enabling such activities remains invisible to practitioners themselves.[27] These ideas can be seen to be reflected in the basic tenets of ethnomethodology, which holds that while the skills of repair are largely invisible to participants when immersed within the flow of social activity, they can nonetheless be rendered visible and therefore accessible to analysis through disruption, a heuristic method famously embodied in Garfinkel's 'breaching experiments'.

Today, (in)visibility continues to form a central theme within scholarship on maintenance and repair. However, invisibility is a multifaceted theme within MRS literature which not only reflects the influence of ethnomethodology but also intersects with a variety of other themes and schools of thought in sociology and closely related fields which touch upon social issues and issues of power. For example, invisibility formed a central theme in the historical and sociological studies of housework which began to emerge during the 1980s. By exploring ways in which the gendered labour of housework failed to gain recognition or valuation in different contexts, these studies gave rise to what has become an enduring theme within feminist sociology – invisible labour.[28] While housework encompasses a variety of activities that include the production of items such as food, some of the activities most readily identifiable as forms of housework, such as cleaning or clothing repair also represent forms of maintenance. For this reason, many well-known studies of housework, such as Ruth Schwartz Cowan's *More Work for Mother* (1983) which focused specifically on the use of domestic technologies, are also considered to represent early works within the field of MRS (Russell & Vinsel 2018). By exploring how these forms of household maintenance have often been performed outside the formal economy and have struggled to gain adequate recognition, studies of housework contributed to establishing invisibility as a dominant theme from the very origins of MRS.

The theme of the invisibility of maintenance in MRS was further reinforced through the influence of studies of 'dirty work', a research line pioneered by sociologist Everett C. Hughes that explores forms of labour which are accorded a low social status or are considered to be degrading or undignified.[29] Invisibility forms a common theme in studies of dirty work, many forms of which are deliberately arranged to be performed out of sight, through the scheduling of cleaning activities 'after hours', for example, or in spaces such as private homes or workshops. As a central topic within the sociology of work, dirty work bears natural associations with MRS literature insofar as the term is often taken to include forms of maintenance, such as janitor work or street cleaning.[30] The influence of this line of research is reflected in the common focus within MRS literature today on the ways in which the visibility of maintenance practices is managed.[31] As a result, literature in the field of MRS has developed increasingly nuanced conceptions of the relationship between the visibility and invisibility of maintenance work alongside examining the different ways in which maintenance and repair may be concealed or downplayed. Certain kinds of artefacts, such as smartphones or laptops, are often designed in ways that prevent the average user from being able to open them so that accessing internal components for repair or maintenance is achieved primarily by technicians in workshops with specialist equipment. In other cases, the successful practice of maintenance is considered to depend on the erasure of traces of the act of maintenance itself (Denis & Pontille 2018) Yet, while invisibility remains a dominant concern within MRS literature, it is important to note that maintenance practices are not always invisible. Some forms of maintenance are displayed openly, such as cultural heritage conservation practices which often display or inform visitors of processes by which artefacts are maintained over time (Jones & Yarrow 2013) or public displays of infrastructure maintenance through which a government may seek to reassert authority and control over structures that have been modified by citizens to meet their needs (Barnes 2017). Furthermore, insofar as maintenance practices are often not distributed evenly through society or the global political economy (Jackson 2014; Russell & Vinsel 2018; Edgerton 2006, p. 80), the visibility of maintenance often depends on social or geopolitical location. For this reason, certain maintenance practices may be visible to some but not to others.

1.1.3 Vulnerability, Fragility and Breakdown

Many MRS scholars understand the relative invisibility of maintenance and repair to be at least partly responsible for cultivating widespread misconceptions concerning the inherent stability of the different objects that comprise our material culture. Within MRS literature, this insight often finds its most sustained elaboration in relation to infrastructure, which as Steven Jackson notes is commonly experienced in many places of the world to

> …*mostly* work, for *most* people, *most* of the time. Because of this we have tended to regard enduring function as a natural and more or less permanent feature of systems, rather than as the ongoing, frequently artful and often fraught accomplishment that it is.
>
> (Jackson 2016)

A common goal in many MRS studies is therefore to move beyond the apparent givenness of functional technology by restoring an appreciation of the inherent fragility of the material world we have constructed around ourselves. This often involves challenging the perceptions of what aspects of technology should be considered typical, such as breakdown and failure which are argued to be normal rather than anomalous occurrences in the lifespan of technologies (Graham & Thrift 2007; Henke 2000; Young 2021b), or fragility or decay which is suggested to represent a normal property of matter itself (Graham & Thrift 2007; Mattern 2018; Denis 2020). These approaches move us in the direction of what Steven Jackson calls 'Broken World Thinking', an analytic orientation to the study of technology which

> asks what happens when we take erosion, breakdown, and decay, rather than novelty, growth, and progress, as our starting points in thinking through the nature, use, and effects of information technology and new media. Broken world thinking is both normative and ontological, in the sense that it makes claims about the nature of technology and its relationship to broader social worlds, some of which may differ from deep-rooted cultural assumptions. But it is also empirical and methodological, an argument and provocation toward doing new and different kinds of research, and new and different kinds of politics, in media and technology studies today.
>
> (Jackson 2014)

Focusing on the inherent vulnerability of our technological environments inevitably propels maintenance to the centre of our thinking about technology insofar as it prevents us from taking the stability which characterises our experience of many technologies for granted. Instead, to account for the apparent durability and permanence of those environments, we are compelled to acknowledge the constant work of rebuilding, repairing and adapting through which such stability arises. At the same time, focusing on the inherent tendencies of artefacts towards decay or breakdown also reinforces connections with common theoretical approaches to the analysis of maintenance such as care ethics, where an attentiveness to dependence and vulnerability has long been considered a central component of relationships of care by feminist theorists.[32]

Thinking through the consequences of fragility has proven to be a fruitful analytic strategy among the variety of different subfields comprising MRS and now forms a central theoretical orientation within contemporary STS and studies of material culture.[33] While such a shift has yet to occur within the philosophy of technology, a focus on vulnerability nonetheless bears important but as yet unexplored implications insofar as it encourages us to invert the dominant line of questioning by resisting the common tendency to focus primarily on how technology affects societies and examining rather than the different ways in which societies affect technology.

1.1.4 Maintenance and Repair as Transformative

An increasingly wide range of scholarship within MRS literature has sought to undermine traditional dichotomies between innovation and maintenance or repair work by emphasising the transformative potential of the latter. This approach strikes a contrast both to traditional conceptualisations of maintenance as consisting exclusively in the attempt to hinder the changes artefacts undergo over time and common understandings of repair as aiming only to restore an artefact to a previous state. While such accounts acknowledge that some forms of maintenance or repair work may indeed seek the preservation or restoration of artefacts in these ways, they are at the same time concerned to illustrate how many of the goals which motivate the practice of maintenance, such as prolonging the time an object can stay in use (Weber & Krebs 2021, p. 11) or preserving physical or technical orders (Russell & Vinsel 2018), themselves often demand the transformation of the underlying technologies. This is apparent in studies of restoration practices which underscore the extent to which artefacts themselves must be changed in the process (Rubio 2016; Rubio 2020), alongside studies of infrastructure maintenance which draw attention to the practices of retrofitting and upgrading by which structures are adapted to emerging interests or needs (Sims 2016; Young 2021a).[34] By acknowledging the creative potential of maintenance, these approaches problematise intuitive distinctions between making and maintaining which align creativity exclusively with the former and cast the latter as a derivative form of technical activity which consists of merely preserving the products of prior creative activity. In doing so, they encourage us to understand the activity of making not as confined to the period in which an artefact is initially created but rather as extending throughout the histories of artefacts. Such approaches also call for a reconceptualisation of the nature of artefacts themselves, not as things with fixed and stable properties, but rather as processes that are open-ended and dynamic, constantly responding to emergent features of the different environments in which they are located.[35]

As a number of scholars have pointed out, acknowledging the creative potential of maintenance and repair opens up novel lines of enquiry by revealing new ways in which political processes come to manifest themselves in the artefacts which surround us. In many cases, changes to artefacts that occur under the guise of maintenance and repair serve the interests of those in possession of political power, such as when infrastructure is modified to facilitate the growth of commercial activity to the detriment of local communities. However, the transformative potential of maintenance and repair also underscores the capacity of these forms of activity to serve as forms of political action through which power may be exercised by marginalised groups. In their study of the history of San Diego's Coronado Bridge, for example, Henke and Sims document how activist work succeeded in enabling members of the local community to influence decision-making processes surrounding seismic retrofits of the bridge itself to ensure that modifications served their interests in addition to those of the California Department of Transportation (Henke & Sims 2020). Henke and Sims' example illustrates how 'transformative repair may also be championed by those who are excluded from existing power structures or otherwise see repair as an opportunity to advance their interests' (Henke & Sims 2020, p. 22).

1.2 Maintenance and Philosophy of Technology: Preliminary Issues

In sketching the key themes and analytical orientations in MRS literature, the foregoing sections illustrate the extent to which there is now an expansive body of empirical research on the topic of maintenance and repair. As we have seen, the analysis of much of this material has been achieved through the employment of a variety of theoretical lenses common in sociology and STS. In contrast, this volume aims to approach the topic of maintenance and repair by employing a different set of theoretical resources and approaches that can be found in contemporary philosophy of technology while at the same time demonstrating how maintenance holds the potential to advance a wide variety of ongoing discussions and themes within the philosophy of technology. While it is common for MRS theorists to draw on philosophical concepts in the elaboration of their arguments,[36] a sustained philosophical examination of many basic themes regarding maintenance remains lacking. For this reason, a variety of basic philosophical questions regarding maintenance remain unanswered. In this section, we sketch a number of preliminary issues related to maintenance which we regard to be particularly suited to philosophical analysis.

1.2.1 The Meaning(s) of Maintenance

In relation to technology, the term 'maintenance' is used to capture a wide and diverse range of human practices. We maintain artefacts through the act of using them in particular ways, cleaning components, applying protective coatings, replacing parts and fixing what is broken.

As highlighted in Section 1.1.1, maintenance is sometimes understood to extend beyond material practice altogether, encompassing activities such as the adjustment of user expectations or the provision of training. In light of this heterogeneity, the question naturally arises concerning the extent to which the term 'maintenance' can be understood to correspond to a coherent and unitary form of technical practice. This is no easy question. For it is not only the activities of maintenance that exhibit significant diversity but also the goals motivating their performance. In some cases, maintenance might be geared towards keeping something in existence, keeping something in use or maybe instead keeping something operating optimally. In other cases, maintenance may apply to the aesthetic features of objects, preserving a particular look and condition. Sometimes maintenance aims to change a state of events which has occurred, such as when we clean away grime that has accumulated on the lens of a camera. Other times, maintenance involves attempts to guard against things that *may* happen, such as when we install protective barriers around windows in advance of a coming storm. Given this diversity, it is not surprising to find a wide variety of different ways of conceptualising maintenance within existing MRS literature. This includes definitions of maintenance as 'all the work that goes into preserving technical and physical orders' (Russell and Vinsel 2018) or as an activity which aims to 'guarantee stability, to provide things with a level of stability they do not possess "naturally"'(Denis and Pontille 2014). Other authors suggest we understand maintenance and repair as the 'myriad forms of activity by which the shape, standing and meaning of objects in the world is produced and sustained' (Jackson 2014, p. 234), or as 'represent(ing) interventions with the temporal aim of prolonging the time an object can stay in use' (Weber and Krebs 2021, p. 11).

One way to gain further insight into the nature of maintenance is by examining its relationship to other forms of human engagement with technology, such as making or using. Maintenance bears deep connections to use, insofar as the proper maintenance of many technologies is often considered to require using things in particular ways. For example, maintaining a car engine may be understood to involve refraining from activities which are known to damage engines over time, such as accelerating in a high gear or shifting from drive to reverse without stopping. As Carl Mitcham notes, maintaining represents a distinct form of using which does more than merely 'put an object to use in the straightforward sense of engaging its technical or end uses' (Mitcham 1994, p. 233). Instead, this form of using is motivated also by the additional goal of prolonging the time a particular artefact may be used. For this reason, maintaining can be understood as a form of using that looks beyond the satisfaction of our interests and goals in the present and actively attempts to influence the future states of a technology. While his account acknowledges the creative significance of maintenance practices, Mitcham is at the same time careful to distinguish maintenance from making, noting that whereas maintenance

> 'might even be said to extend or prolong making, since it protects or retains the object made.... (maintenance) is subordinate not to making but to using, and is a kind of using in that its aim is to keep something usable. Dusting the books and changing the oil in the car are repetitive processes like making the bed or washing the dishes. Indeed, one is almost tempted to say that beds are not just made to be slept in but also "made to be made", that the second making is itself using, and that using dishes entails not only eating off them but also washing and drying them. In this sense maintaining can be seen as a natural prolongation of using, and adjunct or secondary using, certainly a preparation for reusing'.
>
> (Mitcham 1994, p. 233)

Yet, whereas Mitcham is concerned with establishing a distinction between making and maintaining, more recent scholarship has sought to undermine the very same distinction, arguing that maintenance should be considered a form of making. Young (2021a), for example, examines how the distinction between making and using is premised upon the idea that the creation of an artefact ceases at the end of production, a moment philosophers have traditionally understood in terms of the material realisation of a designer(s) intention. Yet by drawing attention to how the ongoing maintenance and repair of artefacts often involve the attempt to realise intentions that differ from those of the original designer(s), Young argues that rather than understanding making to be confined to the initial production of an artefact, we should understand the activity of making to extend throughout the lives of artefacts and to cease only when we no longer intend to continue using them.

Another way of gaining insight into the nature of maintenance is by examining its relationship to repair. Despite the two terms often appearing in tandem throughout MRS literature, questions concerning the similarities and differences between maintenance and repair as forms of technical activity have received surprisingly little attention. However, in a variety of professional contexts, repair is often differentiated from maintenance by emphasising its association with breakdown. According to this perspective, maintenance consists in monitoring and addressing developments in a technology which may impact its ability to function such as the planned replacement of parts or the renewal or replenishment of material such as lubricants or protective coatings. Whereas this perspective casts maintenance as routine work that addresses problems which typically occur throughout the service life of an artefact, repair, on the other hand, is aligned with unexpected developments which result in breakdown and loss of function and is understood in terms of returning an artefact to some prior condition (Perzanowski 2022, p. 15). Maintenance and repair are therefore differentiated according to their temporal relation to breakdown: maintenance is understood to take place before breakdown occurs, as opposed to repair which happens afterwards. Yet, while this way of distinguishing maintenance from repair may be common, as Aaron Perzanowksi notes, drawing distinctions in this way leaves us unable to distinguish maintenance or repair in a wide variety of cases:

> Say your car drifts slightly from right to left on the freeway. Is a wheel alignment from your local auto shop a repair or mere maintenance? Functionality isn't binary. Your car still works with misaligned wheels, although not as well as you might prefer. The line between maintenance and repair is malleable, but the distinction is still helpful.
>
> (Perzanowski 2022, p. 16)

The problem with distinguishing maintenance from repair in this way, as Perzanowksi correctly identifies, lies in the difficulty of drawing clear distinctions between cases in which a technology should be understood as functional or non-functional. The challenges facing such a task are well established within the philosophy of technology where the successful functioning of an artefact is commonly considered to be a matter of degree or perspective and something that depends on the skills and expectations of the user.[37] These difficulties underscore the limitations of conceiving the difference between maintenance and repair as taking place before and after failure, respectively.

A more nuanced conception of the relationship between maintenance and repair can be found in common distinctions in engineering between preventative and corrective maintenance. According to this approach, preventative maintenance encompasses activities performed in advance of the failure of a system, artefact or component. Corrective maintenance, on the other hand, addresses failures that have already occurred that threaten the overall functionality or longevity of an artefact or system. According to this perspective, examples such as wheel alignment, where attempts are made to repair damage which does not prevent an artefact from fulfilling its desired function entirely, are to be considered forms of corrective maintenance. Like maintenance, repair may also take place before breakdown. However, rather than simply conflating maintenance with repair, the corrective preventative distinction may be understood to preserve an exclusive category of repair activities that correspond to cases that are not commonly classified as maintenance.[38] This applies to examples in which technologies are restored to working order after a failure has led to a break in their usage and includes cases such as the act of restoring an old and abandoned car to working condition or rebuilding artefacts or systems after catastrophic failures. Yet while acknowledging the difference between preventative and corrective maintenance may provide a more intuitive way of distinguishing between maintenance and repair, this distinction is nonetheless susceptible to problems surrounding the ambiguity of failure that we noted in Section 1.1.1. The slow, processual nature of many forms of failure makes it difficult to pinpoint a precise moment where failure occurs that could serve to distinguish preventative from corrective interventions. As a result, the distinction between corrective and preventative maintenance is widely acknowledged to be ambiguous in many contexts.[39]

Finally, further insights concerning the nature of maintenance may be revealed by exploring how it has been conceived in relation to practices of retrofitting or rehabilitation in different contexts. A wide variety of different kinds of artefacts undergo modification long after their production or construction has ended in order to align them better with changing interests, needs or values. Examples of these kinds of modifications include the widening of bridges or roads to facilitate greater levels of traffic or the modification of vehicle engines to reduce carbon emissions. Among engineers, these kinds of interventions are termed either retrofitting or rehabilitation and are often distinguished from maintenance and repair based on the notion that the latter forms of activity are limited to restoring or preserving the original designed form of an artefact. However, this distinction does not always hold in practice. As Alampalli notes, in bridge engineering, 'the line where maintenance ends and rehabilitation begins is very subjective. In most cases, the degree of repair, funding source, and agency policies dictate if it is called maintenance activity or rehabilitation activity' (Alampalli 2014). As we have already seen, by emphasising the transformative potential of maintenance, MRS scholarship has tended to downplay distinctions between retrofitting and maintenance. In his study of seismic retrofits of California Bridges, Benjamin Sims, for example, specifically identifies retrofitting as a form of repair (Sims 2016). Akhil Gupta, on the other hand, claims that 'maintenance and repair often result in innovations that enable infrastructures to function for periods far beyond their designed life span and take on new functions that were not even envisioned when the project was built' (Gupta 2021).

The foregoing sections illustrate how the meaning of maintenance itself is contested and inadequately conceptualised, leaving a number of basic questions unanswered: What is maintenance? What are the motivations and goals driving this form of technical activity? Does the term represent a coherent category of technical practice, much like making? Or does the term instead correspond to a heterogeneous collection of different forms of activity? While existing literature on MRS has indeed taken steps towards addressing these questions,[40] a sustained discussion is yet to emerge.

1.2.2 *Surveying the Landscape of Maintenance and Philosophy of Technology*

The range of philosophical issues related to maintenance and repair can be understood to fall into three broad categories: ontological, epistemological and normative. In Section 1.2.1, we already touched upon issues related to the *ontology* of maintenance by exploring questions surrounding whether maintenance represents a coherent and unified category of technical practice or whether the term instead corresponds to a heterogeneous collection of different forms of activity. A different set of ontological issues relating to maintenance concerns the objects of maintenance and repair itself. Philosophers have long debated questions surrounding the ontological status of artefacts and the kinds to which they are understood to belong. As is well known, technical artefacts such as cars or screwdrivers straddle the boundary between human culture and the material world insofar as they cannot be adequately understood in terms of their physical characteristics alone but must also be conceived in relation to the purposes they were created or designed to achieve (Kroes & Meijers 2002). The nature of the relationship between the teleological and physical character of artefacts is itself the subject of debate among philosophers as is the extent to which artefacts should be considered to depend on human minds for their existence. Existing literature on the philosophy of technology has tended to explore these questions by focusing on the way in which artefacts are designed, produced and used. What has received less attention however is the way both the material and teleological characters of artefacts are sustained and very often altered through the practice of maintenance and repair. Exploring how artefacts are maintained over time can reveal new insights into the role social phenomena play in the constitution of artefact kinds and the nature of malfunction, and by doing so, raise a range of new questions that the wide body of philosophical literature on technical functions has yet to address in detail.[41] Furthermore, by focusing attention on the temporality of artefacts, the study of maintenance encourages us to consider the role and significance accorded to change in the underlying ontological frameworks we employ in our research.[42]

Many of the *epistemological* issues explored in the philosophy of technology have focused on the knowledge and processes by which new things are created in engineering contexts. One common and longstanding line of enquiry explores the extent to which knowledge involved in engineering design can be understood to differ from scientific knowledge. Yet questions remain concerning the nature of the engineering knowledge required to maintain artefacts over time and the extent to which this differs from other forms of technical knowledge. In what ways, for example, is the knowledge required for maintenance and repair distinct from the knowledge employed in the design of artefacts? Despite common estimates which suggest the majority of engineers engage in the practice of maintenance rather than design,[43] the knowledge engineers employ in tending to things over time has received little attention. In the philosophy of engineering, epistemological issues pertaining to methodology, such as modelling and simulations, experimentation, heuristics and processes of verification have been examined primarily in relation to design. The same can be understood to apply to the social epistemology of engineering which explores technical norms and standards, collective decision-making and processes of justification. Outside the context of engineering, questions remain concerning the nature of maintenance knowledge and its employment or acquisition by technicians or users. In what ways, for example, should we understand the relationship between the situated character of maintenance and repair and the transmission of repair knowledge through textbooks or online forums and videos? Or to what extent does tacit knowledge, connoisseurship and one's own body play a role in understanding and diagnosing problems during maintenance and repair?

The third category of philosophical issues relating to maintenance and repair are *normative*. Existing literature on MRS has tended to approach questions concerning the ethics of maintenance from the perspective of care ethics or environmental ethics. However, work remains to be done elaborating the connection between maintenance and ethics in these fields, alongside situating ethical issues surrounding maintenance within the landscape of contemporary philosophical ethics. While existential approaches to the topic have emphasised the value of establishing active relationships of engagement and care with technical artefacts, MRS theorists have warned against the common tendency to romanticise maintenance, drawing attention to the social conditions under which such work is often performed (Mattern 2018). Similarly, whereas much scholarship in MRS strives to counter the political and cultural neglect of maintenance by drawing attention to the importance of the practices through which we care for technologies over time, others warn us against uncritically valorising maintenance practices. Henke and Sims, for example, remind us that 'repair can serve to prevent needed social change, exclude or remove certain groups of people from power and stabilise problematic infrastructures and institutions, often at great social cost' (Henke and Sims 2020, p. 15). Exploring the values and interests motivating specific instances of maintenance and repair in connection to social and environmental justice, therefore, represents an important opportunity, not only for deepening our current understanding of the relation between the politics and ethics of technology but also for contributing to a more nuanced discussion surrounding the value of maintenance and repair.

The foregoing sections provide by no means an exhaustive list of issues; however, they do give a sense of the breadth of possibilities that emerge when philosophers of technology turn their attention to maintenance. By representing the first foray into the topic of philosophy of technology, this volume takes the first steps towards addressing these questions and provides the beginnings of what we hope will be an ongoing discussion.

1.3 Structure of the Volume

Unlike some themes in the philosophy of technology which are aligned exclusively with certain traditions or schools of thought, the topic of maintenance cuts across such divisions. By showcasing a variety of different philosophical styles, topics and approaches, the chapters in this volume outline the breadth of the unrecognised significance of the topic of maintenance to a diverse variety of subfields in the philosophy of technology, from continental to anglophone traditions. In addition to illustrating the potential of maintenance to advance thinking on a wide variety of philosophical discussions, this volume therefore also functions to provide a snapshot of the current diversity of methods, topics and approaches within the field of philosophy of technology itself.

The volume is divided into two parts that cover different subfields in the philosophy of technology. Part I, *Metaphysics and Epistemology of Maintenance* begins with a chapter by Steven Vogel 'Maintenance and the Humanness of Infrastructure' which explores the nature of the relationship we share with the many different forms of infrastructure which frame urban life. Examining the Heideggerian notion that technologies in use recede from view when functioning correctly, Vogel draws upon Marx in suggesting that what becomes hidden is not so much their equipmentality but rather their nature as products of human labour. However, against common tendencies to conceptualise this labour as something that occurred in the past, as an historical fact, Vogel argues instead that we must recognise infrastructure as continually produced through ongoing processes of maintenance. By doing so, Vogel draws attention to inherent continuities between the practices of making and maintaining.

In Chapter 3 'Technology in Process: Maintenance and the Metaphysics of Artefacts', Mark Thomas Young examines existing work on the metaphysics of artefacts in order to demonstrate how the underlying framework of substance ontology has hindered our understanding of the relationship between technologies and time. Whereas the Western metaphysical tradition has traditionally approached questions concerning the persistence of artefacts by attempting to demonstrate how they remain the same despite undergoing change, Young examines how attending to maintenance reveals instead the way in which a wide range of artefacts persist only by adapting to changing environments. In order to account for maintenance, Young argues for the need to adopt a new underlying ontology centred around processes rather than substances. In the final sections of his chapter, Young sketches the outlines of such an approach and illustrates how a process-based metaphysics of artefacts tends inevitably towards the empirical analysis of social and political processes which determine the trajectory of artefacts through time.

In Chapter 4, 'There, I Fixed It! On the Status of Meaning and Repair', Tiago Mesquita Carvalho engages in a search for a philosophical theory of the nature of technological repair. In doing so, he takes advantage of two distinct sets of literature within the philosophy of technology: the metaphysics of artefact functions and Ihde's theory of postphenomenology, revealing how important parallels between both approaches provide the resources for a general theory of repair that recognises the capacity of artefacts to evolve over time in conjunction with the interests and desires of their users. By examining how artefacts often derive their meaning from contexts of use, Carvalho argues that repair may not necessarily seek to restore proper function as intended by the designers of an artefact and may instead seek new artefact forms and configurations. In the second section of the chapter, Carvalho turns his attention to the epistemology of repair, situating the knowledge and skills required to perform successful acts of repair in relation to Aristotelian categories of *techne* and *phronesis* alongside modern conceptions of tacit knowledge.

In 'A Standpoint Epistemology of Repair?', Cristina Bernabéu and Jesús Vega-Encabo respond to Steven Jackson's well-known calls for the development of a standpoint epistemology of repair. After exploring the promise and potential of such an approach to yield a deeper understanding of our wider technological environments, Bernabéu and Vega-Encabo turn their attention to examining the central features of feminist standpoint theories in order to ascertain the feasibility of developing such a theory in relationship to maintenance and repair. While they identify several specific difficulties facing the prospect of a standpoint epistemology of repair, Bernabéu and Vega-Encabo nonetheless conclude their contribution by reaffirming the broader epistemic potential of shifting our focus to the ongoing activities of maintenance and repair through which our artefactual worlds are sustained and reproduced.

Jochem Zwier's chapter 'Sustainability as Planetary Maintenance' examines common analogies between maintaining technology and maintaining the environment and illustrates how modern conceptions of sustainability can be understood to arise from the application of conceptions of technological maintenance to the practices by which we care for the Earth. In order to problematise this understanding of sustainability, Zwier traces the metaphysical and ontological roots of the concept of technological maintenance back to Aristotle, revealing essential hylomorphic features that fail to be reflected in modern understandings of sustainability. According to Zwier, these limitations ultimately undermine the meaning of the concept itself, rendering sustainability a nihilistic form of striving that lacks an objective. In the search for a more robust conceptualisation of sustainability, Zwier explores the possibility of alternative understandings based not on the conceptual resources of Aristotelianism, but rather on the work of French philosopher of technology, Gilbert Simondon. Drawing upon Simondon's conception of technological individuation as an exploratory and open-ended process of becoming which arises through establishing new correspondences between different, previously unrelated aspects of the world, Zwier examines how this theory holds the potential to inform valuable new understandings not only of sustainability but also the nature of maintenance itself.

In 'Towards a Realist Metaphysics of Software Maintenance', Keith Begley examines the extent to which computational entities should be considered abstract objects which are therefore immaterial and unchanging and whether this might complicate their status as potential objects of maintenance. Drawing on the account of abstract objects outlined by Jerrold Katz, Begley argues that computational artefacts should be considered composite objects comprised of both abstract and concrete constituents and examines how this accounts for the widely discussed 'dual nature' of such artefacts. In the second section of his chapter, Begley turns his attention to examining what the status of computational artefacts as realisations of their creators' intentions means for how we understand the nature of their maintenance over time. In contrast to conservative conceptions which understand maintenance to consist of the preservation of an original form intended by a maker, Begley illustrates how software comprises a class of artefacts which are explicitly intended by their creators to evolve over time, providing a rich example of the transformative nature of maintenance.

In 'Maintaining Perpetual Actuality in the Digital Age? Simondon's Conception of Maintenance and the Networked Era', Johannes Schick examines the role maintenance plays in the account of technical objects developed by Gilbert Simondon and explores to what extent this account remains relevant in our current digital age. As Schick demonstrates, Simondon understood maintenance to be a mode of human engagement with the internal structure of a technical object and its networked relations to other objects through which we continually adjust it to the present moment. However, unlike the mechanical and electrical objects around which Simondon developed his theories, technical objects in the digital age, such as smartphones, appear to resist these forms of engagement in different ways. In response to this concern, Schick illustrates how the use and development of smart devices nonetheless conform to the processes of concretisation and technicity which form the central elements of his account of technological maintenance. In the final sections of his chapter, Schick examines how Simondon considered maintenance to operate in a temporal space whereby the history of a technical object is related to a possible future and draws from Simondon's pedagogical theories a notion of storytelling which he suggests may help us in overcoming the difficulties we face surrounding the histories of technical objects in the digital age.

Part II of the volume *Ethics, Politics and Aesthetics of Maintenance* begins with a chapter by Steffen Steinert titled 'Maintenance of Value and the Value of Maintenance', which examines the relationship between technology, value and maintenance practices by focusing specifically on socio-technical systems and their lifecycles. Steinert's contribution addresses the lack of attention to the practice of maintenance in current discussions surrounding the relationship between technology and values and expands beyond current literature in looking beyond how values come to be embedded in technologies during the design phase to also include how values are sustained and transformed through practices of maintenance. In elaborating what he terms the 'maintenance perspective', Steinert examines the complex interplay between the preservation and transformation of values in relation to three key aspects of socio-technical systems: agents, institutions and hardware. In contrast to much current work on values and technology that often assumes the nature of technologies to remain fixed in the face of evolving values, Steinert instead foregrounds how the stability and change of values over time are reflected in the historical evolution of socio-technical systems themselves.

In his chapter 'An Eco-Ethics for the End of the Anthropocene: Finding Ethical and Sustainable Paths through Consumerism, Disposability and Planned Obsolescence', Simon Penny examines broad questions concerning the relationship between personal practices of consumption in consumer societies and the impact of global cycles of production, consumption and waste on the environment. By examining how planned obsolescence manifests itself in the use of synthetic and composite materials that decompose within planned timeframes into toxic residues, Penny draws attention to the way consumer products often resist repair, recycling or reuse. Reflecting on the accumulation of waste and environmental consequences that result from such practices, Penny argues that each of us has a responsibility to subvert the cycles of consumption and disposal which typically characterise life in contemporary consumer societies. Drawing on his broad experience upcycling materials as an artist and craftsman, he sketches the outlines of a salvage ethics – a form of life centred around repurposing rather than the disposal of artefacts which no longer serve their purpose.

In 'Aesthetic Values in the Maintenance of Urban Technologies', Sanna Lehtinen approaches the topic of maintenance from the perspective of urban aesthetics. Departing from common conceptions of urban technologies in infrastructure studies as an invisible substratum upon which life in the city plays out, Lehtinen instead examines how the way in which different technologies make cities look and feel is a question of crucial importance in determining how they should be maintained over time. By examining cases of deliberation and disagreement concerning the evolution of urban technologies, Lehtinen's chapter illustrates how aesthetic values often intersect with social and ethical values and argues for the importance of directly addressing aesthetic values in decision-making processes regarding urban maintenance.

In 'Negotiating Visions of Waste: On the Ethics of Maintaining Waste Infrastructures', Joost Alleblas and Ben Hofbauer explore the complex relationships people share with waste in high-income consumer societies. Their contribution illustrates how conceptions of waste as valueless are reinforced by the design and operation of municipal waste systems which render the flow and management of waste invisible in urban settings. By exploring how values are embedded, afforded and resisted through the use of socio-technical systems, Alleblas and Hofbauer illustrate how the preservation of these systems hinders the development of positive affirmations of waste as a resource that is necessary for the establishment of more sustainable practices of municipal waste management. The first step towards achieving such a goal, they argue, lies in rethinking the nature of infrastructure maintenance as a dynamic and reflexive process rather than the activity of caring for a stable and rigid system.

In the final chapter 'Repairing AI', Taylor Stone and Aimee van Wynsberghe investigate how attending to practices of maintenance and repair holds the potential to enrich existing discourse on sustainable AI in the field of AI ethics. Specifically, their chapter focuses on the way current attempts to employ artificial intelligence to address environmental problems are complicated by the underlying environmental problems generated by the ongoing production of the technology itself. In order to address this problem, Stone and van Wynsberghe argue for the need to subvert the dominant tendency in AI Ethics to search for solutions at the level of software design by reorienting our focus towards the global networks of material infrastructure and human labour from which the technology itself emerges. By approaching artificial intelligence as an evolving and dynamic process rather than a fixed and stable entity, Stone and van Wynsberghe elaborate on a conception of repair which extends beyond merely restoring the functionality of a technology and captures the reflexive processes through which the development of technological systems are managed over time. From this perspective, they argue, breakdown presents an opportunity not only to restore the status quo but also to transform technologies in order to address emerging interests and concerns and by doing so, attain more ethical and sustainable configurations.

Together, these chapters signal the arrival of maintenance as an emerging topic in the philosophy of technology. By shifting attention to what happens after technologies are produced, they represent the first steps within the field towards a more complete understanding of the development of technologies over time while at the same time addressing a range of contemporary concerns surrounding the technological cultures we currently inhabit. It is our hope that this volume will contribute to and inspire further philosophical and interdisciplinary work on a phenomenon and aspect of technology and technological practice that is still not fully understood or valued, both in academic literature and beyond.

Notes

1 The few exceptions include Young (2021a, 2021b); Karakas (2020). See also the brief account of maintenance in Mitcham (1994, pp. 233–235).
2 For an account of how our cities would fare in the absence of constant maintenance, see Weisman (2007, chap 3).
3 See, for example, Russell & Vinsel (2019); Mattern (2018); Edgerton (2006, p. 100). For a discussion of this point in relation to software development see Begley (this volume).
4 Examples abound of technologies that defy their producer's expectations, see, for example, Tenner (1996); Huesemann and Huesemann (2011).
5 See Rosenberg (1982; esp chap 6). For a discussion of the implications of this on policy-making regarding technology development, see Collingridge (1980).
6 For an overview of the theme of care within maintenance and repair studies, see Jackson (2016).
7 See, for example, Mattern (2018); Denis and Pontille (2015); de la Bellacasa (2011); Jackson (2014).

8 Interestingly, Pirsig identified a relationship of care as the essence of maintenance already in 1974, before the emergence of care ethics in feminist theory during the 1980s. Reflecting on a mechanic's failed attempt to repair his motorcycle, Pirsig noted that

> The mechanics in their attitude toward the machine were really taking no different attitude from the manual's toward the machine or from the attitude I had when I brought it in there. We were all spectators. And it occurred to me that there *is* no manual that deals with the *real* business of motorcycle maintenance, the most important aspect of all. Caring about what you are doing is considered either unimportant or taken for granted.
>
> (Pirsig 1974, p. 25)

9 See, for example, Tronto (1993, pp. 127–31; 1995).

10 Situatedness represents a key theme in maintenance and repair literature, where many emphasise the situated character of repair work specifically. As Steven Jackson notes, repair work is deeply embedded

> in contexts that structure the need and possibility of repair, and furnish the resources available for its accomplishment. In this view, repair work is deeply contingent, dependent on social and material configurations grown and evolved in particular times and places. Much of the diagnostic work embedded in repair is in fact about calling these configurations to light, both to understand the nature of breakdowns and problems encountered and to craft effective and doable responses from the resources that context affords.
>
> (Jackson 2019)

11 As stated by Willie in Douglas Harper's *Working Knowledge*,

> If you rush things, you can't enjoy them. And it's a challenge – no job's the same. If you had a thousand jobs in a year, not two of those thousand jobs would be the same. Even the ones that are supposed to be the same aren't. Things are broken or worn in different ways – they each have their own characteristics.
>
> (Harper 1987, p. 169)

12 See Penny (this volume).

13 See (Kroes & Meijers 2000; Achterhuis 2001).

14 For an overview of these approaches see (Van den Hoven, Vermaas & Van de Poel 2015; Brey 2022).

15 See, for example, Stone & van Wynsberghe (this volume) and Zwier (this volume).

16 See, for example, MRS studies which draw on the work of Dewey (Schubert 2019; Tironi 2019), Hans Jonas (Dant 2019) or Heidegger (Jackson 2014; Graham and Thrift 2007).

17 Among the earliest studies to explore practices of maintenance were (Cowan 1983; Harper 1987).

18 See, for example, (Orr 1996; Henke 2000; Dant & Bowles 2003). For an influential work of architectural history from this period which focuses on maintenance see (Brand 1994).

19 For criticisms of the dominant tendency within the history of technology to focus on invention at the expense of use, see (Pursell 1995; Edgerton 1999;

Oldenziel 2001). For studies criticising the common tendency in social theory to understand users as passive consumers of technology, see (de Certeau 1984; Oudshoorn & Pinch 2003).

20 Among the most influential studies on the topic from this period are Jackson (2014), Denis and Pontille (2015), Russell and Vinsel (2018).

21 For an account of the establishment of this community, see (Russell & Vinsel 2018; Vinsel & Russell 2020, p. 16).

22 In addition to the classic ethnographic accounts of maintenance and repair (Orr 1996; Henke 2000; Harper 1987; Dant & Bowles 2003; Denis & Pontille 2015), see also the recent collection (Strebel, Bovet & Sormani 2019).

23 For an overview of ethnomethodology see (Sharrock 1989; vom Len 2016). For studies of maintenance and repair which engage explicitly with ethnomethodology, see (Suchman 1987; Henke 2000; Bovet & Strebel 2019).

24 For differences between conversation analysis and ethnomethodology, see (Lynch 1993, chap 6).

25 See, for example, (Orr 1996; Downey 1998; Henke 2000).

26 For a thorough discussion of this idea in Heidegger's work, see (Harman 2002).

27 For statements of this view in relation to the case of language use, see (Wittgenstein 1980, §911, 161e; (Wittgenstein 2009, §122, 54e).

28 For an overview of the topic of invisible labour in feminist theory and sociology, see (Crain, Winifred & Cherry 2016).

29 For overviews of the theme of dirty work, see (Hughes 1971; Simpson et al. 2012).

30 See (Dant & Bowles 2003) for an important study in MRS that draws explicitly on this tradition.

31 See, for example, (Denis & Pontille 2017; Denis & Pontille 2018).

32 For an overview of the theme of vulnerability in feminist theory, see (Mackenzie, Rogers & Dodds 2014).

33 See, for example, (Connolly 2013; Ramakrishnan, O'Reilly & Budds 2021) See also discussions of network fragility in actor–network theory (Callon 1986; Higgins & Kitto 2004; Lovell & Smith 2010).

34 See also Young (this volume).

35 This particular view finds its fullest articulation in infrastructure studies. See also recent work in the philosophy of technology on the nature of artefacts as processes (Young 2021a; Coeckelbergh 2023).

36 See above n.15.

37 See, for example, (Radder 2009; Karakas 2020).

38 Compare this interpretation to that of Jonathan Chapman, who notes that the distinction between preventative and corrective maintenance 'is important, as notions of maintenance and repair are often incorrectly assigned as oppositional ideas. Rather, maintenance and repair are aspects of the same thing, occupying different ends of a "spectrum of care", with preventative maintenance (e.g., washing and ironing a shirt) at one end, and corrective maintenance (e.g., replacing a broken button) at the other' (Chapman 2021, p. 38).

39 Despite this conceptual ambiguity, distinctions between preventative and corrective maintenance remain an important distinction in practice in many contexts, especially those where the costs of failure are high. For a brief historical account of the emergence of conceptions of preventative maintenance, see (Vinsel & Russell 2020, chap 3).

40 For an excellent recent study examining conceptualisations of maintenance see (Denis & Pontille 2022).

41 For an overview of this literature, see (Preston 2009; Franssen 2013).

42 See Young (this volume).
43 See, for example, Edgerton (2006, p.xv); Russell and Vinsel (2019); Vinsel (2019).

Bibliography

Achterhuis, Hans (ed) Robert P. Crease (trans) (2001) *American Philosophy of Technology: The Empirical Turn* Bloomington: Indiana University Press.

Alampalli, Sreenivas. (2014) "Bridge Maintenance" in Wai-Fah Chen & Lian Duan (eds) *Construction and Maintenance* (Bridge Engineering Handbook) 2nd ed. Boca Raton: CRC Press pp. 269–300.

Barnes, Jessica. (2017) "States of Maintenance: Power, politics, and Egypt's irrigation infrastructure" *Environment and Planning D: Society and Space* 35(1) pp. 146–164.

Brand, Stewart. (1994) *How Buildings Learn: What Happens After They're Built.* New York: Penguin Books.

Brey, Philip. (2022) "Understanding Engineering Design and Its Social, Political, and Moral Dimensions" in Shannon Vallor (ed) *The Oxford Handbook of Philosophy of Technology*. New York: Oxford University Press pp. 395–416.

Bovet, Alain & Strebel, Ignaz (2019) "Job Done: What Repair Does to Caretakers, Tenants and Their Flats" in Ignaz Strebel, Alain Bovet & Phillipe Sormani (eds) *Repair Work Ethnographies: Revisiting Breakdown, Relocating Materiality.* Singapore: Palgrave Macmillan pp. 89–125.

Callon, Michel. (1986) "Some Elements in a Sociology of Translation: Domestication of the scallops and fisherman of St. Brieuc Bay" in John Law (ed) *Power, Action and Belief: A New Sociology of Knowledge?* London: Routledge & Kegan Paul pp. 196–233.

de Certeau, Michel. (1984) Steven Rendall (trans) *The Practice of Everyday Life.* Berkely: University of California Press.

Chapman, Jonathan. (2021) *Meaningful Stuff: Design That Lasts*. Massachusetts: MIT Press.

Coeckelbergh, Mark. (2023) "Technology as Process" in Glenn Miller, Helena M Jerónimo & Qin Zhu (eds) *Thinking Through Science and Technology: Philosophy, Religion and Politics in an Engineered World.* Maryland: Rowman & Littlefield pp. 55–68.

Collingridge, David. (1980) *The Social Control of Technology*. London: Frances Pinter.

Connolly, William E. (2013) "The 'New Materialism' and the Fragility of Things" *Millennium: Journal of International Studies* 41(3) 399–412.

Cowan, Ruth Schwartz. (1983) *More Work For Mother: The Ironies of Household Technology from the Open Hearth to the Microwave.* New York: Basic Books.

Crain, Marion G., Winifred, Poster R & Cherry, Miriam A. (eds) (2016) *Invisible Labor: Hidden Work in the Contemporary World.* Oakland: University of California Press.

Crawford, Matthew B. (2009) *Shop Class as Soulcraft: An Inquiry into the Value of Work*. New York: Penguin Press.

Dant, Tim (1999) *Material Culture In The Social World: Values, Activities, Lifestyles*. Buckingham: Open University Press.

Dant, Tim (2019) "Inside the Bicycle: Repair Knowledge for All" in Ignaz Strebel, Alain Bovet & Phillipe Sormani (eds) *Repair Work Ethnographies: Revisiting Breakdown, Relocating Materiality*. Singapore: Palgrave Macmillan pp. 283–312.

Dant, Tim & Bowles, David. (2003) "Dealing with Dirt: Servicing and Repairing Cars" *Sociological Research Online* 8(2). Available at http://www.socresonline.org.uk/8/2/dant.html (accessed January 20, 2021).

De la Bellacasa, Maria Puig. (2011) "Matters of Care in Technoscience: Assembling Neglected Things" *Social Studies of Science* 41(1) pp. 85–106.

Denis, David J. (Denis, Jérôme & Pontille, David) (2020) "Why do Maintenance and Repair Matter?" in Anders Blok, Ignacio Fárias & Celia Roberts (eds) *The Routledge Companion to Actor-Network Theory*. New York: Routledge pp. 283–293.

Denis, Jérôme & Pontille, David. (2014) "Maintenance work and the performativity of urban inscriptions: the case of Paris Subway Signs" *Environment and Planning D: Society and Space* 32(3) pp. 404–416.

Denis, Jérôme & Pontille, David. (2015) "Material Ordering and the Care of Things" *Science, Technology & Human Values* 40(3) pp. 338–367.

Denis, Jérôme & Pontille, David. (2017) "Object Ontologies and the contrasted in/visibilities of maintenance work" Maintainers II conference, April 2017, Hoboken (New Jersey). Available at: http://themaintainers.wpengine.com/wp-content/uploads/2021/04/DenisPontilleObjectOntologiesMaintenanceWork.pdf (accessed July 22, 2023).

Denis, Jérôme & Pontille, David. (2018) "The Multiple Walls of Graffiti Removal. Maintenance and Urban Assemblage in Paris" in A.M. Brighenti & Kärrolm, M. (eds) *Urban Walls: Political and Cultural Meanings of Vertical Structures* New York: Routledge pp. 215–235.

Denis, Jérôme & Pontille, David. (2022) *Le soin de choses: Politiques de la maintenance*. Paris: La Découverte.

Denis, Jérôme, Mongili, Allessandro & Pontille, David. (2015) "Maintenance and Repair in Science and Technology Studies" *Tecnoscienza: Italian Journal of Science and Technology Studies* 6(2) pp. 5–15.

Downey, Gary Lee. (1998) *The Machine in Me: An Anthropologist Sits Among Computer Engineers*. New York: Routledge.

Edgerton, David. (1999) "From innovation to use: ten eclectic theses on the historiography of technology" *History and Technology: An International Journal* 16(2) pp. 111–136.

Edgerton, David. (2006) *The Shock of the Old: Technology and Global History Since 1900*. London: Profile Books.

Edgerton, David. (2010) *Technology and Culture*. 51(3) pp. 680–697.

Fisher, Berenice & Tronto, Joan C. (1991) "Towards a Feminist Theory of Care" in Emily Abel & Margaret Nelson (eds) *Circles of Care: Work and Identity in Women's Lives*. Albany: State University of New York Press.

Franssen, Maarten. (2013) "Artifacts" in D. Pritchard (ed) *Oxford Bibliographies Online: Philosophy*. Available at: https://www.oxfordbibliographies.com/display/document/obo-9780195396577/obo-9780195396577-0145.xml (accessed June 10, 2021).

Graham, Stephen. (2010) "When Infrastructures Fail" in Stephen Graham (ed) *Disrupted Cities: When Infrastructure Fails*. New York: Routledge pp. 1–26.

Graham, Stephen & Thrift, Nigel. (2007) "Out of Order: Understanding Maintenance and Repair" *Theory, Culture & Society* 24(3) pp. 1–25.

Gupta, Akhil. (2021) "Infrastructure as Decay and the Decay of Infrastructure" in Ghassan Hage (ed) *Decay*. Durham: Duke University Press pp. 37–46.

Harman, Graham. (2002) *Tool Being: Heidegger and the Metaphysics of Objects*. Illinois: Open Court Publishing.

Harper, Douglas. (1987) *Working Knowledge: Skill and Community in a Small Shop*. Chicago: Chicago University Press.

Henke, Christopher R. (2000) "The Mechanics of Workplace Order: Toward a Sociology of Repair. *Berkeley Journal of Sociology* Vol. 44 pp. 55–81.

Henke, Christopher R. (2019) "Negotiating Repair: The Infrastructural Contexts of Practice and Power" in Ignaz Strebel, Alain Bovet & Phillipe Sormani (eds) *Repair Work Ethnographies: Revisiting Breakdown, Relocating Materiality*. Singapore: Palgrave Macmillan pp. 255–282.

Henke, Christopher R & Sims, Benjamin (2020) *Repairing Infrastructures: The Maintenance of Materiality and Power* Massachusetts: MIT Press.

Higgins, Vaughan & Kitto, Simon. (2004) "Mapping the Dynamics of New Forms of Technological Governance in Agriculture: Methodological Considerations" *Environment and Planning A* 36(8) pp. 1397–1410.

Van den Hoven, Jeroen, Vermaas, Pieter E. & Van de Poel, Ibo (eds) (2015) *Handbook of Ethics, Values and Technological Design: Sources, Theory, Values and Application Domains*. Dordrecht: Springer Netherlands.

Huesemann, Michael & Huesemann, Joyce. (2011) *Techno-fix: Why Technology Won't Save Us or the Environment*. Gabriola Island: New Society Publishers.

Hughes, Everett C. (1971) *The Sociological Eye: Selected Papers*. Chicago: Aldine Atherton.

Jackson, Steven J. (2014) Rethinking Repair. In T. Gillespie, P. Boczkowski and K. Foot (eds.), *Media Technologies: Essays on Communication, Materiality and Society*. Cambridge: MIT Press, pp. 221–240.

Jackson, Steven J. (2016) "Speed, Time, Infrastructure: Temporalities of Breakdown, Maintenance and Repair" in Judy Wajcman & Nigel Dodd (eds) *The Sociology of Speed: Digital, Organizational and Social Temporalities*. New York: Oxford University Press pp. 169–185.

Jackson, Steven J. (2019) "Repair as Transition: Time, Materiality and Hope" in Ignaz Strebel, Alain Bovet & Phillipe Sormani (eds) *Repair Work Ethnographies: Revisiting Breakdown, Relocating Materiality*. Singapore: Palgrave Macmillan pp. 337–347.

Jones, Siân & Yarrow, Thomas (2013) "Crafting Authenticity: An Ethnography of Conservation Practice" *Journal of Material Culture* 18(1) pp. 3–26.

Karakas, Alexandra. (2019) "Maintenance, Function and Malfunction in Technology" *Informácios Társadalom* 19(4) pp. 40–49.

Kroes, Peter & Meijers, Anthonie (eds) (2000) *The Empirical Turn in the Philosophy of Technology*. Amsterdam: JAI-Elsevier.

Kroes, Peter & Meijers, Anthonie. (2002) "The Dual Nature of Technical Artifacts – presentation of a new research programme" *Techné: Research in Philosophy and Technology* 6(2) pp. 4–8.

Lovell, Heather & Smith, Susan J. (2010) "*Agencement* in housing markets: The case of the UK construction industry" *Geoforum* 41(3) pp. 457–468.

Lejeune, Christoph (2019) "Interruptions, Lunch Talks, and Support Circles: An Ethnography of Collective Repair in Steam Locomotive Restoration" in Ignaz Strebel, Alain Bovet & Phillipe Sormani (eds) *Repair Work Ethnographies: Revisiting Breakdown, Relocating Materiality*. Singapore: Palgrave Macmillan pp. 221–251.

Lynch, Michael. (1993) *Scientific Practice and Ordinary Action: Ethnomethodology and Social Studies of Science*. Cambridge: Cambridge University Press.

Mackay, Hughie & Gillespie, Gareth. (1992) "Extending the Social Shaping of Technology Approach: Ideology and Appropriation" *Social Studies of Science* 22(4) pp. 685–716.

Mackenzie, Catriona, Rogers, Wendy & Dodds, Susan (eds) (2014) *Vulnerability: New Essays in Ethics and Feminist Philosophy*. New York: Oxford University Press.

Mattern, Shannon. (2018) "Maintenance and Care" *Places Journal*. Available at https://placesjournal.org/article/maintenance-and-care/ (accessed June 20, 2021).

McHoul, Alex. (1998) "Can Ethnomethodology be Heideggarian?" *Human Studies* 21 pp. 13–26.

Mitcham, Carl. (1994) *Thinking Through Technology: The path between Engineering and Philosophy* Chicago: University of Chicago Press.

Orr, Julian E. (1996) *Talking about Machines: An Ethnography of a Modern Job*. New York: Cornell University Press.

Oudshoorn, Nelly & Pinch, Trevor. (eds) (2003) *How Users Matter: The Co-Construction of Users and Technology*. Cambridge: MIT Press.

Oldenziel, Ruth. (2001) "Man the Maker, Woman the Consumer: The Consumption Junction Revisited" in N. Creager, E. Lunbeck & L. Sheibinger (eds) *Feminism in Twentieth-Century Science, Technology and Medicine*. Chicago: Chicago University Press.

Perzanowski, Aaron. (2022) *The Right to Repair: Reclaiming the Things We Own*. Cambridge: Cambridge University Press.

Pirsig, Robert M. (1974) *Zen and the Art of Motorcycle Maintenance*. New York: Bantam Books.

Preston, Beth. (2009) "Philosophical Theories of Artifact Function" in Anthonie Meijers (ed) *Philosophy of Technology and Engineering Sciences (Handbook of the Philosophy of Science Vol 9)*. Burlington: Elsevier pp. 213–233.

Pursell, Carroll (1995) "Seeing the Invisible: New Perspectives in the History of Technology" *Icon*, 1, 9–15.

Radder, Hans. (2009) "Why Technologies are Inherently Normative" in Anthonie Meijers (ed) *Philosophy of Technology and Engineering Sciences (Handbook of the Philosophy of Science Vol 9)*. Burlington: Elsevier pp. 887–921.

Rosenberg, Nathan. (1982) *Inside the Black Box: Technology and Economics*. Cambridge: Cambridge University Press.

Ramakrishnan, Kavita, O'Reilly, Kathleen & Budds, Jessica (2021) "The Temporal Fragility of Infrastructure: Theorizing decay, maintenance and repair" *EPE: Nature and Space* 4(3) pp. 674–695.

Rubio, Fernando Domínguez. (2016) "On the discrepancy between objects and things: An Ecological Approach" *Journal of Material Culture* 21(1) pp. 59–86.

Rubio, Fernando Domínguez. (2020) *Ecologies of the Modern Imagination at the Art Museum*. Chicago: University of Chicago Press.

Russell, Andrew L. & Vinsel, Lee. (2018) "After Innovation, Turn to Maintenance" *Technology and Culture* 59(1) pp. 1–25.

Russell, Andrew L. & Vinsel, Lee. (2019) "Make Maintainers: Engineering Education and an Ethics of Care" in M. Wisnioski, E. Hintz & M. Kleine (eds) *Does America Need More Innovators?* Cambridge: MIT Press pp. 249–269.

Schubert, Cornelius. (2019) "Repair Work as Inquiry and Improvisation: The Curious Case of Medical Practice" in Ignaz Strebel, Alain Bovet & Phillipe Sormani (eds) *Repair Work Ethnographies: Revisiting Breakdown, Relocating Materiality*. Singapore: Palgrave Macmillan pp. 31–60.

Sharrock, Wes. (1989) "Ethnomethodology" *Berkeley Journal of Sociology* 40(4) pp. 657–677.

Simpson, Ruth, Slutskaya, Natasha, Lewis, Patricia & Höpfl, Heather (eds) (2012) *Dirty Work: Concepts and Identities*. New York: Palgrave Macmillan.

Sims, Benjamin. (2016) "Seismic shifts and retrofits: scale and complexity in the seismic retrofit of California bridges" in Mike Hodson & Simon Marvin (eds) *Retrofitting Cities: Priorities, Governance and Experimentation* New York: Routledge, pp. 13–33.

Spelman, Elizabeth V. (2002) *Repair: The Impulse to Restore in a Fragile World*. Boston: Beacon Press.

Staudenmaier, John M. (1989) *Technology's Storytellers: Reweaving the Human Fabric*. Baskerville: MIT Press.

Strebel, Ignaz, Bovet, Alain & Sormani, Philippe (eds) (2019) *Repair Work Ethnographies: Revisiting Breakdown, Relocating Materiality*. Singapore: Palgrave Macmillan

Suchman, Lucy A. (1987) *Plans and Situated Actions: The Problem of Human-Machine Communication*. Cambridge: Cambridge University Press.

Tenner, Edward. (1996) *Why Things Bite Back: Technology and the Revenge of Unintended Consequences*. New York: Vintage Books.

Tironi, Martin. (2019) "Inquiring and Experimenting with Urban Ecology: Pragmatist Lessons from Public Bicycle Repair" in Ignaz Strebel, Alain Bovet & Phillipe Sormani (eds) *Repair Work Ethnographies: Revisiting Breakdown, Relocating Materiality*. Singapore: Palgrave Macmillan pp. 187–219.

Tronto, Joan C. (1993) *Moral Boundaries: A Political Argument for an Ethic of Care* New York: Routledge.

Tronto, Joan C. (1995) "Women and Caring: What Can Feminists Learn About Morality from Caring?" in Virginia Held (ed) *Justice and Care: Essential Readings in Feminist Ethics* New York: Routledge.

Weisman, Alan. (2007) *The World Without Us*. New York: Thomas Dunne Books.

Verbeek, Peter-Paul (2005) *What Things Do: Philosophical Reflections on Technology, Agency and Design*. Robert P. Crease (trans). University Park: Pennsylvania State University Press.

Verbeek, Peter-Paul (2022) "The Empirical Turn" in Shannon Vallor (ed) *The Oxford Handbook of Philosophy of Technology*. New York: Oxford University Press pp. 35–54.

Vinsel, Lee & Russell, Andrew L. (2020) *The Innovation Delusion: How Our Obsession With The New Has Disrupted The Work That Matters Most*. New York: Currency.

Vinsel, Lee (2019) Researching "What Do Engineers Do All Day?": Innovation, Maintenance and Everyday Engineering. The Maintainers.Org. Available at: https://themaintainers.org/researching-what-do-engineers-do-all-day-innovation-maintenance-and-everyday-engineering/. Accessed August 10, 2023.

vom Len, Dirk. (2016) *The Creation and Development of Ethnomethodology*. New York: Routledge.

Weber, Heike & Krebs, Stefan (2021) "The Persistence of Technology: From Maintenance and Repair to Reuse and Disposal" in Stefan Krebs & Heike Weber (eds) *The Persistence of Technology: Histories of Repair, Reuse and Disposal*. Bielefeld: Transcript Verlag pp. 9–26.

Wittgenstein, Ludwig. (1980) *Remarks on the Philosophy of Psychology*, vol.1. Oxford: Basil Blackwell.

Wittgenstein, Ludwig. (2009) *Philosophical Investigations*, rev. 4th ed. Oxford: Wiley Blackwell.

Young, Mark Thomas. (2021a) "Maintenance" in N. Doorn & D. Michelfelder (eds) *The Routledge Handbook of the Philosophy of Engineering*. New York: Routledge.

Young, Mark Thomas. (2021b) "Now You See it (Now You Don't): Users, Maintainers and the Invisibility of Infrastructure" in M. Nagenborg, T. Stone, M. Woge & P. Vermaas (eds) *Technology and the City: Towards a Philosophy of Urban Technologies*. Springer, pp. 101–120.

Part I

Metaphysics and Epistemology of Maintenance

2 Maintenance and the Humanness of Infrastructure

Steven Vogel

I

An important characteristic of urban infrastructure seems to be its invisibility. It's *infra*, 'below', very often literally, and therefore hidden: our life in cities depends in great measure on the water mains, electrical conduits, fibre optic cables, steam tunnels and so forth that are constantly there working beneath our feet but which we rarely consider. Even when elements of infrastructure are not so literally buried below us (e.g., roads, bridges, sidewalks, cell phone towers, train tracks), we still hardly ever notice them, and instead simply *use* them, in a way that takes them for granted as a part of urban life that seems to take place beyond (or 'above') them. Our life takes place in a certain sense *with* them without as it were ever being *about* them.

They are *tools*, that is to say: they make urban life possible (it certainly would grind to a halt without them) but are not themselves what that life focuses on. In that sense, we could call them 'invisible', although not without some hesitation: we can and do of course *see* them – even the underground ones, sometimes – but in an odd sort of unseeing sight, a vague awareness (like our awareness of our clothes, say, there when we think about them and, of course, there even when we don't – and yet when we don't nonetheless somehow invisible).[1]

I would like in this essay to investigate the nature of this invisibility or quasi-invisibility of infrastructure, this curious hiddenness of that upon which we rely for our lives in cities (which means, for most of us: our lives), asking about what exactly is hidden about it and in what that hiddenness consists. In doing so, I want to raise in particular the questions of what infrastructure actually *is* and what its relationship is to the human beings without whom it would not exist at all: the ones who use it, first of all, but also the ones who build it and finally and most importantly – and here is the connection to the broader concerns of this volume – those who so crucially and necessarily maintain it.

DOI: 10.4324/9781003316213-3

The alert reader with some familiarity with continental philosophy will already have thought of Heidegger and the well-known account he provides in *Being and Time* of what he calls *Zuhandenheit*. These important passages point out that the 'things' with which we most directly have to do in our everyday lives are tools, equipment, *Zeug*: the chairs, screens, lights, etc., that are all being employed right now as I write, as you read, but all with the same character just mentioned of not being noticed, of being as it were taken for granted. It is only when such objects break, Heidegger argues (or get in our way or turn out to be missing), that we suddenly notice them and bring them – typically in frustration or even in panic – into the centre of our attention. Their mode of Being, as he puts it, switches from *Zuhandenheit* to *Vorhandenheit*: from being (as one translation puts it) 'ready-to-hand' to being 'present-at-hand' (Heidegger 2008 ss. 15–6).

It is tempting to use this analysis to talk about the sort of infrastructure that I have been discussing.[2] Infrastructure, one might say, is precisely the sort of *Zeug* that Heidegger has in mind in these passages: reliable, constantly there, but never thematised or examined – the lights that are on that we don't notice, the heat dependably emitted from the radiators, the Wi-Fi we use to get our email, etc. We count on it working without thinking about it, and yet of course, we know it sometimes also fails: the power goes out, the heating system runs out of fuel or the Internet goes down. And everyone knows what happens at such moments: a sharp and sudden switch of attention as we recognise first the very existence of the systems that have failed but, second, also both their vulnerability and indispensability, followed by a concern, indeed a demand, that they be repaired as soon as possible. *Now* the systems are not hidden or invisible at all: they jump to the centre of our notice, as we (often impotently) stare at them and try to figure out what's wrong. This is clearest of all at moments of generalised failure, as occurs, for example, in large-scale blackouts, where it is not simply the electric lines in one's house or under one's feet that one comes to notice but rather the massive grid upon which large regions so systematically depend. Like the *Zuhanden* in general, one might want to say, infrastructure is constantly there, taken for granted and reliable, yet at the same time is invisible to us – and only comes to visibility when it fails.

Heidegger acknowledges as well that 'invisibility' is not quite the right way to describe the way such infrastructure appears or rather fails to appear: the activity of using and manipulating equipment, he says, is 'not a blind one' but rather 'it has its own kind of sight, by which our manipulation is guided'; the name he gives this sight is *Umsicht*, translated (rather misleadingly) into English as 'circumspection' (Heidegger 2008, 98). The objects with which we interact in our everyday use of them are not strictly speaking *absent*, but rather are simply there for us in that use without our remarking on them. *Umsicht* is a special *sort* of 'sight', associated with the kind of skilful coping or unthematised know-how a trained artisan has with respect to her craft – the kind of coping we all exhibit in our abilities to walk, to write with a pen and to use a mouse without having to think at all about what we are doing.

Heidegger's purpose isn't simply to distinguish between two different sorts of sight: the direct staring-at associated with the present-at-hand and the circumspection with which we use everyday tools, but also, importantly, to assert the priority of the latter over the former. That's the real point of his discussion of failure or breakdown: the direct visibility of the present-at-hand only arises when circumspection *fails*, which is to say that it always and only appears *after* and *on the basis of* prior circumspection. Indeed, even as we carefully examine the broken tool by focusing our vision upon it, circumspection remains operative in our use of the *other* tools that must be employed for that focusing to be possible – the flashlight we use to peer into the stubbornly silent car engine, the clothes we wear as we do so or the garage floor on which we stand. The pure theoretical looking associated with situations of breakdown depends upon circumspection, not the other way around – and the objects with which that looking is concerned are, by definition, *broken* and therefore *deficient* objects, useless objects, which is to say they are *not* the sorts of objects with which we normally have to do. Thus, the moment of breakdown does not provide an opportunity to see the tool as it 'really is', shifting it from an unfortunate invisibility into the shining light of visibility; rather in such a moment, we *lose* our grasp of it, failing to understand it as the very tool it is. Indeed, Heidegger writes,

> The *less* we just stare at the [tool], and the more we seize hold of it and use it, the more primordial does our relationship to it become, and the more unveiledly is it encountered as that which it is.
>
> (Heidegger 2008, p. 98 emphasis added)

The broken object that we stare at is 'veiled'; only in our use of it do we see it – circumspectively – for what it actually is.

These last considerations suggest that it would be a misunderstanding to think of situations of breakdown – the blackout, the broken water main, the Internet outage – as somehow providing special insights into what urban infrastructure really is.[3] Rather, in such moments, we see not what it is but rather precisely what it *isn't*: a complex collection of present-at-hand objects connected neither to each other nor to us and doing nothing at all. This Heideggerian insight is an important and valuable one, suggesting as it does that what's 'invisible' about infrastructure (or rather, what's 'visible' only circumspectively) is *not* the mere fact that it *exists* (it's not 'hidden', i.e., in the way that buried treasure might be, or for that matter a clog in a pipe underground that prevents water from getting to a faucet) but rather *what it is* – and in particular its status as something that human beings *use*. Yet, there's something that this Heideggerian discussion misses, something hidden I would suggest from that account as well.

There's a passage in Marx's *Capital* that's striking both for its similarity to the Heideggerian account of breakdown and for its important difference from that account. Marx notes in it that the objects built by workers in a factory are never built *ex nihilo* but rather always out of *other* objects (Marx 1977, p. 289). In order to produce new things, labourers must work both *on* other things (the materials they transform with their labour) and *with* other things (the tools they use to effect those transformations). Those materials and tools are themselves the product of earlier labour (this is so even of what we call 'raw material', Marx points out: coal must be mined before it is burned to provide energy, cotton or corn must be grown before being turned into fabric or ground into cornmeal and even the horses and cattle employed in farm labour are themselves the result of prior processes of domestication). The labourer's activity, that is, depends upon the activity of previous labourers. But in the passage I'm pointing to, Marx adds that this dependence on previous labour is 'irrelevant' to the labour process itself, just as the baker's labour in producing the bread we eat is irrelevant to the process of digestion. It only becomes relevant, he writes, when something about the labour process goes awry: it is only

> by their imperfections that the means of production in any process bring to our attention their character of being the products of past labor. A knife which fails to cut, a piece of thread which keeps on snapping, forcibly remind us of Mr. A, the cutler, of Mr. B, the spinner. In a successful product the role played by past labor in mediating its useful properties has been extinguished.
>
> (Marx 1977, p. 289)

The resemblance of this last idea to the Heideggerian one about breakdown shattering the 'invisibility' of the *Zuhanden* is clear. But what's revealed in the moment of breakdown, according to Marx, isn't the *vorhandene* object at which we simply stare, but rather the *worker* whose prior acts of production show themselves as having been incompetent. I want to focus on this idea – that what's 'hidden' about our tools is not so much their status as ready-to-hand as the fact that they are *built by human labour*. I'd suggest that this is true not simply when the tools are themselves being used in the sorts of commodity-building labour processes Marx is concerned with in *Capital*, but rather (nowadays) in *all* our use of tools, which is to say – and here Heidegger is perhaps more helpful than Marx – in all our activities in general. As I raise the hammer or wear the clothes or turn on the faucet, I am depending not simply on the reliability of the tool I employ but at what is really a deeper level on those other humans whose labour made it possible for that tool – and therefore that reliability – to come to be. In the city, in particular, it's not simply the specific tool I might be using but rather an entire system of infrastructure upon which I am depending: the wires, the pipes, the roadways, the buildings and the sidewalks that form the taken-for-granted environment within which my life takes place and which indeed makes it possible: all of these are built by human labour.

Yet, there's something wrong with Marx's account too. The idea that a knife that fails to cut reminds us of 'Mr. A, the cutler' seems to refer back anachronistically to a world of individual craft production, where individual cutlers (with various degrees of skill) sell their wares to neighbourhood buyers. Nowadays, the labour process that produces a knife is almost unimaginably more complex, and the labourers involved in that process are too multifarious to even start to consider: those in the factory who produced the blade, others who built the handle, others who assembled the final knife, others who packaged it (and those who made the packaging) not to speak of those who transported it over what are likely vast distances to the store at which it's purchased or the warehouse from which it is delivered to the user's home. We're *not* reminded of all these people, under current conditions of production and distribution, and so under these conditions, Heidegger is right, although for reasons that might be more historically contingent than ontological: when it fails we think instead of the thing in front of us, the useless knife, that frustrates us and leaves us angrily looking about for another one (or ordering it from Amazon). We blame, if one could put it this way, the thing itself and not the human process by which it was (incompetently) produced.[4]

The point I'm trying to make about Heidegger and Marx, it will be clear, is a complicated one. Heidegger is right, I want to say, to criticise the traditional view whereby the world is made of present-at-hand objects (let's call them Things) that exist as static objects independent of us at which we stare, but he's wrong – or rather he does not go far enough – when he says that what we mostly have to do with in our everyday experience (but which are hidden in that experience or seen only circumspectively) are ready-to-hand 'tools' because he fails to notice that those tools themselves are almost always *built* ones. *What's hidden isn't their equipmentality but their builtness* – or rather, the fact that their equipmentality is made possible *by* their builtness. To put it another way, he fails to notice the role of *concrete physical activity* – in Marx's term, labour – in the everyday experience of Dasein that he describes in Sections 15 and 16 of *Being and Time*. Tools aren't just things we 'encounter' (*begegnen*) but rather things we *use*, and that use itself is a *physical action*. Furthermore, and equally importantly, the tool itself is the product of *prior* physical action, which action in turn made use of *prior* tools. What we fail to notice in our use of tools – or what we notice only circumspectively – is the role of *human labour* in their coming to be. It is this replacement of concern for and recognition of human labour that builds the world around us with a view of that world as consisting merely of Things, I am trying to suggest, that explains what is hidden in infrastructure.

II

Infrastructure, I'm arguing, is best thought of not as a Thing, not some present-at-hand object or group of objects or system of objects that exist somehow independently of human activity. Rather it is something that we *use* and that we have *built*. Not a Thing nor the motion of a Thing nor a tool-Thing that humans manipulate to achieve various ends, rather *it is itself a human doing*, a form of human practice. And if this is so, then its hiddenness will not be the hiddenness of a Thing either. A Thing is hidden when it is occluded by another one and is thereby prevented from being (directly) seen. But what's hidden about infrastructure is exactly its non-Thingness, its status as a human doing – as something humans use and have built. (In *this*, literal, sense – and only this one – we could say that infrastructure is a 'social construct'.) To overcome this hiddenness would be to understand that status, to see infrastructure in its humanness and its builtness and to stop seeing it as an environment that, although in principle man-made, nonetheless appears as something separate from and independent of us and our use of it.

If what's hidden in infrastructure is its humanness, its role as a human doing, it follows that it need not be 'invisible' in the ordinary sense – need not be, for example, literally underground. The sort of hiddenness I mean can be operative, for instance, in elements of infrastructure that seem quite visible, even monumental. The George Washington Bridge that spans the Hudson River and connects Manhattan with New Jersey is in no ordinary sense of the word something hidden: rather it is a visible marvel, a piece of sublime architecture that serves as an exemplar of human imagination and ingenuity and the human ability to use reason to increase human mobility in new and extraordinary ways. But whether we stare at it (as present-at-hand) in admiration from afar or simply drive over it (as ready-to-hand) during our regular commute, we rarely if ever notice or consider those who actually *built* it – those thousands of people without whose hard and dangerous work it would never have come to be. We know the bridge to be built, of course – we don't think of it as a piece of 'nature' – but neither its visibility (when we stare at it) nor its reliability (when we drive across it) in any way reminds us of its builtness. It is in this kind of fact, I think, that the real hiddenness of infrastructure consists (Figure 2.1).

When it opened in 1931, the George Washington Bridge was the longest bridge in the world. Construction had taken four years and had been astonishingly complex.[5] Foundations had to be built into the Hudson River on each side, which required (on the New Jersey side) two steel cofferdams to be constructed and then positioned underwater. When one of the cofferdams failed, drowning several workers, a third had to be built to replace it. A total of 32 tower footings were bolted to the foundations and concrete columns more than 600 feet high were floated down the river on barges and set on top of each footing. Anchorages were built on land to anchor the ends of the suspending cables; the cables themselves had to be manufactured, strung over the river in a difficult and dangerous set of operations and then fed over the towers and attached to gigantic eyebars in the anchorages. Suspender cables were then hung down vertically from the main cables. Only after all this was the roadway itself added to the bridge. It had been built offsite, and like the foundation columns, had to be brought downriver on barges to a position just under the bridge, at which point cranes were used to lift it into place (Rockland 2008, pp. 53–62).

I have been able to find no record of how many people worked on the bridge to build it, although we do know that at least 12 people died during

Figure 2.1 The George Washington Bridge.

the process. Still one might simply try to imagine how many people must have been involved: those needed to construct the cofferdams, to build the foundations, to fabricate the tower columns, to place them on the barges, to pilot those barges, to construct the anchorages, to manufacture the cables, to build and then operate the machines that strung them, to check their placement, to attach the suspender cables, to assemble the roadway – and, of course, one could go on, thinking about the painters, the electricians, the security guards, the safety inspectors, plus those who built the barges, built the eyebars, built the hardhats the workers wore to protect themselves, cooked the meals they ate at the end of the day and sewed and repaired the clothing they wore. Then there were the lawyers, the accountants, the architects and furthermore those who cleaned their offices, who typed their contracts, who filed their documents, who answered their telephones, and so forth. The numbers must surely have run into the hundreds of thousands. But although the labour of those myriads of workers is in the bridge – one might even say in a way it *is* the bridge – it remains, and they remain, *hidden*, as we look at it or drive across it: unseen, unthought about and invisible (Figure 2.2).

Of course that the George Washington Bridge was built, and was built by humans, is in one sense obvious, as I have already said – no one would deny it. Yet, in our everyday experience of it, we rarely give any thought at all to the actual people who built it and the actual labour in which they engaged to do so. Those whose work caused the bridge to exist remain in

Figure 2.2 George Washington Bridge Construction Workers.

this sense hidden to us. In addition, there is an important temporal character to this hiddenness: when we experience the bridge (again, either the present-at-hand bridge at which we gaze or the ready-to-hand one upon which we drive), we experience it as a *contemporary* fact, something in front of us *now*, while on the rare occasions when we consider its builtness the labour involved appears, if it appears at all, as a *historical* fact, memorialised occasionally on anniversaries or with plaques, mentioned briefly on the bridge website and written about in books by those with a professional or an enthusiast's interest in the bridge's story (but the plaques, actually, are few; the website is more focused on current traffic conditions than on bridge history and the books seem much more interested in the bridge's physical characteristics – the massiveness of its columns, the strength and number of its cables – than in those who actually did the work). Yet, to treat the building of the bridge as something that *happened*, something that took place once in the past but is now finished, is still to treat the bridge itself as a Thing – not a Thing of nature, to be sure (because it was made by humans), but an artificial Thing and so a Thing nonetheless. To treat it this way, I want to suggest, is still to misunderstand what building is and what the role of human beings and human labour is in producing and making possible ('socially constructing') the world in which we live.

Infrastructure is not built once and then left alone. Its character as a human doing is not a fact simply of its past but also, crucially, of its present.[6] The labour involved in constructing the George Washington Bridge is not simply something that took place in a heroic past of bridge-building, but rather is *ongoing*. Just like every piece of urban infrastructure, whether aboveground or below, the bridge requires constant maintenance. Dozens if not hundreds of people work on the bridge daily.[7] A team of painters is always at work somewhere on the span, engaging in work that is as important as it is dangerous: rust and corrosion are a constant threat and painting helps to prevent them (Posse & DiLella 2018).[8] Structural maintenance workers are constantly inspecting elements of the bridge for signs of failure, using various specialised trucks and other machinery (Ascher 2005, p. 50). Rivets need to be replaced regularly, concrete needs to be repaired, expansion joints need to be emptied of debris and lane markers need to be repainted (Weisman 2007, p. 35). Linseed oil must be periodically injected into the cables to keep them pliable and resist rust (Rockland 2008, p. 22).

In what follows, I want to focus on *ongoing maintenance* as a key and often overlooked element in infrastructure.[9] Ongoing maintenance and repair are a crucial part of all urban infrastructure; without such activities, the infrastructure would quickly collapse.[10] The water mains underneath New York must be regularly inspected for leaks, clogs, faulty valves and bad gaskets and when found each must be repaired or replaced. New York's subway system, which is actually located beneath the water table, is constantly threatened with flooding (as happened catastrophically during Superstorm Sandy in 2012). More than 700 pumps operate night and day to keep it dry, removing 13 million gallons of water *a day* (Weisman 2007, p. 24), while hundreds of workers are constantly inspecting for leaks and plugging those that are found, with more than 150 more employed simply unclogging drains in the system – and, of course, the pumps themselves require regular maintenance, oiling and repair. A 'geometry car' travels daily around the tracks employing complex instrumentation and numerous analysts to look for dangerous changes on the tracks (Estes 2014). Weeds such as the ailanthus tree that sprout in subway tunnel walls and ceilings have to be noticed and removed before they grow so large as to threaten those tunnels' structural integrity (Weisman 2007, p. 26). The city's underground electrical infrastructure requires constant inspection and frequent low-level repairs; since both inspection and repair typically mean some segment of the network must be taken offline, the problem of scheduling such work and using redundancies in the system to avoid local loss of service turns out to be both an important and computationally quite difficult one. Under New York is also a complex system of tunnels for the distribution of steam, used for heat in many buildings but also for other services as well, and it too must be monitored on a 24-hour basis; valves in the steam tunnels that filter out condensed water (which can cause explosions when it comes into contact with the hot steam) have to be inspected regularly and replaced annually (Brown 2018).[11] The manholes workers use to perform these tasks themselves have to be kept clear of debris and water (Ascher 2005, p. 186). The need for constant monitoring and maintenance of the various nodes of the Internet is, of course, well known – and not always very well carried through.

It's important to resist thinking of 'building' (in the past) and 'maintenance' (in the present) as two distinct sorts of activities. Rather they are continuous: building takes account of future maintenance and maintenance itself involves building. The George Washington Bridge's design from the beginning included catwalks whose only purpose was to allow workers to access various parts of the structure so as to make future inspection, maintenance and repair possible. And when such activities take place today, they cannot help but change and therefore (re)build the bridge: a massive repair project currently under way, for instance, will involve among other things replacing all of the suspender cables.[12] The bridge was originally built with the thought that someday a second, lower level would be added, and was designed to allow such an addition, which finally took place in the early 1960s: did that addition count as building, or as repair (of something missing) or as necessary maintenance to prevent the legendary traffic jams the upper level was experiencing? No piece of infrastructure is ever complete; it's always 'under construction'. Design, building, maintenance and repair are all parts of a single process – a process in which human beings work on the world around them to attempt to shape elements of that world into forms that then make up the environment that they inhabit and that help to structure their lives. This is the sense in which I want to call urban infrastructure a (ongoing) human doing, rather than treating it as a once-built but now independent Thing.

It is a human doing in another sense too – one that again is superficially obvious but has implications that are rarely acknowledged. For if we say that infrastructure (and tools in general) comes to be what it is in its use, we need to notice that 'use' is always *active*, always a physical doing. Even we 'end-users' of infrastructure, the ones who take it for granted, treat it as simply and reliably there without noting it, fail to recognise its complexity and its builtness and the ongoing maintenance needed to keep it functioning – even *we* have to act, to do: the infrastructure we use does not function entirely on its own. In order for water to flow from the faucet, I have to turn the handle; in order to get light in the room, I have to flip the switch; in order to change the temperature of my home, I have to adjust the thermostat. These actions are small ones – and, of course, a significant part of the benefit infrastructure brings is the way it gears down, so to speak, the amount of work the user must perform to achieve a task, turning the handle on a sink being much less difficult than going down to a stream to fill up a jug and then bringing it back to one's home – but they are actions, doings, all the same. We users engage in maintenance and repair as well: recharging our phones, washing clothes, replacing lightbulbs and unclogging drains.

Thus building, maintenance, repair and (end-) use, although analytically distinguishable, blur into each other in practice. The crucial point is that *every step in the functioning of infrastructure involves human activity*: this is what I mean by calling it a 'human doing'. I've emphasised the importance of recognising the builtness of infrastructure, but because building turns out to be continuous with maintenance and repair and even with use we can say that building actually never ends, but is rather ongoing. Indeed, 'building' might just be a word for human activity in the world – I am tempted to say here simply that building is *what we do*. Humans are active ('doing') creatures, who transform the world around us at every moment whether this be through constructing new items in that world, working to change the items that exist or using those items to make or change other ones. So the things we build – our infrastructure or one might say our environment in general – are always *being built*, always under construction.

III

Maintenance as an ongoing activity (and one often carried out by someone other than the end-user) fits only uneasily into the Heideggerian schema of readiness-to-hand and presence-at-hand. This is true first of all because as I drive across the bridge its readiness-to-hand, its 'withdrawal' from my attention, turns out to depend upon the bridge maintainers' direct attention to it as the present-at-hand object of their work.[13] *They* have to work on it in order for it to be able to withdraw from *me*. But, second, even that latter presence-at-hand is only imprecisely described as the result of failure or breakdown because of course it's exactly in order to *prevent* breakdown that the work of maintenance takes place. Even what could be called repair – the removal of debris from the expansion joints on the bridge roadway, say – may be so common as to be planned for and scheduled, suggesting that it occurs not in response to failure but rather as something expected, and so merely an element of regular maintenance. And of course, if the failure is *expected* it starts to be unclear in what sense it can be said to be 'failure' at all. One might ask whether the rusting of the steel, against which so much of the activities of bridge maintenance are directed, is correctly described as a *failure* of the steel – or is it simply part of what steel *is*, part of what it is for steel to function as a tool? (What would steel that did *not* rust be like? It surely wouldn't be *steel*.)

In truth, the necessity of ongoing maintenance suggests that 'failure' is not an occasional event interrupting the otherwise successful functioning of the equipmental totality within which our day-to-day life takes place, but rather is in a certain way constant, indeed is *part of what it is to be equipment*. It is a fact of the world – one that engineers and architects are fully aware of but that often escapes the notice of philosophers – that *things break*, and are always breaking. Pipes leak, walls crack, fasteners crumble, adhesive bonds weaken and joints split apart. Wires overheat. Weeds grow. Electrical circuits short out. Software faces situations its designers never anticipated. These are not unfortunate and surprising accidents but are rather intrinsic characteristics of the real world (arguably consequences of the second law of thermodynamics). To build something in the world is to create something that from the very start is already collapsing, crumbling, falling apart, succumbing to entropy – already, that is, 'failing'. Again, that's part of what building *is*: it's not something that building 'confronts' or is 'constrained' by. The necessity for constant maintenance of the things that we build follows directly from this fact and means as I have suggested that building and maintaining are not distinct activities.[14]

When we think of a piece of technology as a 'tool' that we 'use', we tend to imagine it as an independent item in the world that the user manipulates to achieve some particular goals and one that in the ordinary case does so smoothly, flawlessly and invisibly. But what I have been arguing is that this apparent smoothness – the 'withdrawal' of the ready-to-hand – is itself a *difficult (human) achievement* and one the production of which requires an enormous amount of work: both that needed to construct the tool in the first place but also crucially the ongoing work of maintenance required to keep it functioning normally. I drive over the bridge without incident; but that drive would not be possible were it not for the work of those who paint the bridge, inspect and repair the cables, inject the linseed oil, etc. Indeed, as already mentioned, my own relationship with the tool is not free of work either, and so not *entirely* smooth: as I cross the bridge, I have to engage in an effort too, even if it's merely a matter of pressing my foot down on the accelerator or maintaining my grip on the steering wheel. *These too are forms of maintenance*, responding to the tendency of the automobile on its own to 'fail' by moving in directions other than I intend or by succumbing to friction by stopping entirely.[15]

But at this point 'failure' no longer seems like the correct word. To build an object in the world, whether it be a bridge or an automobile or a sewage system, is to produce something subject to the laws of physics, including those having to do with friction, momentum, entropy and so on. That such objects obey those laws and thus are inevitably and from the very beginning always breaking, crumbling and collapsing is simply a fact about what it is to be an object in the world: why would one call this a 'failure'? (What would it be to be a 'success'? *Not* to be subject to the laws of physics? But that would mean: not to be real!). It is a fact of building that the objects we build never fully do what we expect them to – they never look nor function nor age the way we intended; they always have surprising qualities and always produce results we did not anticipate. 'Failure' in this sense does not occur occasionally or only when our building somehow goes awry: these rather are simply unavoidable facts about what it is to build and about what it is to be a real object in the world.[16]

That the objects we deal with are *real* means that they never fully do what we expect, but it also means that our dealings with them always involve activity, 'doing', work on our part. The world where effort is unnecessary, where what we build is exactly what we want and where failure never occurs and so maintenance is never needed is the world of fantasy; in the world of reality, our doings are always difficult, inevitably incomplete and never-ending. Building cannot help but require maintenance and repair; maintenance and repair are themselves kinds of building; and even 'use' requires physical action, which is to say effort, labour.

To recognise the builtness of everything around us in the urban world – where building is understood to involve maintenance and use as well – is to recognise, perhaps paradoxically, that everything around us escapes our control and that our labour does too. To say we build the world around us is not to say that we control or 'dominate' it, although surely we try to do the best we can to produce the results we wish our actions to have. That lack of control is evidenced, of course, when tsunamis destroy nuclear power plants or blackouts bring down large cities or regions, but it is evidenced too by the fact that we cannot get water from our faucets without turning the knobs and cannot magically produce light from our lamps just by imagining it.[17] We have to *act* in the world to do anything, and to act in the world is immediately, and constantly, to be subject to forces beyond our control.

One might be inclined to find in these last considerations the justification for the idea that there is in the world something beyond or beneath or before human action whose 'otherness' from that action (and from us) we need to recognise and before which a certain humility is required: something that *produces* the difficulty of work, the inevitability of collapse and the constancy of 'failure' in our buildings. I have been arguing that what's hidden in infrastructure – and indeed what is hidden in our relationship with our environment in general – is its humanness, which is to say its character as the product of human labour. But perhaps something else needs to be acknowledged as well: the moment of resistance that our labour faces, the entropic and unexpected elements that I'm insisting arise in all our buildings and that explain the need for constant maintenance. Perhaps, it could be argued, that moment deserves a name like 'nature', meaning something independent of the human, a fundamental otherness to the world.

Yet, this would be mistaken, I think, and would evince a misunderstanding of what labour is. For the concept of labour *contains within itself* this moment of 'resistance': the latter is not something that labour *faces* or *responds to* but rather it is part of what labour *is* – which means in turn that that moment is not correctly described as a moment of 'otherness' at all nor as something beyond the human. To work is to do something *in the (real) world*, as opposed to thinking about it or planning it or desiring it. Those latter things might be said to 'face' resistance or otherness, but work does not. To ask about the 'source' of the difficulty humans encounter when they interact with their environment, and then to identify that source with nature or with otherness, is to view humans as at first somehow distinct from that environment and then *only subsequently* 'encountering' it – but to do this is once again to divide humans from the rest of the world and to imagine them as (at first) outside it. Such a view subtly betrays the sort of anthropocentrism that it thinks of itself as leaving behind. Rather we are *in* the world, always already, and so our labour in the world does not 'encounter' resistance or difficulty but rather simply *is* difficult. The infrastructure we build 'fails', and so requires ongoing maintenance, not because a nature separate from us sadly always thwarts us, but because our building of infrastructure – our labour – is *real* and takes place within the real world. That there is no confrontation with 'otherness' here becomes clear if one starts to think about what labour would be *without* this moment of so-called resistance – how could there be labour without effort, without doing something *in the world*? When I turn the knob on the faucet to get the water, just as when I place the jug in the stream for the same purpose, I don't face anything 'other': doing things like that is *all I do* in the world – it's the most familiar, non-other, sort of thing there is. And so the claim that infrastructure is a human doing, precisely because the 'doing' here is meant as *real* doing, as *labour*, does not require supplementation by the notion of a 'nature' independent of or standing below or prior to the humanly built world. Dualism needs to be rejected *all the way down*, including the dualism between human activity and an 'otherness' such activity is said to 'encounter'.

IV

I want to conclude with some brief remarks about those who engage most directly with the infrastructural labour I have been discussing: the workers themselves, that is, who are constantly building, rebuilding, maintaining and repairing infrastructure. Among the benefits infrastructure provides, of course, is a reduction in the effort involved in various daily tasks – I've used the example above of getting water from a faucet rather than from a stream. But if one counts the labour involved in constructing the water system itself – the aqueducts, the pipes, the pumps, etc. – and adds to it the maintenance and repair work constantly required to keep all the elements of that system running as well, one realises that part of the labour saved has simply been shifted elsewhere, away from the end-user to others. The building of infrastructure does not only add efficiency or save labour with respect to achieving various human goals but it also serves as a means of organising and distributing the labour needed to achieve those goals – which is why the question of what infrastructure to build (and how to maintain it) is a *political* one. Surely, furthermore, the hiddenness of the human labour involved in the building, maintenance and repair of infrastructure that I have been discussing is not unconnected to the relative invisibility in our society of those who actually engage in that labour – labour that is generally poorly paid, not well respected, associated with marginal populations, etc.[18] The hiddenness of infrastructure and the hiddenness of those workers are two sides of the same phenomenon.

But there is no necessity to either side, no intrinsic reason why infrastructure's builtness or why those whose labour in fact builds it must inevitably be hidden. Rather it follows from certain facts – political ones – about the social arrangements under which we live. One could also imagine alternative systems where those whose ongoing work makes it possible for the rest of us to enjoy heat, water and electricity and travel over bridges with relatively low expenditures of labour would be highly honoured by society, or one where that work was shared among all capable citizens. Such systems would be ones where infrastructure would not be so hidden (in the sense I have been discussing), because the fact that the building and maintenance of our environment is always a *social* effort, requiring the work of many different members of the community, would be a matter of everyday awareness and acknowledgement. The system we actually live under, of course, is not like this, but rather one where the built environment we inhabit appears as the result of a series of individual projects, each of which is engaged in for private monetary gain: there is little occasion under that system for a sense of social solidarity to develop that might make it possible for our deep reliance on each other's labour to become something we daily notice and appreciate, and that might therefore also make possible an overcoming of what I have been describing as infrastructure's hiddenness.

Still, there is a sense in which the very workers I have just mentioned – the ones who build the infrastructure we depend upon and whose ongoing work to maintain and repair that infrastructure is crucial to its continued functioning – are in what might be called a privileged position with respect to the possibility of such an overcoming of infrastructure's hiddenness. Whereas for the rest of us infrastructure does function as something simply there, simply reliable and barely noticed – the taken-for-granted background condition that makes possible the ordinary activities of our own daily lives (which is to say, of our own labour) – for those who actually work directly on that infrastructure itself, that simplicity can likely never seem so simple. For them, the fact that infrastructure is humanly built and humanly maintained, and furthermore is constantly subject to 'failure' and in danger of full-scale breakdown, is quite obvious: infrastructure is not hidden from *them*. Nor too can they fail to notice that their fellow citizens, as they drive across bridges or turn on lights or enjoy the heat in their homes, depend upon the labour of others in order for their own lives and work within the city to be possible. One might therefore find in those infrastructural workers themselves – poorly paid and marginalised as they often are, in many cases literally hidden underground in their work, and rarely noticed by those whose lives they so significantly support – a group with a kind of special knowledge of the real character of the world we all inhabit, a knowledge to which we might be well advised to pay attention.[19]

Notes

1 And this is so even for that important element of urban infrastructure that does require citizens to venture – eyes open – underground, the subway (sometimes named the Underground itself), seen and heard and smelled for sure and yet somehow rarely itself the object or goal of their activities. They are focused, one might say, on what will happen when they reach their station and emerge above the Earth, but hardly ever on the fact that *right now* they are travelling under that 'ground' on top of which their ordinary lives take place and which they normally take to be solid.

2 See, for example, Star (1999). For this citation and an introduction to some of the issues connected with the maintenance of infrastructure, I'm indebted to Mark Thomas Young and his paper 'Now You See It (Now You Don't): Users, Maintainers and the Invisibility of Infrastructure' (Young 2021).

3 There is a tendency to do this, for instance, in various studies of infrastructure in the philosophy of technology and urban studies. Young makes this point in Young (2021, pp. 103–5). Thus, Stephen Graham writes in the introduction to his edited volume *Disrupted Cities: When Infrastructure Fails* that 'studying moments when infrastructures cease to work as they normally do is perhaps the most powerful way of really penetrating and problematising those very normalities of flow and circulation to an extent where they can be subjected to critical scrutiny' (Graham 2010, p. 3). The 2003 blackout in the Eastern United States and Canada is often used as an example – see, for instance, Timothy Luke's essay 'Power Loss or Blackout' in the same volume (Luke 2010) or Jane Bennett's *Vibrant Matter* (Bennett 2010, esp. chap. 2).

4 There is a name for this process, in the history of Marxist thought: the fetishism of commodities. Georg Lukács called it reification (*Verdinglichung*), the process whereby human labour appears in the form of apparently independent Things. See his *History and Class Consciousness* (Lukács 1971).

5 See 'George Washington Bridge Across the Hudson River at New York, NY' reprinted from the *Transactions of the American Society of Civil Engineers* vol 97 (New York: The Port of New York Authority, 1933), as well as Michael Aaron Rockland, *The George Washington Bridge: Poetry in Steel* (New Brunswick, NJ: Rivergate Books, 2008).

6 In *The Poverty of Philosophy*, Marx writes that for bourgeois economics 'there has been history but there is no longer any' (Marx 1973, p. 105).

7 I have been unable to find anything like an exact number which in itself indicates how little awareness or interest this sort of labour excites.

8 It is however an urban myth – which is also apparently told about the Golden Gate Bridge and the bridge over the Firth of Forth in Britain – that painting begins at one end of the bridge and then when completed at the other simply starts all over again.

9 See Young (2021, pp. 109–110). Stephen Graham and Nigel Thrift in their essay 'Out of Order: Understanding Repair and Maintenance' (2007, p. 10) speak of 'the remarkable neglect' in contemporary discussion of urban infrastructure 'of the massive and continuous work that is necessary to sustain' it. The current volume is a welcome attempt to overcome this neglect.

10 Alan Weisman's strange and moving book *The World Without Us* (Weisman 2007) describes in startling detail what would happen if human beings suddenly one day simply disappeared, which means among other things the end of that ongoing maintenance: the results would not be pretty.

11 In the middle of the last century, a series of billboards in Times Square advertising cigarettes used the steam from this system to provide realistic-looking puffs from the cigarettes of larger-than-life smokers.

12 See 'Restoring the George: 10 Year Capital Plan Construction Program', *Port Authority NY NJ*. Available at: https://www.panynj.gov/bridges-tunnels/restoring-the-george.html (accessed June 15, 2022).

13 And note that Heidegger actually never points out that work always does have such an object – because Heidegger, unlike Marx, never really focuses on work itself. Despite all that hammering in *Being and Time*, it's never really made clear what is *being built*.

14 See, along these lines, Steven J. Jackson's essay 'Rethinking Repair' (Jackson 2014) which argues for what he calls 'broken world thinking', a thinking that he describes as what happens 'when we take erosion, breakdown, and decay ... as our starting points in thinking through the nature, use, and effects' of things like infrastructure (Jackson 2014, p. 221). He suggests, as I would too, that such an approach might lead us to 'a deep wonder and appreciation for the ongoing activities by which stability ... is maintained', leading to a kind of 'admiration and even reverence [for] the sheer magnitude of the work represented in the ongoing maintenance and reproduction of the established order' (Jackson 2014, p. 222). See also Graham & Thrift (2007, pp. 5–6).
15 And this is true of 'self-driving' cars as well, which continue to require human drivers prepared to take over when (inevitably) 'failure' threatens.
16 This holds for so-called 'digital' objects too, which are often mistakenly discussed as if they were non-physical. 'Software' is only *relatively* soft: it takes the form of a physical pattern of transistors and resistors conducting currents in various complex ways, and it only functions as it does (and 'fails' when it does) because of something about that pattern.
17 Yes, 'smart' homes are making it possible to turn lights on remotely, or to program them to turn on 'by themselves': but *still* work has to be done to make those things happen. Nothing happens in the world that doesn't involve – well, something happening in the world.
18 See Young (2021, p. 12) and Graham and Thrift (2007, p. 18).
19 Jackson, for instance, talks of the idea of something like a 'standpoint epistemology of repair'. See Jackson (2014, p. 10).

References

Ascher, Kate (2005) *The Works: Anatomy of a City*. New York: Penguin Press.

Bennett, Jane. (2010) *Vibrant Matter: A Political Ecology of Things*. Durham: Duke University Press.

Brown, Nicole. (2018) 'NYC steam system: Why we have it and how Con Ed maintains it' *AMNY*. Available at: https://www.amny.com/news/nyc-steam-system-1.20146953/ (accessed on April 27, 2022).

Estes, Adam Clark. (2014) 'This Superheroic Train Keeps New York City's Subway Safe' *Gizmodo*. Available at: https://gizmodo.com/this-superheroic-train-keeps-new-york-citys-subway-safe-1571987376 (accessed June 20, 2022).

Graham, Stephen. (2010) 'When Infrastructures Fail' in Graham, S (ed) *Disrupted Cities: When Infrastructure Fails*. New York: Routledge. pp. 1–26.

Graham, Stephen & Thrift, Nigel. (2007) 'Out of Order: Understanding Repair and Maintenance', *Theory, Culture and Society* 24(3) pp. 1–25.

Heidegger, Martin. (2008) *Being and Time*, translated by J. Macquarrie and E. Robinson. New York: Harper Perennial.

Jackson, Steven J. (2014) 'Rethinking Repair' in Gillespie, T., Boczkowski, P. & Foot, K. (eds) *Media Technologies: Essays on Communication, Materiality and Society*, Cambridge, Mass.: MIT Press. pp. 221–240.

Lukács, Georg. (1971) *History and Class Consciousness*, translated by Livingstone, R. Cambridge, Mass.: MIT Press.

Luke, Timothy W. (2010) 'Power Loss of Blackout: The Electricity Network Collapse of August 2003 in North America' in Graham, S (ed) *Disrupted Cities: When Infrastructure Fails*. New York: Routledge. pp. 55–68.

Marx, Karl. (1973) *The Poverty of Philosophy*. Moscow: Progress Publishers.

Marx, Karl. (1977) *Capital Volume 1*, translated by Fowkes, B. New York: Vintage Books.

Posse, Erica & Dilella, Chris. (2018) 'These workers have one of the scariest jobs in the world' *CNBC Make it*. Available at: https://www.cnbc.com/2018/06/20/this-crew-of-bridge-painters-has-one-of-the-most-dangerous-jobs-in-the-world.html (accessed on 28th April 2022).

Rockland, Michael A. (2008) *The George Washington Bridge: Poetry in Steel*. New Brunswick: Rutgers University Press.

Star, Susan Leigh. (1999) 'The Ethnography of Infrastructure', *American Behavioral Scientist* 43(3) November/December pp. 377–391.

Weisman, Alan (2007) *The World Without Us*. New York: St. Martin's Press.

Young, Mark Thomas. (2021) 'Now You See It (Now You Don't): Users, Maintainers and the Invisibility of Infrastructure', in Nagenborg, M., Taylor, S., Woge, M., & Vermaas, P. (eds.), *Technology and the City: Towards a Philosophy of Urban Technologies*. Cham: Springer Nature Switzerland. pp. 101–119.

3 Technology in Process

Maintenance and the Metaphysics of Artefacts

Mark Thomas Young

In this chapter, I aim to demonstrate that despite the breadth of existing work on the metaphysics of persistence, we're yet to properly understand the relationship between technology and time. In order to address this problem, I'll argue for the need to begin exploring a topic which to date has attracted little attention from metaphysicians: maintenance. However, as I'll aim to show, doing so requires us to flip the script in the metaphysics of artefacts, by taking leave of the substance ontology which has traditionally provided the framework for western metaphysics.

Before we begin, it is worth clarifying what I mean by substance ontology because there are at least two senses in which philosophers use the term. The first is used to refer to a class of approaches within ontology that employ philosophical conceptions of substance. According to this definition, substance ontology was bequeathed to the western philosophical tradition by Aristotle and has remained dominant to the present day where it currently exists in different variations.[1] The second sense in which philosophers use the term substance ontology is broader and applies less to specific theoretical commitments than it does to a set of implicit presuppositions guiding ontological analysis. According to this second sense, substance ontology is more properly understood as a research paradigm, one which includes theories that not only predate Aristotle but which often don't involve philosophical conceptions of substance at all (Seibt 1997; Seibt 2004). It is this latter conception of substance ontology in particular, which is often contrasted with rival ontologies which understand change and process to be fundamental. Central among the assumptions which unite the different theories that are considered to fall under the research paradigm of substance ontology is the idea that at the most fundamental level, reality consists of static, enduring entities (Rescher 1996 p. 29; Dupré 2018).

Increasingly, philosophers have been drawing attention to the way in which the paradigm of substance philosophy forms an implicit framework that lurks beneath the apparent diversity of philosophical thinking in the western tradition. According to Johanna Seibt, for example, because the

DOI: 10.4324/9781003316213-4

principles of substance ontology are located within the 'presuppositional depth structure' of debates in metaphysics,

> they have hardly received any attention so far. Even those contemporary ontologists who drop the traditional talk about substances still import the presuppositions of the traditional paradigm; even in ontologies whose basic entities are events, tropes, or state of affairs, the presuppositions of substance ontology are still operative.
>
> (Seibt 1997)

As Anne Sophie Meincke points out, theories which are often viewed as standing in opposition to one another, such as Democritian Atomism and Aristotle's account of substance, for example, or even the more contemporary metaphysical debate between Perdurantism or Endurantism, nevertheless all share a commitment to the central insight of the substance philosophy paradigm – that reality is fundamentally composed of static enduring entities (Meincke 2019, 2021). For philosophers concerned with uncovering the research paradigm of substance ontology, the goal is very often to show how, by forming an implicit framework within which discussions themselves take place, it has constrained the development of alternative approaches which understand change or process, rather than static entities as the basic building blocks of reality.

While a number of recent studies have attempted to illustrate the potential of such process-based approaches to enable progress in areas as diverse as the philosophy of biology (Nicholson & Dupré 2018) and philosophy of psychology (van Geert & de Ruiter 2022), the benefits of adopting a process-based approach to questions surrounding the passage of artefacts through time has yet to be elaborated. My goal in the current chapter is to do precisely that: by outlining how the paradigm of substance philosophy has constrained philosophical thinking surrounding questions concerning the relationship between technology and time. As I'll attempt to show, artefacts represent a particularly unique area of research for which process perspectives hold the potential not only to solve existing problems but also to reveal new avenues for research.

The first section of this chapter provides a brief outline of the various ways in which the western metaphysical tradition has approached questions concerning the persistence of artefacts. Here I aim to show that underlying the diversity of approaches in this tradition, are a number of shared assumptions which align closely with the research paradigm of substance philosophy. In the second section, I draw upon existing literature in maintenance and repair studies to illustrate how a wide variety of artefacts can be understood to undergo continuous change after production through the act of maintenance. My goal here is to show that insofar as the underlying framework of substance ontology faces a variety of limitations in accounting for these forms of change in the metaphysical analysis of artefacts, the study of maintenance highlights the need for an alternative ontological framework. As I argue that this alternative can be found in process philosophy, the final section of the chapter explores the form a process-based metaphysics of artefacts might take by drawing on resources central to the recent 'process turn' in the philosophy of biology. In contrast to existing methods and approaches in metaphysics, I demonstrate how a process philosophy of artefact maintenance tends inevitably towards the empirical analysis of social and political processes which determine the trajectory of artefacts through time.

3.1 The Problematic Status of Change in Western Metaphysics

While it is not often recognised as such, maintenance forms the topic of one of the earliest and most enduring metaphysical puzzles in the western philosophical tradition: The Ship of Theseus. Consider, for example, how the original formulation of the problem by Plutarch depicts the replacement of materials as occurring specifically within a context of maintenance.

> The ship wherein Theseus and the youth of Athens returned from Crete had thirty oars, and was preserved by the Athenians down even to the time of Demetrius Phalereus for *they took away the old planks as they decayed, putting in new and stronger timber in their places*, insomuch that this ship became a standing example among the philosophers, for the logical question of things that grow; one side holding that the ship remained the same, and the other contending that it was not the same.
> (Plutarch 1859, p. 21 emphasis mine)

Over time, however, the practice of maintenance has come to be excluded from discussions surrounding the Ship of Theseus. In modern philosophy, the version of this paradox which attracts most discussion stems from the work of Thomas Hobbes. In Hobbes' modified retelling, instead of being simply discarded, the planks composing the original ship (which we must assume have not suffered decay), are reassembled into another ship (Hobbes 1656, p. 100). Hobbes' version introduces a case of fission, whereby one artefact appears to have evolved into two: the ship consisting of replaced but new planks and the ship consisting of the reassembled original planks, prompting the question of which, if any, should be considered identical with the original ship. As a result, most philosophers today will recognise the Ship of Theseus paradox more immediately as a puzzle concerning identity rather than maintenance. Yet, this downplays the deep connections which exist between conceptions of identity and maintenance which the original formulation of the paradox succeeded in highlighting. After all, to ask whether the ship remains the same throughout the replacement of materials is at the same time to ask what it means to *maintain* a technology. For this reason, conceptions of how artefacts persist through change always imply particular ways of conceiving maintenance. In the latter sections of this chapter, we'll return to the question of the relationship between identity and maintenance, but for now, I want to dig deeper into questions surrounding the persistence of artefacts.

Regardless of the formulation we consider, the Ship of Theseus paradox can be understood to represent a version of a perennial metaphysical dilemma known as the *problem of change*. Put simply, the problem is this: if an object can be said to undergo change, then it must become different from what it was before. Yet, if it is not the same both before and after change, on what basis can we say that one object has actually undergone change? (Wasserman 2006). Among classical philosophers, the problem of change was often viewed as the central difficulty underlying metaphysical theories that posited change as the fundamental basis of reality. The problem stemmed from the concern that the presence of change meant that things possessed contradictory properties – that the ripening of a banana meant that it could be *both* green and unripe *and* yellow and overripe, for example. Insofar as this was widely understood among classical philosophers as threatening the possibility of knowledge, language and even thought itself (Lee 2010), resolving the problem of change has been considered a task of crucial philosophical importance since antiquity. In fact, this problem continues to draw the attention of philosophers today. There is now a wide variety of responses to the problem of change, among which it is possible to distinguish two main orientations:

Milder approaches seek to resolve the problem of change by maintaining that not all aspects of an object undergo change. Instead, these approaches seek to demonstrate how certain features, specifically those that grant identity to an object, remain the same over time. The most well-known example of such an approach appears in the work of Aristotle, who drew a distinction between two different kinds of change things could undergo: accidental change, such as the painting of a house, during which the identity of the object is preserved, and essential change, such as the bulldozing of a house, during which the thing in question ceases to exist. Coupling the identity of things tightly to their essential properties, Aristotle sought to segregate those parts of an object that underwent change from those which determined the identity of the object, making it possible to say that an object was changing while still at the same time remaining the same. Modern variants of this approach employ the notion of identity criteria which are understood to determine the kinds of changes a particular object can tolerate or not. For artefacts, identity criteria have been suggested to stem from a variety of sources including collective beliefs about kind membership (Bloom 1996; Pearce 2016), sortal concepts (Hilpinen 1993) or even stipulations from the makers of artefacts themselves (Thomasson 2003).

Stronger Approaches seek to provide accounts of identity over time which satisfy the requirements set out by Leibniz's law of the indiscernibility of identicals. According to such approaches, in order to justifiably assert that an object has undergone change, it must be demonstrated that the object both before and after the change are identical, understood in terms of sharing *all* their properties. Among responses that satisfy this more stringent requirement, the most successful arguably stem from the perspective of four dimensionalism which considers artefacts to be extended not only in space but also through time. Repudiating the common and intuitive notion that an object wholly exists at any one moment in time, four-dimensionalism instead considers an artefact to exist as a series of temporal parts extended throughout time. What this means is that an object is not wholly present before or after any change, instead what we find at each moment are simply temporal parts of a larger process which is always identical to itself.

The foregoing passages highlighted how existing responses to the problem of change within the western philosophical tradition exhibit significant variation. However, what is striking is that despite their variety, these different responses nonetheless share at least two assumptions concerning the nature of change and identity. Perhaps, not surprisingly, these shared features, which following Johanna Seibt can be understood to be located in the 'presuppositional depth structure' of the different theories, point towards a particular kind of ontology – one based around things rather than processes. The first assumption concerns the *ontological priority of substance over change*. In neither of the two main approaches to resolving the problem of change that we considered above, does change occupy a position of ontological priority. While milder approaches recognise the existence of change, by aligning the identity of objects with those aspects which remain the same, they also tend to shuffle it into the corner. As Anne Sophie Meincke notes, this tendency to deny ontological significance to change represents a fundamental assumption underlying substance ontology which

> crucially rests on the idea that change – if there is any – must have a non-changing basis. Whatever appears to change is believed to have an unchanging core. Change, if it exists at all, is taken to be parasitic upon

> stasis; it does not have an independent ontological status. It resembles the fluctuating images on a mirror surface, leaving the mirror itself the same whether it shows microbes on a petri dish, a murder, or passing clouds.
>
> (Meincke 2019)

According to this perspective, change is understood as something that happens *to* things and for this reason, it depends on the existence of stable, enduring entities. Stemming from the intuition that for change to be possible there must be some *thing* that changes, these theories typically understand change as the 'doings' of stable entities (Rescher 2000, p. 6) or as something that happens to them (Dupré & Nicholson 2018, p. 12) and for this reason, accord it a derivative and secondary ontological status. The denial of the ontological significance of change emerges even more forcefully from the stronger approaches considered above. One of the most common criticisms of four-dimensionalist solutions to the problem of change, for example, is that they succeed in resolving the problem only by denying the existence of change altogether.[2] In order to explain how the same object can instantiate contradictory properties, such as the banana which is green and unripe at a particular time and then yellow and overripe at a later time, four dimensionalism encourages us to view the green and yellow bananas as different temporal parts of a larger, timeless whole. By explaining the development of the banana in terms of a succession of unchanging temporal parts, this approach eschews the possibility of becoming itself. As Simons notes, four dimensionalism offers 'not an explanation of change, but an elimination of it, since nothing survives the change which has contrary properties' (Simons 2000).

The ***second assumption*** underlying the two approaches concerns the *separation of identity from change*. As Johanna Seibt has noted, the tendency to conceive of identity as static is a shared feature of substance-based approaches to questions of persistence (Seibt 1997). The idea that only that which does not undergo change can be a candidate for identity has informed discussions of persistence in western metaphysics since antiquity. Aristotle's distinction between essential and accidental properties, for example, posited a division at the heart of objects which demarcated those aspects which undergo change from those which confer identity. A similar opposition between identity and change continues to inform modern approaches within the metaphysics of identity. As Peter Simon has noted, the ship of Theseus paradox draws out two opposing tendencies within the sortal concepts we employ to track the identity of objects such as 'ships'. On the one hand, there is a tendency to link the identity of an artefact with the matter composing it. On the other hand, an opposing tendency considers identity to reside in an artefact's form. Yet, regardless of the conception we employ, however, the identity of the ship is nonetheless conceived as being unable to tolerate even the smallest amount of change.

> Form-constant objects cannot survive a disruption of continuity of their form: even the shortest intervention which interferes with the capacity to function counts as destroying the artifact. Matter-constant continuants cannot survive the slightest change in their constituent matter. On the other hand, a matter-constant continuant does not require a continuously maintained form. It could survive dismantling and reassembly, provided it is put back together without any matter being lost or gained.
>
> (Simons 1987, p. 200)

In discussions of persistence then, an identity changed is very often an identity lost. This common tendency to strictly separate change from identity in Western metaphysics emerges as a necessary consequence of the fact that for the vast majority of theorists, change is understood to *corrupt* identity (Meincke 2019). This is reflected in the nature of the dilemma which these approaches are intended to address, which understands change to be the problem that must be solved, in order to rescue the possibility of a world which can be understood, thought and communicated of. The underlying strategy western metaphysics has hitherto employed to address the challenge of explaining persistence therefore consists of demonstrating how a changing thing has remained the *same* despite undergoing change. Despite the diversity of views concerning exactly how objects should be understood to remain the same, this is one assumption which is rarely subjected to critical scrutiny or even considered a topic for discussion at all. In the following sections, we will turn our attention to examining the processes by which artefacts develop through time, in order to begin exploring reasons why this apparently self-evident way of approaching questions surrounding the persistence of artefacts may not be so obvious after all.

3.2 Maintenance and Modification

In relation to technology, maintenance encompasses an extremely wide variety of different kinds of activities. We replace parts, apply lubricant or protective coatings, clean away grime or salt, polish, sharpen and shine, all in the name of maintenance. Insofar, as we perform these activities deliberately in order to achieve specific goals, maintenance can be understood to be inherently normative and inextricably linked to specific expectations concerning how particular artefacts *should* be. The particular activities we perform under the guise of maintenance therefore reflect our underlying intuitions that our knives should be sharp and our skillets seasoned, our cars free from rust and our bike chains well oiled. Despite the diversity in the kinds of activities we regard as maintenance, focusing on the goals we aim to achieve by maintaining allows us to see how, at the broadest level, maintenance may be characterised as an activity which aims to ensure coherence between our understanding of the state of an artefact and a particular set of normative expectations.

In philosophy, the normative expectations we seek to realise through the maintenance of artefacts are often identified with the intentions of those responsible for producing the artefacts in the first place. This view corresponds to a particular conception of how artefacts are made which has remained dominant in the Western philosophical tradition since antiquity. According to this view, which I call the 'intentional conception of making' (Young 2021), artefacts are produced through the material realisation of the intentions of one or more human agents. These intentions, which may include specifications regarding function, form or material composition, are understood as determining the identity of the artefact that is created, alongside establishing persistence conditions which determine the kinds of change the artefact in question can tolerate and those which result in its destruction.[3] An important corollary of this view, and one which again highlights the pervasiveness of the research paradigm of substance ontology, is that the intentions which determine the identity of artefacts are not themselves considered to undergo change over time.[4] For this reason, the proponents of the intentional conception of making tend to understand the goals of maintenance to consist of the preservation or restoration of the original intentions an artefact's creator sought to realise in the act of production. A prominent example of this approach is found in the work of Simon Evnine, who considers various maintenance activities, such as the replacement of parts, as a form of restoration:

> a restorer of a sculpture may replace its lost nose; the restorer of a car may replace a failed battery....The process of replacement mimics the original making: it involves an agent who supplies the matter to do what the original matter did, acting with an intention to repair an object of the kind originally made...Let us call this kind of replacement *restoration*.
>
> (Evnine 2016, p. 76)

Insofar as this perspective understands the persistence of artefacts to depend on the continued possession of identity-granting features intended by their designers, maintenance has an important role to play. By ensuring the coherence between the state of an artefact and the intentions of its creators through activities such as the replacement of parts or the protection of certain features from damage, maintenance can be understood to be crucial in enabling the persistence of artefacts. At the same time, this account also determines what operations count as maintenance for any particular artefact because it draws an implicit distinction between acts of maintenance and modification. Consider again the Ship of Theseus. According to this view, certain operations performed upon the boat, such as the replacement of planks, may count as maintenance only to the extent that they succeed in restoring the original intentions of the maker (assuming of course that the maker intended to create an artefact which could tolerate the replacement of the material constituting it). Other forms of activity, such as converting the ship to a DUKW by adding wheels and a chassis to the hull, would represent a form of modification that would count as a new act of making because the intentions we aim to realise in performing these operations differs from what we can assume the maker of the ship originally intended.

Despite forming the standard view on the processes by which artefacts are made and maintained, this view encounters various difficulties when we attempt to apply it to actual histories of artefacts in use. As philosophers have long recognised, the normative expectations users hold about how their artefacts should be do not always match the intentions of those who produced them. There are two main reasons for this: the *first* is that normative expectations differ among users depending on their individual interests and expertise. Whereas one user may regard a heat pump as performing adequately, for example, another user who is a heating, ventilation, and air conditioning (HVAC) engineer may consider the same heat pump to require modification in order to match their expectations for optimal function (see Figure 3.1). Still other users may seek to modify the same heat pump in order for it to be operated in an environment for which the artefact was not optimised during the design process. The *second* reason concerns the capacity for our expectations regarding artefacts to change over time. We may be satisfied with our phone, for example, until we learn that a similarly priced model takes far better pictures or is able to run much larger programs smoothly. The capacity of normative expectations for particular artefacts to evolve over time is seen not only in relation to personal devices but also to artefacts that we own or use collectively. Consider, for example, safety codes for engineered structures. Many structures which were originally regarded as safe when they were constructed in the early twentieth century eventually came to be viewed as unsafe as our understanding of structural mechanics and the risk of collapse during seismic events increased.[5] The evolution of engineering standards provide a clear indication of the capacity for our normative expectations to change over time.

Yet while philosophers have long recognised that users come to hold ideas about how particular artefacts should be that differ from what was

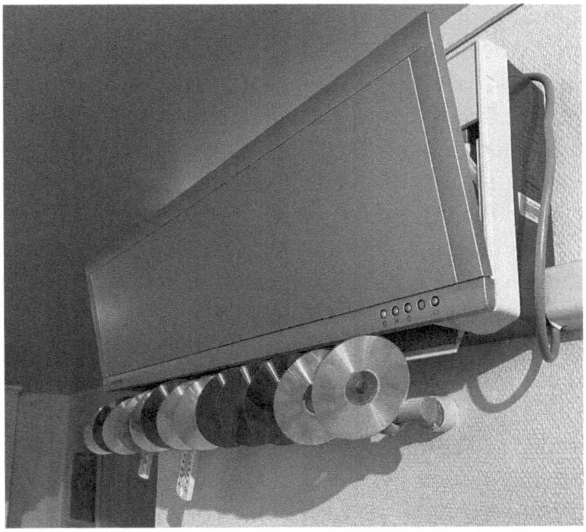

Figure 3.1 Heat pump modified by the author's father-in-law to direct airflow and monitor temperature.

intended by the producers of those artefacts, what has not received adequate attention is the extent to which this diversity leads artefacts themselves to change over time. Examining the histories of artefacts in use reveals how a wide range of different kinds of artefacts undergo change through the activities of retrofitting, repair and modification. These processes of transformation occur for different reasons. Artefacts are often transformed, for example, when they change hands between owners who possess different normative expectations. When buying a house, for example, we are often interested in how it can be altered in order to better align with our interests: whether an office can be converted to a kid's room, for example, or whether there is enough space to build a patio in order to host barbeque parties in the summer. In other cases, artefacts transform when our collective normative expectations evolve over time. This is particularly apparent for long-lived engineered structures such as bridges, whose histories often reveal a wide range of changes that were undertaken in order to adapt the structures to evolving standards for safety. Consider the Golden Gate Bridge in San Francisco as an example. During the 1950s, the bracing on the underside of the bridge deck was modified to prevent torsional instability during high winds which had recently led to the collapse of the Tacoma Narrows Bridge. After the collapse of the Cypress Viaduct during the Loma Prieta Earthquake in 1989, the Golden Gate Bridge District decided to fund a series of seismic retrofits which have been ongoing for the last two decades and involved the modification of almost all parts of the structure, including the approaches, pylons and bridge deck (GGBHTD n.d.). Further changes made to bring the structure in line with current safety standards include the installation of pedestrian barriers (Bateson 2012, p. 70) and concrete median barriers to prevent cars from crossing into the path of oncoming traffic (Josephson 2017, p. 76).

Until only recently, the processes by which artefacts continue to undergo development after their production had received only scant attention in scholarship on technology.[6] The recent emergence of the interdisciplinary field of maintenance and repair studies, however, reflects an increasing interest in exploring these processes and there is now a growing body of literature examining how technologies undergo adaptation after production. However, exactly how many artefacts undergo processes of adaptation throughout their histories still remains an open question and one which can only be addressed through further empirical research. It is nonetheless clear already that for some artefact kinds, such as houses and boats, transformation represents a normal feature of their histories. For other artefact kinds, however, transformation depends more on the particular context in which they will be employed. Consider cars as an example. While a small minority of car owners in the Global North may be interested in modifying their vehicles, most cars undergo very little transformation throughout their histories insofar as maintenance routines typically consist of the replacement of parts ordered directly from the manufacturer. Outside this context, however, transformation is much more common. In many African countries, for example, the lack of access to cheap genuine replacement parts means that vehicles are often modified through the activity of repair. In their biography of a 504 Peugeot that had been imported from the Netherlands to Ghana in the early 90s, anthropologists Verrips and Meyer recount the modifications made to the vehicle by its owner Kwaku:

> he replaced the two-chamber carburettor of his old Peugeot by a one chamber one in order to save petrol. The air inlet was not original any longer, but consisted of a piece of pierced tin. A lot of gaskets and rubber parts, such as bushings were 'indigenous', that is, cut out of old tubes and tyres...at several places nails were used a lock pins. Some rubber tubes were fixed with iron wire, whereas others which evidently were out of use were closed with old spark-plugs, butterfly nuts or even pieces of wood. In order to prevent the knocking of some worn-out main bearings in the engine block, Kwaku had put pieces of greasy paper between them and the crank-shaft. Instead of using a special spring, his distribution chain was held in its proper place with a piece of copper pipe. He had raised the oil level in his shock-absorbers beyond normal so that he could drive more comfortably on roads full of potholes.
>
> (Verrips & Meyer 2001)

Transformation is also common for artefacts that move *between* contexts. This might include cases where objects trade hands between owners who intend to employ them in different ways or for different purposes, such as when a new owner converts a fishing vessel to a recreational vessel by removing a crane and winch assembly in order to make space for relaxing on the back deck. Or it might include cases of artefacts travelling between different geographical areas, such as when cars imported from Europe to Eritrea required specific modifications in order to better handle the local terrain and high altitudes (Belluci & Zaccaria 2012). Transformation which occurs as a result of shifting contexts also includes cases where artefacts exist over long time periods through which new patterns of use or social values emerge. Examples of such cases include the modification of buildings to include wheelchair ramps or retrofitting cargo ships with wingsails.

A common claim in the literature on maintenance and repair is that modifying artefacts in order to adapt them better to the contexts in which they will be employed itself constitutes a form of maintenance. In their seminal paper on the topic, Stephen Graham and Nigel Thrift, for example, emphasise the transformative potential of maintenance practices by claiming that:

> maintenance and repair can itself be a vital source of variation, improvisation and innovation. Repair and maintenance does not have to mean exact restoration. Think only of the bodged job, which still allows something to continue functioning but probably at a lower level; the upgrade, which allows something to take on new features which keep it contemporary; the cannibalization and recycling of materials, which allows at least one recombined object to carry on, formed from the bones of its fellows; or the complete rebuild, which allows something to continue in near pristine condition. And what starts out as repair may soon become improvement, innovation, even growth.
>
> (Graham & Thrift 2007)

Despite forming a common theme in the literature on maintenance and repair,[7] the idea that maintenance can be transformative in this way appears to conflict with dominant metaphysical approaches to the analysis of artefacts which recognise sharp distinctions between the practices of maintenance and modification. As we noted above, such distinctions are understood to stem from the nature of the intentions guiding each form of activity; only when an operator's intentions are identical to those of an artefact's creator does their activity count as a form of maintenance. In contrast, many scholars of maintenance and repair studies appear content to recognise cases in which artefacts are deliberately modified as examples of maintenance. In the remaining part of this section, I want to outline some reasons why I think we should be inclined to relax distinctions between maintenance and modification and follow suit in considering operations performed to realise intentions that are distinct from those of the original producers of an artefact, as forms of maintenance.

The *first* reason emerges from a closer examination of the kinds of practices, such as replacement, that we often identify with maintenance. In philosophical discussions surrounding the Ship of Theseus paradox, it is often assumed that the new planks involved in the replacement of the original material differ only to the extent that they are not composed of the same wood as the original planks. Yet, replacement of this kind, where replacement occurs with the same yet numerically different parts, constitutes the exception rather than the rule when it comes to replacing parts on a boat. It is when parts are due for replacement, that owners often take the opportunity to upgrade their vessels by replacing the part with one which expands the capacity of the boat in different ways. When a motor is due for replacement, for example, an owner may decide to replace it with one which includes the possibility for autopilot functions, leading them to modify the steering mechanism and install new components at the helm. Often, the service lives of different parts of a boat make replacement with the *same* part either impossible or extremely costly. After all, it is often unlikely that owners will be able to find the exact same engine model that was produced 40 years before. Exploring the histories of actual boats provides further insights into the way in which the replacement of parts is very often undertaken with a view towards modifying vessels for new purposes. This may be best illustrated by the case of battleships. In his study of the maintenance of twentieth-century battleships, historian David Edgerton, for example, notes how

> A ship was not necessarily a stable entity. Ships were often radically changed, often more than once, in the course of their lives.... What a Queen Elizabeth battleship was and looked like in 1918 was pretty clear, but by 1939 they were different ships. Between 1924 and 1934 they were given substantial refits, a process which included trunking the funnels into one, and fitting huge anti-torpedo bulges on their sides, as well as changes in the smaller armament. Then, in the 1930s....they were given new engines, major changes were made to the guns and mountings, and much of the superstructure was rebuilt.
>
> (Edgerton 2006, p. 93)

As the central example through which philosophers have approached the question of maintenance, the Ship of Theseus paradox therefore exhibits serious limitations insofar as it *vastly* understates the capacity of boats to change over time. And it's only boats that undergo transformation through the replacement of parts. As most people who own cars or bikes will immediately realise, routine maintenance activities very often involve replacing components such as tires or seatposts with versions better suited to the conditions in which these artefacts will be employed and the specific goals and interests of their owners. The customisation that occurs through such routine maintenance practices reflects how the replacement of parts is often undertaken with a view towards upgrading artefacts.

The *second* reason for embracing a conception of maintenance as transformative connects with broader meanings of maintenance as enabling something to continue existing over time. Put simply, the modification of artefacts in ways that were not intended by their designers is very often what enables them to avoid destruction. To see why, consider again the case of the Golden Gate Bridge. After reviewing vulnerability studies for the bridge in the wake of the Loma Prieta Earthquake, the Bridge District judged the risk of collapse in a future earthquake to be unacceptably high. In order to align the structure with current expectations surrounding seismic integrity, the district was faced with two options: either replacing the entire structure with a new bridge or initiating a major retrofitting project which would lead to the modification of nearly all parts of the existing bridge over a period of three decades (GGBHTD n.d.). As we have seen already, the district chose the latter option and began the first of three phases of seismic retrofits in 1997. What these deliberations illustrate, however, is something that existing discussions on the persistence of artefacts appear to have missed entirely. That is, for many artefacts, it is only because they have been able to change and adapt to emerging interests that they continue to exist today. When we examine the histories of a variety of different kinds of artefacts, this simple fact is found to apply widely. Houses, for example, often face demolition when they can no longer be changed to match our interests. It is when our phones can no longer support updates that we begin to consider replacing them. For many artefacts then, it is the capacity to evolve and adapt which determines how long they remain in use before they are disposed, decommissioned or destroyed. Yet, while this simple insight most likely will not strike us as particularly surprising, it nonetheless represents a challenge to deeply held assumptions in the existing discourse on the metaphysics of artefacts. In stark contrast to dominant approaches that understand artefacts to persist by *remaining the same*, what we find when examining the histories of artefacts in use is precisely the opposite conclusion – that the continued existence of a wide variety of different kinds of artefacts is premised on their capacity to *change*. In Section 3.3, we will begin exploring what it might mean for the metaphysics of identity over time, if we take seriously the insight that artefacts persist by changing. In order to do so, we'll turn our attention to the recent 'process turn' in the philosophy of biology, where similar insights concerning the dynamic nature of biological entities have motivated a shift towards an ontology based on processes rather than substances. In seeking the outlines of a process philosophy of technology, our starting point will be an examination of the extent to which some of the basic concepts and approaches central to this growing body of literature might also be applicable to the analysis of the passage of artefacts through time.

3.3 From Biology to Technology: Towards a Process Philosophy of Artefacts

A growing number of philosophers of biology now suggest that the nature of the processes by which biological entities move through time demands a different kind of ontology to that which has remained dominant in western metaphysics since antiquity. Instead of an ontology based on substances, they argue that the capacities of organisms to grow and adapt to their environments necessitates a shift towards an alternative perspective which takes processes as the fundamental constituents of reality.[8] This perspective, known broadly as process philosophy, inverts the traditional ontological hierarchy of western metaphysics by taking change rather than stasis to be primary. Process philosophy is a broad category used to capture a diverse range of philosophical theories, spanning antiquity to the present, which together share a number of basic tenets. These shared features include the idea that time and change are the basic categories of metaphysical understanding, that processes should be regarded as the fundamental components of an ontological theory and accordingly that entities which we experience as static and enduring are best understood in terms of processes and change (Rescher 2000, p. 5). Contemporary proponents of process philosophy often consider this broad theoretical orientation to represent a return to the philosophical perspectives of antiquity which characterised the nature of reality in terms of constant change and against which classical philosophers reacted through the development of substance ontology.[9]

Among the features of organisms that philosophers have identified as being particularly suited to a process-based metaphysics are their capacity to 'develop progressively over time, acquiring certain properties and capacities and losing others along the way' alongside their possession of a metabolism through which they 'constantly exchange energy and matter with their surroundings' (Dupré & Nicholson 2018). In previous sections of this chapter, we explored how the histories of a wide range of artefacts reveal similar processes at play. As the Ship of Theseus paradox so aptly illustrates, artefacts also exchange matter with their surroundings, through processes which challenge the dominant ways of understanding the relationship between technology and time in metaphysics. In his discussion of the material composition of artefacts, for example, metaphysician Simon Evnine suggests that the capacity of many artefacts to exchange their matter over time warrants the ascription of a 'metabolism' to artefacts in a metaphorical sense (Evnine 2016, p. 16). Similar observations for large socio-technical systems can be found in the burgeoning literature on 'urban metabolism'. Yet, despite broad similarities between the processes by which organisms and artefacts persist through time, a process philosophy of artefacts nonetheless has yet to be developed. One of the main reasons why relates to a widespread and longstanding resistance among philosophers to recognise the extent to which these characteristics can also be understood to apply to technologies. Since antiquity, metaphysicians within the Aristotelian tradition have been more interested in exploring the differences between natural objects and artefacts than they have been in investigating their similarities (Preston 2022).[10] However, this tendency to focus on their dissimilarities has led many to overlook the various ways in which the passage of artefacts and organisms through time share a variety of important similarities, and by doing so, has helped to hinder the development of a deeper appreciation of the relationship between technology and time in western metaphysics. In arguing for the relevance of a process-based metaphysics for biological entities, Daniel J. Nicholsen, for example, draws a comparison between the life cycle developments of machines and organisms and concludes that:

> the form of an organism does not stay fixed during its lifetime. Organisms grow and develop. Some undergo major morphological changes, such as metamorphosis. Nothing comparable happens to a machine...given that the fixity of its internal structure is precisely what allows it to perform the function it was designed for.
>
> (Nicholson 2018)

But machines also undergo growth and development. While it may be true that the internal structures of many machines remain fixed throughout their service lives, there are nonetheless a wide range of machines which undergo major changes while continuing to perform the functions that were intended by their designers. Consider, for example, the common practice of modifying car engines with turbochargers, aftermarket air filters or newer performance chips in order to improve their performance. Furthermore, in some cases, machines *must* undergo change in order to continue performing the functions they were designed for. Examples such as this include industrial machinery, which avoids replacement by being upgraded to match the operational capacity of newer models, or the retrofitting of vehicle engines in order to remain compliant with evolving legal requirements regarding emissions.

In earlier sections of this chapter, we examined how a wide range of artefacts develop and change throughout their histories in order to match evolving expectations concerning how they should be. Furthermore, we noted how for many artefacts, this capacity to undergo adaptation is crucial for avoiding replacement and consequent disposal or destruction. Examining the histories of artefacts in use therefore allows us to see how H.G Wells' famous dictum 'adapt or perish, now as ever, is nature's inexorable imperative' can be understood to apply as much to artefacts as it does to natural organisms. Given such broad similarities between the temporal character of organisms and artefacts, we should be forgiven for suspecting whether a process-based metaphysics might also present a number of benefits for the analysis of technology. Our suspicions may be enhanced further when we recognise the extent to which the basic analytical approaches currently applied in the study of biological entities as processes, bear an immediate relevance and clear applicability to the examination of the histories of artefacts. Consider, for example, how the adoption of process perspectives generates a new range of explanatory tasks that differ from those which have traditionally occupied the focus of metaphysics. As John Dupré notes:

> One way of seeing the difference between a process and a substance ontology is in terms of the explanatory problems they face. Whereas a substance ontology has to provide explanations of change, a process ontology instead faces the problem of explaining stability or the appearance of stability...This perspective profoundly transforms questions about identity over time. First, the persisting individual is not a metaphysical given, in which the individual is identified and investigated to detect the conditions under which it might persist. Rather, it is something the individual does: persisting is, indeed, hard work.
>
> (Dupré 2018)

Rather than attempting to deny the presence of stability altogether, process philosophers seek to show how the semblance of enduring things emerges through the *stabilisation* of processes. In recent work in the philosophy of biology, homeostasis, metabolism and evolution have all been suggested to represent processes of stabilisation from which the stability of species and organisms over time can be understood to derive.[11] Yet, stabilisation also represents a particularly promising way of conceptualising the temporality of artefacts insofar as it provides an analytical framework through which we can explore the different ways in which artefacts continue to change after production. The fact that, unlike organisms, artefacts do not stabilise themselves and instead require this work to be performed by external agents reflects the importance of the practice of maintenance in enabling the persistence of artefacts over time and helps shed light on what maintenance might be understood to mean in a metaphysical context. For artefacts, persistence is also hard work. And that work is maintenance.

Yet, exactly what maintenance represents in metaphysical terms requires further elaboration. Before a process philosophy of artefacts can develop an explanatory framework through which the stability of artefacts can be accounted for, we must first determine what the stabilisation of processes should be understood to mean in relation to technologies over time. The answer to this question is anything but obvious. The literature on process philosophy of biology, for example, offers at least two distinct suggestions for how stability should be understood in relation to biological entities: the *first* is the stability of properties, such as an organism's achievement of a constant temperature through homeostasis. The *second* is the stability of directional processes, such as the sequences of life cycle development individual organisms undergo (Dupré 2018, n.7). In the following, we will consider each conception of stability: of properties and developmental processes, to see whether they might be fruitfully applied within a process philosophy of artefacts.

The first candidate, the stability of properties, represents perhaps the most intuitive option insofar as it reflects common conceptions of maintenance as hindering processes of change which the materials comprising artefacts undergo. Consider, for example, the iron which makes up much of the steel from which engineered structures such as bridges or building frames are constructed. Iron begins as iron ore – rocks and minerals rich in oxides. In order to produce metallic iron which can be used to make steel, iron ore must be refined through a process which requires great amounts of heat to drive off the oxygen, converting the oxide to a metal. The resulting product however is inherently unstable, and if left unprotected, it immediately begins corroding – reacting to water and oxygen and slowly reverting back to an oxide in the form of rust, a similar compound to the ore from which it began. Here, we see how the materials from which engineered structures are composed exist as processes that must be stabilised in different ways, through the application of protective coverings such as paint and vinyl coatings, for example, or through the replacement of corroded elements and materials, in order for the bridge to appear as a stable entity. Yet, while it might be tempting to understand stabilisation in terms of the attempt to hold the material properties of artefacts in stasis, we saw in the last section a variety of reasons why this would not provide a suitable basis for a general process philosophy of artefacts based around the practice of maintenance.

In the *first* place, conceptualising stability in terms of the stasis of properties would severely restrict the range of practices that such an account could consider. After all, very few activities that we identify as forms of maintenance manage to prevent the change of materials from which artefacts are composed without also modifying them in the process.[12] Even activities which explicitly aim at delaying or hindering the transformation of materials often involve *adding* other materials: protective coatings or weather barriers, for example. Furthermore, a wide range of activities we identify as maintenance often deliberately facilitate the transformation of materials composing artefacts over time. For example, proper maintenance of a cast iron skillet allows the build-up of material on the cooking surface, facilitating the development of a superior cooking surface, while at the same time preventing the development of rust. *Second*, understanding maintenance in terms of preventing the change of materials would sever the relation between maintenance and persistence for a wide variety of artefacts. As we noted in the previous section, the continued existence of many artefacts depends crucially on their capacity to adapt and change. For such artefacts, understanding the stabilisation of processes as consisting of preventing change would leave us unable to explain how it is they persist through time.

What we require then is a broader understanding of the stabilisation of processes which is wide enough to encompass activities which both prevent and enable the change of materials. Viewed in this light, directional processes yield a more promising option for an object of stabilisation. As we saw in the last section, a closer examination of the range of maintenance practices that we perform on artefacts reveals their shared characteristic to lie in the attempt to guide change in the direction of certain normative expectations. Here we find a conception of the stabilisation of processes that can serve as a basis for a general account of technological maintenance. According to this account, instead of understanding artefacts to be *comprised* of processes that demand stabilisation, artefacts themselves are understood as processes that can develop in different ways over time. Stabilisation consists of actively intervening in the processual development of an artefact in order to guide it in ways which ensure coherence between our understanding of how the object *is* and our normative expectations for it (i.e., how the artefact *should* be). The central benefit of interpreting stabilisation in this way is that it enables us to explain how artefacts persist through time not only by resisting change but also by undergoing it too. It applies as much to cases in which our normative expectations do not undergo change, as it does to cases in which we seek to adapt artefacts to emergent interests and uses. Maintaining an artefact therefore means different things; in some cases, it may involve seeking to prevent the material change of an artefact over time while in others it may involve their deliberate modification through practices of redesign, retrofitting and customisation.

Understanding stability to pertain to the relation between artefacts and normative expectations in this way also enables us to see how problems maintenance can be understood to address, alongside the ways in which it does so, may extend beyond materiality altogether. For example, by conceiving instability in terms of a divergence between our understanding of the state of an artefact and our normative expectations, this account makes room for the emergence of a instability even in cases where artefacts do not undergo change. In his study of seismic retrofits of Californian bridges, Benjamin Sims examines the problems addressed by practices of repair as forms of 'slippage' which occur between our normative expectations and the material conditions of artefacts (Sims 2016).[13] In addition to identifying slippage in cases where artefacts undergo material change over time through degradation, Sims also highlights how slippage can occur through obsolescence.

> when objective conditions are perceived as stable, but normative expectations have changed: for example, engineering standards and building codes have been updated, there is a movement to add bike lanes to existing bridges, broadcasters want to transmit high definition television signals. Existing technology comes to be seen as outdated or not responsive to social needs.
>
> (Sims 2016, p. 16)

Obsolescence, like all forms of instability which may develop throughout the history of an artefact, requires attention. Left unchecked, obsolescence threatens the persistence of artefacts, drastically increasing their risk of replacement, disposal or destruction. However, obsolescence, like all forms of slippage, can be resolved through stabilisation in which we perform work to restore coherence between understandings of the state of an artefact and certain relevant normative expectations. One of the most important activities through which such coherence is restored is practices of repair and modification which bring the state of an artefact 'back in line' with how it is expected to be. Yet, examining the different ways in which stability can be restored leads inevitably towards a broader conception of maintenance than that can be found in the existing literature on the metaphysics of artefacts. Artefacts can be stabilised not only through material practices but also through discursive work. If you become convinced that the state of an artefact is not as it should be, for example, I can attempt to convince you either that you are mistaken about the state of the artefact or that your normative expectations should be altered.[14] In either case, if I am successful in my attempts to convince you, then coherence will have been restored and the artefact stabilised.

Yet, despite the various options at our disposal, not all cases of slippage are resolved through stabilisation. Failing to stabilise artefacts can bring their processes of development to an end. It is when we fail to see how our bike can be changed to match our expectations that we may decide that it needs replacement or should be discarded, for example, or when a governing body decides that it is too costly to modify a bridge that no longer adequately meets the needs of society that it ends up being decommissioned and replaced. This latter example highlights the extent to which the stability and instability of a wide variety of artefacts often possess irreducible social and political dimensions. Not only does the question of *whose* normative expectations matter but also the wider political structures that determine the power that different actors and groups within society may exercise. Insofar as instability represents the potential for change, cases in which it is primarily actors and groups with little political power who feel that an artefact does not match their expectations, most likely will not result in destabilisation. As an example, consider again the case of the Golden Gate Bridge. Since its construction in 1937, the number of people who have jumped to their deaths from the bridge has steadily increased. Today, the bridge remains the deadliest structure in the world for suicides: to date over 1500 people have lost their lives in this way. It has long been thought that this unusually high number of deaths is at least partly a result of the great height of the bridge above the water (75 metres) combined with a relatively low guard railing (just over one metre). Already by the 1960s, activist groups began campaigning in order to convince the Bridge District to establish safety barriers such as higher guard railings in order to help prevent further deaths. However, 30 years of lobbying the board directly ultimately failed to destabilise the artefact insofar as members of the Bridge District remained convinced that the bridge matched their normative expectations. The cause of the high rate of suicides, they argued, lay not in the design of the bridge but rather in wider political developments such as the defunding of mental health services in California. After years of failed attempts, activist groups realised the need to switch tactics and began attempting to cultivate broader support for change among members of the public. This new strategy eventually proved successful in convincing enough members of the board to change their normative expectations regarding the safety features of the bridge.[15] This slippage led to attempts to restore coherence, resulting in the installation of safety nets which now span almost the entire length of the bridge.

3.4 Conclusion

Throughout this chapter, I've attempted to outline the rudiments of a wider research program that approaches the metaphysics of maintenance by analysing the nature of artefacts as processes. Such a program would involve, in the first place, empirical work to chart the contours of such processes, not only in the case of individual artefact tokens such as the Golden Gate Bridge but also within and between different artefact kinds. Second, it would involve exploring the structure and nature of those processes by investigating how they are influenced by the range of different environments within which artefacts are located alongside identifying the conditions that enable or constrain material change.

By now, however, the attentive reader will have noted a shift towards concepts and questions which seem more naturally associated with science and technology studies (STS) than metaphysics. Yet, while exploring social and political processes which drive technological change might seem methodologically far from home for metaphysicians, we shouldn't be too surprised to find such tasks at the centre of a metaphysical research program for the study of artefact maintenance. After all, artefacts have long been understood to straddle the boundary between the human and material worlds. As is well known, artefacts cannot be apprehended entirely in terms of their physical shape and dimensions but must also be understood in relation to human purposes, to what they are considered to be *for*. As noted above, existing studies in the metaphysics of artefacts have tended to overlook how this teleological component changes over time.[16] However, attending to the maintenance of artefacts renders such tendencies infeasible. As I have argued in this chapter, the changes that many artefacts undergo through practices of maintenance can only be explained through recourse to changing perceptions regarding what they are for and consequently, how they should be. Examining such changes leads inevitably to a process-based account that takes change to be primary.

Yet, there is a final concern about this account that I have yet to address, one which metaphysicians in the Western tradition have long considered to be of crucial importance. After all, if we cannot specify those aspects of an entity which remain the same after undergoing change, does not such an account fall foul of the same concerns regarding the possibility of meaningful communication about the world which motivated the development of substance ontology in the first place? Not at all – eschewing the idea that reality, at the most fundamental level, consists primarily of static enduring entities does not mean simply accepting the problem of change. Instead, by operating with a different set of basic ontological assumptions, proponents of process philosophy are able to address the problem in ways that are unavailable to those operating within the paradigm of substance ontology. Johanna Seibt, for example, advances a process account which recommends framing questions concerning the ontology of persistence in terms of processes rather than things. Her 'recurrence account of persistence' addresses the problem of change by taking statements about change and identity over time to apply to processes at different levels of specificity. According to Seibt, statements about identity over time, such as the 'the Golden Gate Bridge today is the same bridge that was built in 1937' apply to *general* dynamics: processes understood as dynamic subjectless activities. Such processes are both ontologically fundamental and homoeomerous. What this means is that while these processes are dynamic, they do not themselves undergo change and so are not analysable in terms of temporal parts or phases. Statements about change, on the other hand, such as 'the bridge used to have a concrete deck, now it has a steel orthotropic deck' apply to *specific* dynamics: processes which, because they do possess different parts, exhibit the presence of change from one time to another (Seibt 2008).

Another approach, proposed by Robert Koons and Timothy Pickavance, employs a notion of processes in terms of human practices to explain how artefacts can be understood to persist through time despite undergoing change. What they call the 'continuous history theory' of artefactual persistence

> exploits certain practices of use and maintenance. For example, if a watch persists, it is because there is a certain ongoing history of use of the watch as a watch and of maintenance of the watch as a watch. Let's suppose that these social processes or practices have a kind of unity through time, that the process as a whole is metaphysically fundamental, not the various instantaneous events that make up the process.... If that is so, then the metaphysical unity of the process over time can be used to ground the persistence of the artifact.
>
> (Koons & Pickavance 2017, p. 561)

While the full potential and limitations of process-based accounts remain to be fully elaborated, the approaches above illustrate how a transition to an ontology based on processes rather than substances does not necessarily preclude us from responding to the problem of change. At the same time, the shift towards an ontological framework which understands change to be primary holds important consequences for the topic of maintenance within metaphysical discussions of artefactual persistence. Rather than being a derivative activity tending to the preservation of that which has already been created, process philosophical perspectives reveal how maintenance can be understood as constitutive of the nature of artefacts themselves, as the glue which binds their identity through time. As calls for a process philosophy of technology grow louder,[17] maintenance begins shifting from the periphery to the spotlight, and in doing so, moves closer to emerging as a central topic in the metaphysics of artefacts.

Notes

1 For an overview of these variations, see Dumsday (2019, pp. 69–72).
2 For the original statement of this objection to four dimensionalism, see McTaggert (1927, p. 14).
3 The role played by artefact kinds in helping to determine persistence conditions, alongside their relation to the intentions of the producer, is a matter of debate. See, for example, Bloom (1996); Evnine (2016, p. 72); Koslicki (2018,

p. 221) for different views on the relationship between intentions, artefact kinds and the modal flexibility of artefacts.

4 This is a standard feature of intentionalist theories from antiquity to the present, see, for example, Young (2021); Koslicki (2018, p. 222).

5 For an account of such processes in relation to bridges in California, see Sims (2016).

6 This has often been considered a consequence of the dominant focus on invention and innovation in a variety of fields attending to the historical study of technology, see, for example, Edgerton (2006, p. 81); Russell & Vinsel (2018). For a notable exception, see Brand (1994).

7 See, for example, Edwards et al. (2011); Schaffer (2011); Henke & Sims (2020, p. 21); Gupta (2021); Harper (1987, p. 19); Lejeune (2019); Denis, Mongili & Pontille (2015).

8 See, for example, the contributions collected in Nicholson & Dupré (2018), alongside Dupré (2017, 2018).

9 See, for example, discussions of Heraclitus in Dupré and Nicholson (2018); Winters (2017, chap 1); van Geert and de Ruiter (2022, chap 1).

10 For an overview of the different ways in which this tradition has upheld oppositions between artefacts and natural objects, see Newman (2004).

11 See, for example, Dupré (2017); Nicholson (2018); Meincke (2018).

12 The removal of salt residue from boats or the underside of cars with the use of water or cleaning lenses with fluids are among the few examples.

13 By choosing the term "slippage" for such cases, Sims encourages a shift in discussions of notions of stability in science and technology studies, away from the traditional focus on how stability is established and secured through processes of closure and towards an examination of how stability is *lost*, often repeatedly throughout the lifetime of an individual artefact. See also Denis and Pontille who encourage a similar shift in noting how "even at rest, artefacts are not as sealed and as stable as they may appear, maintenance and repair studies show. The ability for technologies to remain the same and to be taken for granted by most of their users requires a constant work that such terms as 'black box' or 'immutable mobiles' seriously understate'. (2020).

14 Examples of cases of 'discursive repair' such as this are a common feature of ethnographic studies of maintenance and repair. See, in particular, Henke (2000); Orr (1996).

15 For a more detailed recounting of this story, see Bateson (2012).

16 Among the few exceptions are studies exploring cases of functional exaptation, see for example Thomasson (2014); Preston (1998).

17 See, for example, Coeckelbergh (2023); Young (2021); Brunet (2021).

Bibliography

Bateson, John. (2012) *The Final Leap: Suicide on the Golden Gate Bridge*. Berkely: University of California Press.

Brand, Stewart. (1994) *How Buildings Learn: What Happens After They're Built*. New York: Penguin Books.

Belluci, Stefano & Zaccaria, Massimo. (2012) "Engine of Change: A Social History of the Car-mechanics Sector in the Horn of Africa" in Jan-Bart Gewald, André Leliveld & Ive Peša (eds) *Transforming Innovations in Africa: Explorative Studies on Appropriation in African Societies*. Leiden: Koninklijke Brill NV pp. 237–256.

Bloom, Paul. (1996) "Intention, History and Artifact Concepts" *Cognition* 60 pp. 1–29.

Dumsday, Travis. (2019) *Dispositionalism and the Metaphysics of Science.* Cambridge: Cambridge University Press.

Brunet, Tyler David Price. (2021) 'The Machine-Organism Distinction', PHD Thesis, University of Cambridge. Available at: https://www.repository.cam.ac.uk/items/83ecf29a-d462-45e8-88d0-1e4d3729375c (accessed June 10, 2023).

Coeckelbergh, Mark. (2023) "Technology as Process" in Glenn Miller, Helena M. Jerónimo & Qin Zhu (eds) *Thinking Through Science and Technology: Philosophy, Religion and Politics in an Engineered World.* Maryland: Rowman & Littlefield pp. 55–68.

Denis, David J. (Denis, Jérôme & Pontille, David) (2020) "Why do Maintenance and Repair Matter?" in Anders Blok, Ignacio Fárias & Celia Roberts (eds) *The Routledge Companion to Actor-Network Theory.* New York: Routledge pp. 283–293.

Denis, Jérôme, Mongili, Allessandro & Pontille, David. (2015) "Maintenance and Repair in Science and Technology Studies" *Tecnoscienza: Italian Journal of Science and Technology Studies* 6(2) pp. 5–15.

Dupré, John & Nicholson, Daniel J. (2018) "A Manifesto for a Processual Philosophy of Biology" in Daniel J. Nichlson & John Dupré (eds) *Everything Flows: Towards a Processual Philosophy of Biology* Oxford: Oxford University Press pp. 3–45.

Dupré, John. (2017) "The Metaphysics of Evolution" *Interface Focus* 7: 20160148. Available at: http://dx.doi.org/10.1098/rsfs.2016.0148 (Accessed May 20, 2023).

Dupré, John. (2018) "Processes, Organisms, Kinds and the Inevitability of Pluralism" in O. Bueno, R. Chen & M.B. Fagan (eds) *Individuation, Process and Scientific Practices.* New York: Oxford University Press pp. 21–38.

Edgerton, David. (2006) *The Shock of the Old: Technology and Global History since 1900.* London: Profile Books.

Edwards, Paul N. et al. (2011) "AHR Converation: Historical Perspectives on the Circultation of Information" *American Historical Review* 116(5) pp. 1393–1435.

Evnine, Simon J. (2016) *Making Objects and Events: A Hylomorphic Theory of Artifacts, Actions and Organisms* Oxford: Oxford University Press.

van Geert, Paul and de Ruiter, Naomi. (2022) *Toward a Process Approach in Psychology: Steeping into Heraclitus' River.* Cambridge: Cambridge University Press.

GGBHTD (n.d.) *Seismic Retrofit Construction Project* [Online]. Available at https://www.goldengate.org/district/district-projects/seismic-retrofit/ (accessed June 15, 2022).

Graham, Stephen & Thrift, Nigel. (2007) "Out of Order: Understanding Maintenance and Repair" *Theory, Culture & Society* 24(3) pp. 1–25.

Gupta, Akhil. (2021) "Infrastructure as Decay and the Decay of Infrastructure" in Ghassan Hage (ed) *Decay.* Durham: Duke University Press pp. 37–46.

Harper, Douglas. (1987) *Working Knowledge: Skill and Community in a Small Shop*. Chicago: Chicago University Press.

Henke, Christopher R. (2000) "The Mechanics of Workplace Order: Toward a Sociology of Repair. *Berkeley Journal of Sociology* Vol. 44 pp. 55–81.

Henke, Christopher R. & Sims, Benjamin (2020) *Repairing Infrastructures: The Maintenance of Materiality and Power*. Massachusetts: MIT Press.

Hilpinen, Risto. (1993) "Authors and Artifacts" *Proceedings of the Aristotelian Society*, 93, pp. 155–178.

Hobbes, Thomas. (1656) *Elements of Philosophy, the first section, concerning the body: Written in Latine by Thomas Hobbes of Malmesbury and now translated into English*. London: Andrew Crooke.

Josephson, Paul R. (2017) *Traffic*. New York: Bloomsbury Academic.

Koons, Robert C. & Pickavance, Timothy (2017) *The Atlas of Reality: A Comprehensive Guide to Metaphysics*. West Sussex: Wiley Blackwell.

Koslicki, Kathrin. (2018) *Form, Matter, Substance*. Oxford: Oxford University Press.

Lee, Mi-Kyoung. (2010) "Antecedents in early Greek philosophy" in Richard Bett (ed) *The Cambridge Companion to Ancient Scepticism*. Cambridge: Cambridge University Press pp. 13–35.

Lejeune, Christoph (2019) "Interruptions, Lunch Talks, and Support Circles: An Ethnography of Collective Repair in Steam Locomotive Restoration" in Ignaz Strebel, Alain Bovet & Phillippe Sormani (eds) *Repair Work Ethnographies: Revisiting Breakdown, Relocating Materiality*. Singapore: Palgrave Macmillan pp. 221–251.

McTaggert, John M.E. (1927) *The Nature of Existence Vol 2* C.D. Broad (ed) Cambridge: Cambridge University Press.

Meincke, Anne Sophie. (2018) "Persons as Biological Processes: A Bio-Processual Way Out of the Personal Identity Dilemma" in Daniel J. Nicholson & John Dupré (eds) *Everything Flows: Towards a Processual Philosophy of Biology*. Oxford: Oxford University Press pp. 357–378.

Meincke, Anne Sophie. (2019) "The Disappearance of Change: Towards a Process Account of Persistence" *International Journal of Philosophical Studies* 27(1) pp. 1–19.

Meincke, Anne Sophie. (2021) "Processual Animalism: Towards a Scientifically Informed Theory of Personal Identity" in Anne Sophie Meincke & John Dupré (eds) *Biological Identity: Perspectives from Metaphysics and Philosophy of Biology*. New York: Routledge pp. 251.

Newman, William R. (2004) *Promethean Ambitions: Alchemy and the Quest to Perfect Nature*. Chicago: Chicago University Press.

Nicholson, Daniel J. (2018) "Reconceptualizing the Organism: From Complex Machine to Flowing Stream" in Daniel J. Nicholson & John Dupré (eds) *Everything Flows: Towards a Processual Philosophy of Biology*. Oxford: Oxford University Press pp. 139–166.

Nicholson, Daniel J. and Dupré, John (eds) (2018) *Everything Flows: Towards a Processual Philosophy of Biology*. Oxford: Oxford University Press.

Orr, Julian E. (1996) *Talking about Machines: An Ethnography of a Modern Job*. New York: Cornell University Press.

Pearce, David. (2016) "Collective Intentionality and the Social Status of Artifact Kinds" *Design Science* 2(3) pp. 1–28.

Plutarch. (1859) John Dryden (trans) *Plutarch's Lives Vol 1*. Boston: Little Brown and Company.

Preston, Beth. (1998) "Why Is a Wing Like a Spoon? A Pluralist Theory of Function" *The Journal of Philosophy* 95(5) pp. 215–254.

Preston, Beth. (2013) *A Philosophy of Material Culture: Action, Function and Mind*. New York: Routledge.

Preston, Beth. (2022) "Artifact" in Edward N. Zalta & Uri Nodelman (eds) *The Stanford Encyclopedia of Philosophy* (Winter 2022 Edition). Available at: https://plato.stanford.edu/archives/win2022/entries/artifact/ (accessed August 23, 2022).

Rescher, Nicholas. (1996) *Process Metaphysics: an Introduction to Process Philosophy*. Albany: State University of New York Press.

Rescher, Nicholas. (2000) *Process Philosophy: A Survey of Basic Issues*. Pittsburgh: University of Pittsburgh Press.

Russell, Andrew L. & Vinsel, Lee. (2018) "After Innovation, Turn to Maintenance" *Technology and Culture* 59(1) pp. 1–25.

Schaffer, Simon. (2011) "Easily Cracked: Scientific Instruments in States of Disrepair" *Isis* (102(4) pp. 706–717.

Seibt, Johanna. (1997) "Existence in Time: From Substance to Process" in J. Faye, U. Scheffler & M. Urchs (eds) *Perspectives on Time (Boston Studies in the Philosophy of Science Vol. 189)*. Dordrecht: Springer pp. 143–182.

Seibt, Johanna. (2004) "Process and Particulars" in Michel Weber (ed) *After Whitehead: Rescher on Process Metaphysics*. Frankfurt: Ontos Verlag pp. 113–134.

Seibt, Johanna. (2008) "Beyond Endurance and Perdurance: Recurrent Dynamics" in Christian Kanzian (ed) *Persistence*. Frankfurt: Ontos Verlag pp. 133–164.

Simons, Peter. (1987) *Parts: A Study in Ontology*. Oxford: Oxford University Press.

Simons, Peter. (2000) "Continuants and occurrents" *Proceedings of the Aristotelian Society Supplementary Volume* 74(1) pp. 59–74.

Sims, Benjamin. (2016) "Seismic Shifts and Retrofits: Scale and Complexity in the Seismic Retrofit of California Bridges" in Mike Hodson & Simon Marvin (eds) *Retrofitting Cities: Priorities, Governance and Experimentation*. New York: Routledge pp. 13–33.

Thomasson, Amie. (2003) "Realism and Human Kinds" *Philosophy and Phenomenological Research* 67, pp. 580–609.

Thomasson, Amie. (2014) "Public Artifacts, Intentions, and Norms" in M. Franssen, P. Kroes, T.A.C. Reydon & P. Vermaas (eds) *Artefact Kinds: Ontology and the Human-Made World*. Cham: Springer pp. 45–62.

Wasserman, Ryan. (2006) "The Problem of Change" *Philosophy Compass* 1(1) pp. 48–57.

Winters, Andrew M. (2017) *Natural Processes: Understanding Metaphysics Without Substance*. Cham: Palgrave Macmillan.

Verrips, Jojada & Meyer, Birgit. (2001) "Kwaku's Car: The Struggles and Stories of a Ghanian Long-Distance Taxi Driver" in Daniel Miller (ed) *Car Cultures*. Oxford: Berg pp. 153–184.

Young, Mark Thomas. (2021) "Maintenance" in N. Doorn & D. Michelfelder (eds) *The Routledge Handbook of the Philosophy of Engineering*. New York: Routledge.

4 There, I Fixed It! On the Status and Meaning of Repair

Tiago Mesquita Carvalho

4.1 Introduction

What happens during repair? How does one learn to repair an artefact? and what is the relationship between repair and function? And if repair is a skill that requires knowledge, how can it be described? Can there be a general theory of repair capable of reducing knowledge to steps and procedures? To address these matters, we will use the theoretical underpinnings of post-phenomenology as formulated by Don Ihde, as well as other analytical concepts from authors who have put forward a metaphysical theory of artefact function. The similarities with other forms of knowledge in science, technology, design and ethics will also be explored to clarify several issues around repair, mainly its relation with function and the designer's intention, its meaning and the nature of the knowledge it uses.

In Section 4.2, we start by examining the definition of repair as a means to restore the proper artefact function, bridging a discontinuity between its initial state before the breakdown and its final state after the repair. This definition will be explored to determine how one can tell whether a repair has taken place since a good test often means reusing the artefact. Nevertheless, there are artefacts that cannot be tested with use. This leads us to question the nature of the knowledge required to repair an artefact and its relation to the knowledge used to produce and design said artefact.

In Section 4.3, we move forward with Ihde's understanding of the place of technology in the lifeworld. We draw some implications about the nature and meaning of their function and breakdowns. The lifeworld is filled with users' intentions and these interplay with the ambiguity of artefacts. An analysis of repair has to take this into account. Proper function and the designer's intention are not enough to appreciate their selection and variation in reproduction, as well as the dynamics of the transfer between cultures. Scrutinising examples of how the same artefacts are used and repaired in different cultures, rather than a mere comparative exercise, has

DOI: 10.4324/9781003316213-5

analytical value for the concepts surrounding repair and maintenance. Furthermore, an analogy between artefacts and texts is drawn, highlighting relevant affinities. Section 4.4 continues this exploration by framing repair in context and exploring how a technical syllogism regarding repair must be context-specific. The technical syllogism of repair serves to qualify what comprises a good or bad repair and which judgements and actions are valid to achieve the functionality of an artefact. However, the syllogism is highly dependent on the context and its available parts and tools.

Section 4.5 discusses the relationship between repair knowledge and scientific and technological knowledge in the design of artefacts. The similarities with other forms of knowledge are pointed out and their role in the skill of repair is discussed. Here we distinguish between repair knowledge that can be formalised in rules and procedures and that which represents the cultural consequences of the accumulation and transmission of personalised forms of that knowledge. Finally, Sections 4.6 and 4.7 discuss a phenomenology of the repair attitude and its structure.

4.2 About Repair

Preston (2009, p. 217) draws attention to how phenomena like maintenance, repair and rebuilding should be integral to a theory of artefact function. Although there are significant differences between individual and collective repair concerning the division of labour and the organisation of work, at first sight, it does not seem essential for the purpose of this analysis to distinguish types of repairs based on artefact kinds. Repair is either done individually or collectively, and there must be a vision of all the steps required for an artefact to achieve its function. The complexity of the artefact, as well as time constraints and lack of personnel and resources, have a bearing on whether a repair should be accomplished individually or collectively.

The object of repair seems to be artefacts. To cultivate and trim a garden or to heal a patient is to cooperate with an already active process in such entities. It seems inappropriate to say that we *repair* a garden or a person except perhaps when used as a figure of speech. There is something distinctive about the object of repair that means it is more of a machine than part of life or nature. The much-vaunted term 'technological fix' for solving issues denotes a cultural propensity to turn social or political issues into technical ones. Some authors, such as Spelman (2002, pp. 4–6), seem to identify repair on a continuum with other terms like restoration, preservation or even healing, which encompass a recovery of some earlier state before a breakdown, whether this is understood to apply to living entities such as ecosystems, geopolitical entities like nations or things such as relationships.

Following Baker (2009, p. 82), we will consider repair to only be concerned with technical artefacts. However, we are not saying that repair does not play a role in the nexus between material and social realities. As Henke and Sims (2020, p. 19) note, repairing a bridge, for instance, is not only about fixing a structure but also a whole system of artefacts, institutions and organisations. There is a continuum across the social and technological orders, implying that repair is not only about changes in artefacts but also social systems. Furthermore, repair sometimes depends on the best available knowledge. Repair usually occurs after something breaks down, but the causes for the failure may remain hidden while a makeshift solution is adopted. Knowledge, standards and conceptions of maintenance and repair coevolve.

Besides, when we say that repair concerns artefacts, these include artefacts generated by activities such as art, craft, invention or engineering. Repair may apply to artistic, cultural or decorative artefacts. One can repair a painting or a frame, an activity which is often related to the conservation and restoration of cultural property. The renovation and refurbishment of technologies such as infrastructures, buildings and houses are also a type of repair. In addition to being a collective endeavour, they must be carefully planned with tasks assigned and deadlines met. Activities such as amending software code, adjusting the brightness of videos and images or subtitling films are also examples. These cases are not really an exception to the definition so much as a narrowing and specialisation of the meaning and judgements involved.

At the same time, it does not matter whether we are discussing the repair of mobile phones, automobiles or bridges. Each repair skill seems distinct and has its own language shaped by the particularities of the materials and components that usually break down, but there should be something common to the whole structure of the repair skill, that is, the need to restore their functional state. In the first instance, the repair is then defined as restoring an artefact to working order after a breakdown. It appears to be a return to the past, a way of reestablishing continuity.

Repair is therefore teleological, an end-directed activity, towards a final state of the artefact, such that it works and delivers a function. For now, we will assume that such a final state should resemble the initial state, as accomplished by the artefact's proper function (Vermaas & Houkes, 2003, p. 265).[1] In other words, repair must assume, from the outset, the existence of a solution. There is no repair without belief in a solution and the ability to find it; one moves towards a problem not as an end in itself but on the assumption that it is a means to an end, towards the meaning of 'it works!' or 'there, I fixed it', whatever that may be. Repair is thus guided by the aim of providing the artefact with a working state, imagined as reliable, and acting as the regulator of all steps and actions to be taken.

In addition, a more demanding definition of repair requires that the final state is defined as one where the fulfilment of the artefact function can be accomplished without any mishaps, provided that the user is aware of how to use it. A test is needed and use is often what suffices provided one knows what to do. A car out of fuel is not broken but it cannot be driven. Moreover, we do not need to look at a more complex artefact with several subsidiary functions to understand how repair can be about correcting secondary mishaps and not just restoring the proper function. A dull knife still cuts and a wobbly chair can still be sat on. Repair may well contribute to restoring the artefact's proper function, but often it also means perfecting a secondary part of the artefact so the performance of that function can be perfectly accomplished.

In addition, although repair is concerned with the artefact's future use, one who repairs does not necessarily know how to use it. The repaired artefact can be used as a subsidiary or as a part of another one. In collective repair, the purpose of functionality is usually achieved without it being necessary for everyone involved to know how to use the artefact. For example, a team of mechanics can repair a car engine without it being necessary for each of them to know how to drive an F1 car. On the other hand, to accomplish the repair of a wide range of artefacts, the repairer has to know whether they have reached their final state by performing a test to check whether the repair was successful. While it may be tempting to assume that testing requires use, this is not always the case. If the pillars of a metal road bridge start to become corroded by the salt and moisture of an aquatic environment, they have to be repaired, although crossing the bridge is not considered a reliable test of a well-performed repair. The use of the repaired part of the bridge is not sufficient evidence of the appropriateness of the repair. Here, to be validated, repair perhaps requires an instrument to show that the repaired structure now accords with safety guidelines. This then implies that various types of knowledge are required to validate a successful repair.

In the above examples, repairing seems to involve a type of knowledge distinct from the specialised knowledge required to produce the artefact. Repairing requires knowledge, although it does not have to be the one used in production, but rather knowing about the artefact's inner workings. In the bridge example, theoretical and conceptual knowledge about the bridge's corrosion performance in the surrounding aquatic environment is essential to understand how it should be both monitored and repaired. Without this knowledge, it would be impossible to repair the bridge and ensure its safety and structural soundness. For our purposes, specialised knowledge can be considered either scientific or technological. Knowledge about the chemical and mechanical corrosion of the bridge's materials and structures has a bearing on the need for repair and what to repair.

The question arises as to the relationship between the knowledge needed to repair the bridge and the knowledge needed to make it.

4.3 Defining Repair

Building on Ihde's extension of Heidegger's thinking, it is assumed that many, if not all of the artefacts we possess, are not isolated but instead interwoven with other systems and infrastructures (Ihde, 1990, p. 32). In repair, there is often a *ceteris paribus* principle, that is, the assumption that the context in which artefacts work is consistent. Repairing rests on the assumption that the artefact's technological context remains subsidiary to its use. To succeed, it needs stability in the world. Our analysis of repair is based on the assumption that there are no artefacts *per se* in the lifeworld. The artefacts and infrastructures that surround us belong to an 'in-order-to' context where they fulfil their function and provide meaning.

Repair is concerned with how to make an artefact work again. If repair implies knowing how an artefact works, it also implies that the repairer understands the significance of the artefact and how it fits into the interests of *that* everyday user. The point is that an artefact being broken can only be understood in relation to a lifeworld where artefacts already possess a normative function which determines how they are supposed to perform (Houkes & Vermaas, 2010, p. 111). The lifeworld proper to artefact designers, repairers and users is one where intentions, assumptions and skills are built around what artefacts ought to accomplish. An artefact can exist in a world lacking minds with propositional attitudes, but the same artefact is only broken by virtue of being made by human beings. Hence, any given artefact is more than its chemical and physical components arranged in a given structure. Malfunction, proper function and repair are then conceptually linked as they are dependent on human intentions. One cannot conceive an ontology of artefacts' function or malfunction while maintaining that they are mind-independent or that our concepts do not have a bearing on what they are (Baker, 2009, p. 89).

Furthermore, in line with Ihde, the ambiguity of each artefact is revealed according to their contexts of use and the spatial and temporal constraints of each lifeworld (1990, p. 69). This means that the ambiguity of any artefact stems from the way its current proper function is dependent on its social and cultural history.[2] This history hinges on several contingencies that blur the division between proper and non-standard accidental functions, while simultaneously providing them with a meaning. Every artefact is subject to a hermeneutic process that secures, extends or modifies the designers' original intentions. And this, as Ihde notes, is because a technological object, whatever else it is, becomes what it 'is' through its uses by being brought into the range of praxis (Ihde, 1990, pp. 70–71). A philosophical treatment of repair must therefore recognise how it plays out with the ambiguity of technology in general and how malfunction is the nexus where accidental functions can acquire relevance. Repair is never about the object as an object, but about what it is used *for* and its practical purposes in the user's lifeworld. An instrumental and metaphysical view of artefacts which understands the use, behaviour and repair of an artefact to be deduced from the physical and material properties of its components does not suffice. Every artefact has a social and cultural element, and without considering such elements, artefact function cannot acquire the usual everyday meaning in daily practices. The artefacts of the built environment are therefore not just matter arranged according to scientific principles and technological procedures. As Ihde points out, they enter the everyday lives of users through immediate, sensuous and perceptual contacts of the body that foster or curb disclosing possibilities. The artefacts in the lifeworld become interpretable while manifesting their multistability[3] (Ihde, 1990, p. 144), subject to various interests.[4] It is through them that, as users or consumers, we can identify malfunctions.

It is also useful to distinguish two senses of perception, a sensory one and a cultural or hermeneutical one. Imagine two subjects A and B, who while returning to their respective parked bicycles, find that each has a flat tire. While subject A considers repairing the puncture with a patch, subject B reads the situation as requiring a trip to a bicycle shop. The *microperceptions* are the same, but the *macroperceptions* are different. These two senses of perception both belong to the lifeworld (1990, pp. 29–30). A sensory embodied impression about what malfunctions in an artefact is interwoven and linked with a hermeneutic and cultural perception of its meaning. The relationship between these two perceptual features is not one of derivation, but one of figure-to-ground. Any microperception needs a hermeneutic-cultural macroperception field to occur and there is no macroperception without its microperceptual foci. The interaction between these two perceptual features makes it possible to understand issues linked to the transfer of artefacts from one culture to another, as well as to understand why the repair of any artefact depends on the meaning it acquires in a certain culture.

This distinction allows one to see why an artefact's proper function is necessary but not sufficient to understand the reproduction of artefacts inside a culture and their various transformations. One needs to recognise that the designer's intentions regarding artefact function are only part of what defines them. There is a whole field of artefact design that is usually unintended, as well as interactions of the intended proper function with social and cultural realities. The purposes of artefacts are often found by trial and error and cannot be removed from human intentions. In the same way, assuming that the meaning of a literary work lies in its author's intentions is a fallacy (Ihde, 2008, p. 51), the idea that one or more designers are responsible for a given artefact's future uses or proper function is also a fallacy. This would be analogous to the idea that readers must subscribe to a single interpretation of a given literary work. Such a hermeneutic game does not ignore that there are intentions underlying texts and artefact design, but when the texts and the artefact move on to practical contexts, these intentions end up interplaying with users' intentions.

Against the deistic and individualistic claims of the designer who creates a form with a single purpose, the material plasticity of artefacts subjects them to a lifeworld constituted by macroperceptual intentions of users with their own cultural history. Artefacts should not be seen as conditional on transparency and control (Ihde, 2008, p. 58). Design, and repair for that matter, should be regarded more as experimental and material interrogations of possibilities that may result in what is desired and expected. Examples in the history of technology abound. The phonograph, the email (Nye, 2006, p. 44) or the Minitel (Feenberg, 1999, p. 121) have all slipped from their authors' original intentions and were then reconfigured by both users and designers. An accidental function became a proper function once it merged into social reality.

These examples reveal how multistability is the result of several human and social processes and their interaction with changing interests and contexts. Artefacts evolve because of unplanned contingencies of use, driving the complexity of relationships between designers and their own cultural and geographical contexts. Consequently, the original proper function is often subverted, adapted, diminished or even ignored. An artefact is never completely transparent to its designers, as it contains an indeterminate set of unintentional possibilities inscribed in its physical structure. More generally, the relativisation of the designer's intentions vis-à-vis how a given artefact should work highlights the heuristic value of flaws, failures (Petroski, 1996, p. 89) and users' interpretations, ultimately promoting new trajectories and practices of material culture.

To understand both repair and design, one must then account for a theory of culture that recognises how artefact functions are subject to the interplay between practices, materiality and the intentions of both users and repairers. Authors with an analytic background, such as Vermaas and Houkes (2003) and Preston (2003, 2009), have examined non-intentionalist etiological theories about the history of selection and reproduction of artefact functions and their similarities with biological functions. Ihde's viewpoint seems to be in line with Preston's (2003, p. 603) critique of Vermaas and Houkes (2003) concerning the insufficient role played by designers' deliberations and how proper functions often start out as accidental or system functions. Intentionalist theories of function downplay the history of an artefact's uses and contexts and place the onus of function on designers. Like Preston's example of marigold beer, the designer's hopes, expectations and intentions seem to be the criteria that qualify a repair as acceptable or deviant if it restores proper function or introduces an accidental function. On the other hand, users too have a deliberative history (Preston, 2003, p. 606) that repair often helps to realise, given that use itself is creative. We argue that it is the suitability of users' intentions in their lifeworld that ultimately grounds the adequacy of repair and what counts as a function. Users have desires, wishes and expectations concerning what the artefact should accomplish. The reproductive and selective processes of artefact evolution depend not only on designers and their intentions but also on organised patterns of macroperceptual practices.

Artefacts, by belonging to a culture and the users' lifeworld, then become alive. They are quasi-living entities, focal points of cultural and technological networks in which function and meaning are reconfigured through breakdowns, repairs, design, collaborative participation and improvements that may stimulate the development of communities and skills. Many times, an accidental function is instituted through a hermeneutic or material reappropriation of users that will have a bearing on the designer's intentions. This allows repair to be conceived as exploring and establishing new uses and stimulating new practices in which individual creativity may make an artefact functional according to the meaning it acquires in any given culture.

Repair must then restore a function to artefacts and while such a function may be the proper one, *this is not necessary*. It can be about assigning a function that the artefact already allows or that some minor alteration permits. Repair discloses the artefact and lingers in the oscillation between restoring its proper function or modifying it to another one. Hence, comparative exercises about the uses of an artefact and how it is repaired in distinct cultures or socio-economic contexts are more than an exercise of curiosity. It has an analytical scope for the concepts used in the philosophy of technology and engineering. The history of selection and reproduction of a given artefact shows the instability of its proper function and the limited role of the designer's intentions. But it also displays the limits of the idea that there can be a pure or neutral artefact, that is, an artefact that does not carry the designers' intentions and the macroperception about the meaning of its function. The adoption of a particular transferred artefact depends mainly on its multidimensional ability to fit into an existing practice (Ihde, 1990, p. 126).

Most often, an artefact is transferred without the technology, that is, the macroperception that assigns meaning to the artefact, just as when a text is transferred between cultures and ages its reading becomes different. Thus, the transferred artefact engages with its context of use, becoming 'what it is' by regularly fulfilling a function and acquiring its meaning in relation to such a new context. A ready-to-be-scrapped automobile in affluent societies is, in other cultural contexts, a source of materials or may even be considered repairable, for example, if converted to an animal trailer (Houkes & Vermaas, 2010, p. 110). A commonplace glass Coca-Cola bottle, as portrayed in Jamie Uy's 1980 film, *The Gods Must Be Crazy*, appears as a single, detached artefact for an isolated tribe in the Kalahari Desert. But soon, notwithstanding its weirdness, it starts performing functions already of existing artefacts.

We can now broaden our earlier preliminary definition of repair. What we have seen so far has revealed how repair should not be conceived as restoring the proper function of an artefact, but instead about restoring the meaning a particular function acquires in the user's lifeworld. Mechanics like Willie (Harper, 1987, pp. 19–21), for example, practice repair not only by restoring artefacts to their initial states but also by making modifications that go far beyond the designer's intentions and fit his client's standard of living. The tradition of repairing Japanese porcelain used in tea ceremonies with golden lacquer, known as *Kintsugi*, is another example. Such porcelains are valued for having traces of their fragility and displaying a history of use. The goal of repair is bound to show imperfections while ensuring functionality. The golden lacquer mends, but also becomes an integral part of the bowl. Thus, within the tea ceremony macroperception, the artefacts acquire an exquisite aesthetic value that ordinary pottery cannot. On the other hand, as Spelman details (2002, pp. 17–20), a painting restoration is a repair that aims not to *remove* but to *reverse* the effects of time. That is, repair does not only have to restore a function but also to safeguard the integrity of the artwork. In the case of a painting, repair is personalised but it should be minimal. The steps taken matter and may, according to the context, be considered as damage or not, as having an aesthetic value or harming the appreciation of the painting.

4.4 Repair and Context

In Section 4.3, we saw how two factors should be considered when defining repair: the artefact to be repaired and its context, both as a place where micro- and macroperceptions are joined. Repair and malfunction (Houkes & Vermaas, 2010, p. 108) are highly context-specific and, hence, the knowledge of how to repair something is mostly acquired in each context. Existing tools, parts and materials are needed that assist, hinder or enable the success of the repair. The judgement about what is exactly malfunctioning and what is to be done with a broken artefact token cannot be abstracted from the in situ context in which it ought to work again. Repairing an artefact token is neither pre-given nor identical to a universal end state of a given type of artefact.

Repair takes place when the contextual conditions surrounding an artefact token provide the background context in which perceptions, judgements and actions become appropriate. This can be represented by a *technical syllogism* that qualifies what is a good repair and depends on the knowledge exacted in a given context.[5] Abstracting the actions and judgements that compose each context as if they were rules that apply across all repair activities of a given artefact type may lead to flawed approaches. Gerrike et al. (2017, p. 107) have advised that applying methods successfully depends not only on the formal adequacy of the method but also on the skills and judgements utilised by users in applying it to a particular context. However, the conditions of an artefact's context do not always reveal from the outset what can be repaired and what counts as repaired. Often, there are no appropriate tools or materials, which can lead to ingenious solutions. In such cases, a depleted context does not hinder repair, but rather fosters a more hands-on creative activity closer to tinkering and invention: *informal repair*. This is how we often encounter a broken artefact in everyday life. Its analysis is also phenomenologically rewarding. Furthermore, we often observe, especially in more marginal areas of the planet, customised and patched artefacts that reflect their context (Smith, 2010, p. 1). These make-do repairs are not temporary solutions but rather reflect the very way in which repair knowledge is manifested and how its efficiency relates to its context.

Other contextual features surrounding some artefacts are related to their importance for everyday social reproduction, such as in the case of routine, preventive or life-cycle maintenance. Repair can be an ongoing process throughout the life span of an artefact or technological system. Sometimes repair work never really restores a system to the much-envisioned state of its proper function. In the rural countryside of Portugal, for instance, houses are seldom finished. They are often inhabited while being repaired or upgraded existing next to the building materials in a state of stasis, depending on the next capital input. A final solution may not even be available and, while a temporary one can be achieved, it is dependent on variable circumstances.

It may also be the case that successive repairs lead to the replacement of an artefact's parts, in what could be another instance of the Ship of Theseus. Rather than exploring the metaphysical question, however,[6] let us look at examples of how the same broken artefact will be repaired and restored in different ways in different cultural and economic geographies. Different macroperceptions give rise to different interpretations of the artefact's proper function that go beyond designer intentions. In societies where repair is primarily a professional activity and there is an abundance of tools, parts and components, it is easy to overlook how the context of repair dictates what is or is not possible to do. Repair may seem merely a return to the initial proper function of an artefact that appears more or less bound to an already fixed meaning. In other contexts of greater economic scarcity, however, repair appears as a more comprehensive approach that explores the possibility of using different materials and tools as replacements for defective parts.

Ernesto Oroza, a designer, has examined how artefacts in Cuba are prone to fluctuate beyond their proper function. He calls this patched work character of artefacts 'technological disobedience' (Oroza, 2009, p. 20). Since Cubans are subjected to a political regime and an embargo that limits their access to essential commodities, many have become maker-repairers. The Rikimbili (Figure 4.1), for example, are two-wheeled adapted vehicles that utilise chainsaw engines or plastic bottles as fuel tanks. Some are used as movable, pedal-activated sharpeners or grinders.

In keeping with the analogy between technology and text, we can say that informal repair often means a partial rewriting of the original material *inscription* given by engineers, designers and past users (Latour, 1992 p. 236). As Bruno Latour notes, the very *description* of what artefacts do is brought to light once they are removed from their everyday lifeworld

Figure 4.1 A Rikimbili.

Photo by Ernest Oroza. Oroza, E. (2021) Rikimbili. Available at: http://www.technologicaldisobedience.com/wp-content/uploads/2021/07/Rikimbili-55.jpg (Accessed: 19 June 2022).

through an extraordinary event such as a crisis (Latour, 1992, p. 260) that shatters, breaks or interrupts the usual context of use. Disruptive possibilities appear that illustrate what Feenberg (1999, p. 221) called secondary instrumentalisation in which users can assign technology other meanings.

The point is that examining actual cases of repair illustrates how the proper function of artefacts is not fixed, but frequently interpretable. Similarly, after breaking down, the postponement of repair can reveal all the social relations implicit in that artefact as a way of life. This is precisely Winner's argument (1977, p. 331). Breakdowns and dismantling could foster an intended research programme, an 'epistemological Luddism' that can have high educational value by illuminating the 'buried substance' of our civilisation: how artefacts are more than a means to an end and how human relations are transformed by their presence. According to Winner, repair is a confession of our technological addiction that causes severe withdrawal symptoms when consumption is interrupted. The symptoms could, however, provide an explanation for how the structure of human relations is determined by the technology in question, thus making it possible to justify technological choices.

Returning to the analogy between technology and text, informal repair is in many ways similar to the practice of a medieval copyist monk tasked with transcribing an ancient yet incomplete manuscript of a work they had once read. Because the text has gaps, the monk has to 'repair' it in the act of copying. Likewise, any artefact may contain scripts of action and perception (Verbeek, 2006, p. 1113) put forward by engineers and designers and interpreted by its users. The artefact may present defects relative to how it usually performs, just as the medieval monks' attempt to be as faithful as possible to the original meaning encountered through previous contact with the text. He cannot, however, avoid introducing certain expected words or terms where they are absent. The final meaning of the manuscript may or may not be preserved, but the text has undergone some changes. While copying, however, he notices, like the translator of José Saramago's novel 'The History of the Siege of Lisbon' (1998), that his practice allowed him to rewrite a manuscript, allegedly changing its meaning by introducing a simple 'not'.[7]

4.5 Repair Knowledge

How does someone who repairs proceed to be successful? We have already noted that repair is a knowledge-driven activity, but what kind of knowledge is employed and how does it relate to the activity? In many cases, knowledge of repair seems to be acquired in the field, through prolonged contact with the artefact and its context. Someone who repairs seems to possess a knowledge distinct from others, a skill by which they know exactly how to apply that knowledge through a technical syllogism. Nevertheless, this seems to have been changing for several decades. Informal repair linked to the 'shade-tree mechanic' has faded. A case in point is modern bicycles.[8] With the advancement in technology, bicycle maintenance and repair have become increasingly more complex and are now rigidly controlled by manufacturers in an attempt to constrain interchangeability between parts from the same artefact type. The solution to a repair problem is becoming predetermined from the outset.

Many new artefacts contain either scientific or technological state-of-the-art knowledge. However, that knowledge does not necessarily include the knowledge required to repair the artefacts themselves. The scientific and technological knowledge guiding the design of many artefacts does not seem to be of the same nature as the practical knowledge required to repair them. A more general epistemological question that repair raises is whether certain human activities can be defined by another kind of knowledge than the propositional knowledge based on laws, rules and procedures. We will now explore this point.

Informal repair implies a form of perception that does not resort to rules, procedures, concepts and their formal sequencing but is nevertheless operative and highly effective. This idea can be traced back to Aristotle regarding the relationship between *phronesis* and *techne*, both intellectual virtues that although related to what might be otherwise, are also distinguished from each other (Aristotle, 2008, NE 1140a 24–30). For Aristotle, the image of the good is not given beforehand but depends on the circumstances surrounding the agent (Broadie, 1991, p. 198). The right action resists a universal prescription precise enough to be valid in all cases: it is up to the agent to judge and feel what is right in any particular context. Although *techne* requires a universal *logos* for all bodies – the image of the final, functional state of the artefact – informal repair concerns knowledge appropriate to a specific form of making. Whoever makes or repairs something proceeds with an idea of the final state to be achieved. However, this knowledge is not enough to know *how* to achieve that state or *how* to give form to the matter or artefact. Experience is needed. There are many similarities between doing and making, but as Mitcham (1994, pp. 131–4) emphasises, there is a distinction between *techne* and modern technology, in that the former is always driven towards particular goods, while the latter, given the rationalisation of the productive process, allows for mass production and procedural-like rules. The major difference between repairing and science is akin to the contextualism of *phronesis* and *techne*. It lies in how the repair skill based on the knowledge required to fix certain artefact tokens does not imply universal knowledge about all artefacts of that type which hold regardless of time and place. The erosion of time, wear and tear from use, and the flexibility with which users can interpret the artefact, alters both its meaning and what counts as its ideal function. Whereas the natural sciences apprehend a body and its behaviour as an example is generally framed in a lawful pattern or rule valid for all times and places. Although there are analogous cases, each informal repair deals with singular individuals not with examples of artefact kinds.

Thus, put briefly, if *techne* is analogous to art and the personal element in making something is lost in mass production, informal repair still seems to retain an element of being in touch with concrete materials and singularities and with the contingencies of each poietic process. Processes of repair, like craft, are never identical but involve the discovery of what is appropriate. The craftsman and the repairer have to let the artefact speak to their senses in order to work with it. Similarly, to become good practical agents, we have to live with others to 'pick up' and emulate how to act properly while understanding why we do it (MacIntyre, 1999, p. 69). The conclusion is that there appears to be a non-conceptual element (Dreyfus, 2002, p. 398) in both *techne* and *phronesis* since concrete circumstances are what matter in determining the criteria for what to do or what to make. In its similarity to the contextual qualities of *phronesis*, repair is thus able to take on the features of a 'practical labour' that pays attention to the local features where it takes place, with the 'concrete and actual realities of its natural setting' (Harper, 1987, p. 20).

Despite the attraction of rigorous formalisation based on mathematical and scientific knowledge, the above considerations seem to prevent repair, much like *techne* and *phronesis*, from being formalised in a decision theory capable of identifying optimal operative actions. A step-by-step procedure cannot be determined in advance as it depends heavily on the particulars of each item, just as attempting to create a science of ethics or a science of repair seems futile, although the general structure of each can be identified (Aristotle, 2008, NE 1094b11). The skill of repair appears to grow on a 'knack for' or a 'feeling' which employs knowledge that is only revealed in the action itself, that is, in the practice of repair. It is therefore an example of both tacit knowledge and connoisseurship (Polanyi, 1962, p. 54). Each repairer knows more than they can say and tends to express it inaccurately. Harper defines Willie's intuitive knowledge along similar lines to Polanyi's: an embodied, imaginative and explorative action that is attentive to sound and tactile clues and signals and that is familiar with the nature and behaviour of the shape, hardness, plasticity and resistance of the materials that make up the type of artefacts he repairs. To some extent, this is a more refined, objective knowledge. The repair person must have a deep understanding of the materials of the artefact, as well as an operative knack for the laws of physics and engineering that govern its operation, an old 'mélange of hand and theoretical knowledge' (Harper, 1987, p. 20).

Science, guided by truth and ideals of value-neutrality, has a fundamental criterion for judging the adequacy and justification of explanatory theories. However, the criterion of efficiency that guides the production and design of technological objects allows for a whole variety of cases that satisfy the criterion of functionality, of what really works. What makes a technological proposition successful is not its truth value, but its adequacy. The same problem admits different interpretations of what counts as 'being solved'. The multistability of artefacts is demonstrated in repair and the fact that its procedure cannot be put in binding formulas and rules. Moreover, context dependency hinders the possibility of constituting a heritage of repair practices.

The methodical observation and idealisation of reality in mathematics and science are codified in laws and formulas. They foster the historical construction of a cultural and scientific heritage with technical potential that grows from generation to generation and becomes continuously accessible as both truthful and operational. In mathematics and science, what is new is built upon old premises and discoveries which do not lose their relevance and application. On the other hand, the skill of informal repair implies an original personal contact with the meaning of artefacts and a heuristic understanding of how they work without necessarily appealing to first principles. This has many similarities with how actual design proceeds according to the description offered by Bucciarelli (1994, p. 18). Design and engineering are not just applied science (Boon, 2006, pp. 29–30, Kant & Kerr, 2019, p. 687). By exploring the question 'What does this artefact do?', repair becomes akin to a way of thinking and acting in engineering design by which the artefact is a living element of a cultural matrix.

In informal repair, artefact and mind come into play through a hermeneutic game of responses and counter-responses, projections and explanations that form an arc of intentionality about what remains to be done in the technical syllogism. Through the repair attitude, the artefact is made personal, even if it has been designed according to mathematical formalism. A set of models, concepts and principles of scientific or technological knowledge can nevertheless be helpful. They can be included in a hermeneutic game through the tokens they instantiate and the way they are embedded in particular components. It is moreover to be expected that the precision of concepts or instrumental measurements should assist the repair of complex mechanisms, as in the case of the bridge discussed in Section 4.2.

However, in order to be applied, these concepts must be translated into concrete particulars. Formal knowledge at the level of concepts, laws and principles has to be converted into the technical syllogism of repair and by which the particulars, whether they are screws, bolts, metal plates, batteries or electrical circuits, become contextualised and operative. That is, the in situ personalisation intrinsic to tacit knowledge comes into play if it is to be effective and acquire a vital form. This is how the gap between abstract knowledge and personal, tacit and concrete knowledge is bridged. Repair knowledge is therefore personalised knowledge. It is built on the basis of stories, mistakes, attempts, experimentation and readings about what is going on. This accords with the traditional epistemological distinction between the propositional knowledge of 'knowing that' and the praxical knowledge of 'knowing how' (Vincenti, 1990, p. 3; Ryle, 1949, p. 441). Repair, as mentioned previously, blurs such distinctions because knowing that an artefact obeys certain principles and laws is sometimes needed to account for use (Houkes, 2006, pp. 103–4) or to test whether an artefact is working. Propositional knowledge can also point to clues and hints useful for tracing a repair hypothesis. Knowing how thus involves the means by which truth is made operational and even useful.

Nevertheless, artefacts today are more likely to become obsolete due to the increased applicability of natural scientific knowledge affecting innovations that crawl into the built environment of everyday life. As old artefacts become replaceable and are less likely to be used and repaired, they become relics of past forms of life. The car mechanic today cannot fathom how the same cars were repaired in the past while Newton's laws of mechanics have not yet lost their relevance.[9] The conclusion is that for societies bent on reaping the fruits of scientific and technological knowledge, each age has its own forms of repair, as artefacts are either replaced or improved, while craft, maintenance and repair practices are obliterated, precisely because repair knowledge cannot be accumulated or gathered in the same way or pace as scientific and technological knowledge. A scientific or mathematical discovery, on the other hand, does not remain bound to its historical origin, but rather projects forcibly beyond it and is diffused into the artefacts of the world of repair. Repair always remains attached to its age, but on the condition that the renewal of repair skills keeps up with the transformation of the artefacts to be repaired.

There are then two distinct procedures for the accumulation and transmission of knowledge based on the distinct features of theoretical and practical knowledge. Informal repair practices have a testimonial character. Like *phronesis*, practical reasons are context-specific and depend on personal transmissibility due to their having to be learned in concrete and particular situations. Since they are prone not to be formalised, they are more fragile and vulnerable. The transformation of the traditional workshop of old and the practices it harboured was precisely one of the points in Harper's study (1987, pp. 20–24). The separation between the workers and the knowledge surrounding their work is increasingly giving way to *formal repair*. The division of labour and its rational organisation also achieve the unified and continuous character that unites former crafts and informal repair and in which, as we have seen, a living, personal intelligence is attuned to circumstances, materials and tools. The knowledge of informal repair becomes more and more external to the worker, and the condition for this externalisation is its conservation through procedures, rules and formulas.

With the rationalisation of repair, there is however an epistemic transformation. It ceases to be a hands-on experience and becomes more theoretical and detached. As a result, certain tasks become routine, specific and predictable and, therefore, less dependent on the lengthy acquisition of practical knowledge. The solutions for repairing something also become more piecemeal and specific as these practices go through the same process of rationalisation typical to production. As Harper notes, the solutions of rationalised repair become more objective, limited to the replacement of broken parts with functional ones and less 'intuitive' and less dependent on understanding the whole that makes each artefact. Repair becomes a 'by the book' procedure, losing in the process the faculty of the repairer to refer each fault and each part to the artefact's unity: 'the process of repair has become simple, but the knowledge of the repairman in relation to the system he or she repairs is very limited. Repair has become deskilled in much the same way as the work of assembly' (Harper, 1987, p. 23). The repairman loses authority over what is repaired because, in a way, he does not know what he has repaired. Repairing is increasingly seen by consumers as the replacement of parts, granted by procedures, as the authority of the tacit knowledge of repair loses power. Formal or rationalised repair and informal or non-rationalised repair then imply different kinds of skills and competence learning (Harper, 1987, p. 26).

One of Husserl's points (2008, pp. 64–7) in his *Krisis* was to show how the legacy of cultural achievements, such as theorems, concepts and tools, while becoming widely available for use by everyone, conceal the original meaning that once led to their creation, severing the living evidence of their source in the pre-scientific lifeworld. This fosters an obliviousness of the original meaning, as though the condition for formally transmitting this objectified heritage over generations is the forgetfulness of the original evidence. Husserl's quest thus consists of retrospectively questioning the meaning of the geometric and formal methodologies each scientist receives, making them personal and akin to the original meaning. Otherwise, the meaning of the original evidence is emptied by a formal, methodological and operative game where only rules must be followed, creating a scientific life entirely devoted to logical activities, crowned by the success of its practical application and without consideration for the good life.

Repairing, at least in its informal features, is thus a skill that often involves a continual search for meaning to be successful. It proceeds by exploring the purpose of artefact components before any rules and methodological procedures. The dependence of repair on labour is then not merely an economic question but an epistemic condition for the thriving of this type of knowledge. This evidence and its comprehension allow one to demonstrate and communicate to others the 'how it is done' of repair through intersubjectivity. This is why repair cannot be taught but only understood by people who already share the same context.

Following Husserl's critique, we argue that the practice of repair is not as prone as mathematics and science to forget the original grounding in the lifeworld from which all theoretical and practical life proceeds. Like practical reason, informal repair deals with particular entities in different contexts and, thus, can hardly become systematic. Harper's observations about how repair is being transformed highlight how formal rationalised repair inevitably tends toward a procedural theory of replaceability. Although repair practices are subject to recurrences that obscure and pale in comparison to a discovery, they retain the property of never becoming a formal game. Indeed, every 'there, I fixed it!' suggests that the artefact provides evidence of the technical syllogism, an 'it works!' that validates the perceptions, judgements and maxims that mean a repair skill has been successful. Repair may and does in fact work with inherited tools and instruments, but it cannot inherit abstract knowledge and concepts without translating them into embodied practical knowledge.

This is one of Polanyi's most important points (1962, p. 53) made in his defence of the value of epistemic traditions. Motor or cognitive practices based on tacit knowledge, such as informal repair skills, resist being summarised in rules and procedures not only because knowing them as rules does not ensure successful performance but also because these procedures are always insufficient. The possibility of transmitting and accumulating knowledge internal to these practices becomes restricted since this knowledge can only be transmitted by example, between master and apprentice or in convivial settings (Illich, 2009, p. 11). The apprentice only becomes capable when they begin to make their own judgements. Both master and apprentice will then be aware of what is correct or incorrect in their activity without being able to say precisely why. There are striking similarities with the development of practical moral reasoning as a skill (Vallor, 2015, p. 109; Coeckelbergh, 2019, p. 271). The drawback is that much of the tacit knowledge linked to skills, such as those found in arts and crafts, against a backdrop of accelerated changes in the material environment, is prone to being lost.

Following Kant and Kerr (2019, p. 699), informal repair knowledge is then essentially a know-how that goes beyond the know-that about the artefact's physical features. It includes: 1) procedural knowledge about a sequence of actions and 2) operational knowledge to go through the particulars, an embodied practice that involves interventions aiming for an end state. Following such observations and based on the knowledge type proper to repair, some conclusions can be drawn. A first, somewhat common-sense assertion is that, although an artefact can be made either through craft or industrial mass production, there can be no industrial mass repair. Moreover, whether artefacts are handmade or industrially fabricated, repair can be expected. However, in the former, repair knowledge remains generically in the makers' skill set, whereas, at the other end of the spectrum, there are differences between producing and repairing in terms of the intensity and the type of knowledge required. Formal repair, nevertheless, can be gradually expected because advanced everyday artefacts that compose our modern lifeworld – cars, aeroplanes and network systems – experience recurrent malfunctions requiring troubleshooting manuals with rules and procedures to identify breakdowns and swiftly indicate replacement of malfunctioning parts.

Regarding the technical knowledge used in production, it is perhaps useful to distinguish general artefact types, such as commodities, and unique artefacts, such as prototypes, industrial machines, infrastructures or technologies used for scientific instrumentation and calibration. The former are mass-produced, fostering a labour market for informal repair and skills that can, as a rule, ignore the scientific and technological knowledge used in production. In the case of the latter, a failure might be a flaw in the design or production process and thus may require the designers themselves, as possessors of the technological knowledge that supervised their making, to return to the worktable or to actively collaborate with repairers.

The repair of unique artefacts is then extremely restricted. In these technologies, there is usually just one token per type and hence their function must still be restored without a labour market for repair. In other words, because in such cases repair depends on restoring the proper function, it may hinge on scientific and technological knowledge and careful methodological procedures. Nevertheless, as we have noted previously, repair only happens when scientific and technological knowledge are realised in action and disposition. Here repair remains simultaneously linked to the major premise of an abstract understanding, even perhaps to a set of procedures and steps, but as well as to an empirical application that requires the use of tacit knowledge guided by theoretical clues. In many cases, this leads to a division of cognitive labour in which the knowledge required for repair is accessible only to specialists.

Herein lies much of Winner's (1977, pp. 330–2) social critique of artefacts that increasingly demand ongoing, specialised training to be repairable, severing users from engaging with them. This process echoes the rationalisation of the social order through labour division and the very design of mass-produced artefacts, making the appearance of informal repair more unlikely and reducing it to formal repair. As a set of knowledge, formulas, rules and procedures, technology has progressively crept directly into human praxis. However, these examples show that this process is limited because it is anchored to informal, tacit repair skills supported by an ongoing reflective conversation between individuals, artefacts and their context (Schön, 1992, p. 142). Repair may thus be an example of how some skills resist formalisation and calculation. Their epistemic relevance appears ironically within the very process of expanding scientific and technological knowledge to various fields of human activity.

4.6 The Phenomenology of Informal Repair

We now proceed to inspect the basic structure of the informal repair skills that precede the rationalisation of formal repair. Such a basic structure may be informed by scientific and technological principles that underscore judgement and troubleshooting. These skills are exercised in the pre-scientific, intuitable lifeworld in which all life is played out. In such an everyday world, one does not experience artefacts as if they were geometrical or ideal figures. The artefacts of our surroundings are experienced in an actual, usual way, without any need to add the formal possibilities of these subjects to them. They remain equal to themselves, connected to each other and far from having an arbitrary behaviour, exhibiting characteristic and recurrent modes that foster an empirical intuition about their future behaviour. The natural attitude takes the surrounding artefacts for granted, constituting the background context of the lifeworld. An accident or failure can remove them from this everyday character, inviting users to understand how artefacts work. The surprise that artefacts can break down triggers an inquiry, marking the beginning of the repair attitude.

If the theoretical or scientific attitude stems from an *epoche* of the pre-scientific world, the repair attitude is also a modification of praxis, a 'going towards' that signals the artefact as 'repairable', but without abandoning the lifeworld structure. The world of human purposes linked to praxis remains the same in the repair attitude. It changes how some artefacts appear, making them thematic and not just damaged or broken, but also having 'possibilities for' as objects that can be combined, assembled and disassembled. For the repair attitude, there is no point in objectively knowing the artefacts unequivocally and universally as in the way mathematical or physics practices approach their objects. It is enough to deal with the artefacts in how they appear to experience in a regular and empirical style of a general causality with its own essential structure based on the familiarity of the surrounding pre-scientific lifeworld (*Umwelt*). This general causality of the lived, somatic body has its usual truths that suffice for repair and for which the 'true being-in-itself' of the natural sciences don't add much. As Husserl says, our kind of practical knowledge of things as they are for us is perfectly adequate for human purposes and is open to other transformations that possess unity in themselves (Husserl, 2008, p. 352).

The first point to note is the way in which the repair attitude regards artefacts, that is, seeing them according to a problem interpreted in light of the artefact's intended meaning. One should not lose sight of the fact that every artefact, broken or not, may hold other meanings. Artefact function is inextricably linked to meaning, but they are to be considered different. A broken chair may become a source of firewood and a broken stereo may become a bookend. Meaning is related to the users' lifeworld and how the material environment grants a temporary or permanent function to attain that meaning. The interpretation of an artefact as being 'broken' is also the moment when meaning swings and can be pushed to other final states. The temporary suspension of the usual function of things draws limits to technological determinism (Feenberg, 1999, p. 78).

Through the repair attitude, everyday artefacts, despite having their proper function, now appear with various functional shades and various 'possibilities for' that can be explored. Adopting Uexküll's terminology (1988, p. 113), the *funktionale Tönung*, the shading of the aspects and qualities of artefacts takes on a character of 'manipulable by tool' or 'replaceable by another part' that other attitudes do not grasp. Each part or tool now has the shade of 'serving for something' subsidiary to the desired function. Those who repair do not shade the parts of the artefact as objects of scientific knowledge or aesthetic appreciation but rather as objects that may or may not be manipulated by certain tools and have parts that can be removed, replaced or adjusted.

The connection between the artefact and its context, that is, between the artefact and the available tools, parts and the macroperception, forms the action circle of the repair attitude, composed of perception marks and action marks. For example, the availability of a spare part reveals that repair by replacement is achievable. Similarly, the availability of a tool makes it possible to perceive the manipulable character of other aspects of the artefact. Perception and action are not only pre-existing logical or physical possibilities but also possibilities that stem from the world specific to the repair attitude and the repair skill.

Tools and spare parts are also subject to interpretation according to the repair attitude for a given class of artefacts. Spare parts have a specific use content, that they 'work like' the component to be replaced, and tools have a 'coupling content' regarding the artefact. For example, when we want to sit down, we do not have to look for a chair but for a surface that has a 'seating content'. These shading possibilities contribute to establishing the difference between what is possible vis-à-vis what is feasible,

We can now see how the repair attitude seeks replacements for an 'intended shading' that can fulfil the same function of a broken part or an unavailable tool. Consequently, the difference between repair *stricto sensu* and repair as recreation lies in how much of the artefact's initial state should be aimed for and what counts or not as a repair solution. A patch, a make-do or a temporary arrangement restores the artefact to its initial state, recovering its functionality, but does not restore it equally. The informal repair attitude opens semantic possibilities: the artefact can be repaired but it has shifted to another function or even to an object of appreciation. According to its intended meaning and contextual possibilities, an artefact undergoes an intervention that can range from *stricto sensu repair, improvised repair, creative repair or recreation*. Repair can be creative. The repair attitude thus reveals how an artefact's meaning is available to be fixed. A damaged mountain bike can be repaired to make it work again as a city bike, but it can also be transformed into a motor vehicle or even a coffee bean mill. In the latter case, the meaning of the artefact has changed.

We can now try to summarise how repair knowledge is acquired and practised. What distinguishes an amateur repair from a more advanced one is the former's inability to limit a large number of hypotheses to an operative number. In the repair attitude, the learner goes through a period of discomfort and exploration by trial and error. The problem is identified and assessed before they can devote their attention to it. Relevant clues, aspects and patterns are followed until the source of the problem can be pinpointed with relative certainty. A trial use of the broken artefact is then conducted so that it can reveal the origin of the malfunction. Hypotheses and verification tests are put forward to restrict the scope of explanations and repair steps.

In the early stages of learning repair skills, repairers try to detail each component's role in relation to the whole. Each change is followed by observation of the artefact's behaviour, paying attention to the way slight modifications bring about change that is not random, but causally linked to it. After observing the results, each further change may or not continue to cause a more satisfactory result. Error, therefore, has a fundamental heuristic role in improving skill and establishing a good or bad judgement or action. Only those who are capable of perceptions and judgements proper to the repair attitude can say why there is an error. To repair is also to accept the possibility of going back to make amends, in the constant readiness to discover other issues, in an oscillation between 'seeing the whole' and 'seeing the detail'. This process manifests itself as a 'focusing-defocusing' structure. With time and practice, for the same artefact type, repair becomes embodied and a state is reached where 'one knows what one is doing'. It means a tenure, a familiarity with the nature of the recurring problems usually affecting the artefact. It marks a state of readiness to perceive what needs to be done. A skilful repairer has perfected the focusing–defocusing structure to the point where all steps appear to be predetermined, as though they were just following rules. In reality, a repairer follows perceptions and judgements recognisable as significant. It is this personal character of the repair skill that appears not only to be a procedural compliance but potentially describable by an algorithm (Young, 2018, pp. 64–5). As pointed out previously, there is nevertheless a difference between understanding that repair can be described and explaining it as if the repairer was following explicit rules (Dreyfus, 1997, pp. 256–7), mainly because interpretation is involved in the repair attitude. There is no knowledge of repair without a skill whose mastering is fraught with doubts, hesitations and dead ends. It is a lived skill marked by an emotional history bound with frustration, persistence and resolution that enables its embodiment. Its reluctance to be spelt out in rules is the counterpart to not being easily forgotten, as it seems always available to be reinstated.

Moreover, it is possible to assess retrospectively how appropriate a repair may be, not based on the ideal solution for achieving the final state, but on how that repair has successfully enabled certain options in the search for a functional solution. An illustrative scene can be seen in the docudrama *Apollo 13*. After the moon-bound spacecraft suffered an unforeseen explosion, all that mattered was bringing the crew back to Earth. One scene raised the question of how a carbon dioxide leak could poison the astronauts. The mission of the NASA ground team was to show the astronauts on board the Apollo 13 spacecraft how to repair a filter, so they did not asphyxiate. To do this, the ground team had to reproduce materials they knew existed on the space shuttle and interpret them as appropriate for the problem at hand, eventually using duct tape to arrive at the 'filter purifier' solution. As flight director Gene Kranz, played by actor Ed Harris, previously states: 'I don't care what anything was designed to do. I care what it can do' (Apollo 13, 1995). The technical syllogism of the repair enacted on Earth was then described in a set of schematic procedures and transmitted to the astronauts by radio. They reproduced the instructions without ever understanding the meaning of what they were doing. There was never any question of whether another solution was possible, as though the ideal solution could be different from the only solution the ground team was able to find under those circumstances. In short, repair, like any practice, presents context-specific criteria to define the solution that works but which, nevertheless, can define what counts as a good or bad repair.

4.7 Conclusion

This chapter advanced a post-phenomenological framework inspired by the work of Don Ihde in order to understand what repair is about. In the first part, this framework was used to question some assumptions in the metaphysics of artefact function by underlining how artefacts are part of the user's lifeworld. Designers and repairers must acknowledge how artefacts, especially when they break down, are prone to become interpretable. Artefact function theories should then take into account that what an artefact does hinges on how it is used, according to the meaning it may acquire in a lifeworld. The difference between the proper function assigned by the designer's intentions and the accidental function is blurred as the user's intentions and the repairer's context-dependent character play a role: repair should be understood as an activity that restores a function, but not necessarily as restoring the proper function. Social and cultural processes thus have a bearing on the selection and reproduction of artefact function. Consequently, cultural differences and technology transfers gain philosophical relevance: other social relationships, interests and institutions interact to define what exactly repair restores.

Afterwards, an epistemology of repair skill was advanced in order to outline its relationship to different types of knowledge. The skill of informal repair was analysed and contrasted with the emerging formal repair in the context of progressive automation and rationalisation of productive processes. Some conclusions were drawn from the likelihood of informal repair turning into a formal, procedural-like knowledge given the advancement of labour division and work organisation. Additionally, a phenomenology of the structure of informal repair was sketched: it highlighted how its knowledge is akin to current practices of design and engineering, allowing one to see how scientific principles and laws are not necessary for every repair practice but are nonetheless apt to complement it. Informal repair rests on a translation of scientific and technological propositional knowledge into practical, useful knowledge.

Funding Information

This publication is funded with National Funds through the FCT/MCTES – Fundação para a Ciência e a Tecnologia/Ministério da Ciência, Tecnologia e Ensino Superior (Foundation for Science and Technology/Ministry for Science, Technology and Higher Education – Portugal), in the framework of the Project of the Institute of Philosophy – University of Porto with the reference UIDB/00502/2020.

Notes

1 For Vermaas & Houkes, any theory of artefact function should be able to distinguish between its proper and accidental function: 'Proper functions can typically be understood as functions ascribed standardly to the artefact, whereas accidental functions are ascribed only occasionally'.
2 This does not excuse the fact that prototypes or highly specialised artefacts are more rigid or more stable regarding their capacity to acquire accidental

functions. But as Robert Smithson's 'A Tour of the Monuments of *Passaic*' shows, any artefact can become something else under an aesthetic 'seeing how'.

3 Ihde (1990, pp. 144–6) puts forward this term as both a way to question a single or unified trajectory of Technology and its autonomy. Technologies are ambiguous and have a variety of embeddedness across cultures. The polymorphism that visual perception has regarding some objects like the duck-rabbit or the Necker cube is expanded by phenomenological analysis to highlight how objects are also multistable in the relationships that humans establish with them.

4 In an overview of philosophical theories of artefact function, Preston (2009, pp. 214) notes how such theories should account for how artefacts are *multiply realisable* and *multiply utilisable* and that such notions are two sides of the same coin.

5 A technical syllogism is the description of a practical judgement about what is at stake in a certain artefact and what needs to be done and can be done in *that* context. It's important to note that a technical syllogism is not a claim about how in fact human beings think and proceed when dealing with artefacts (for another view, cf. Dreyfus, 1997, p. 233). Its aim is to provide an explanation. It consists of a major and a minor premise followed by an appropriate action in case both premises hold true and possible, respectively. In the case of repair, the major premise contains the general end of restoring a function to the artefact or a step towards it. The minor premise is causally linked to the major premise as a particular means and takes into account the circumstances of the context in which the agent finds herself. The skill of a repairer shows itself upon identifying the relevant features of the minor premise. Deliberation about how to proceed thus ceases when the principle of action is integrated into the agent. The conclusion of the syllogism is the principle of the next step of reparation.

6 For further discussion on the metaphysics of artefact persistence, see Young (this volume).

7 This 'not' completely shifted the historical meaning of the siege of Lisbon by stating that the foreign crusaders have 'not' helped the Portuguese in conquering the city to the moors after all. Saramago's point was to illustrate how history's meaning is open to be interpreted through the very same way used to settle what were the facts.

8 Lindsey, J. (2022). 'Are modern bicycles the end of DIY maintenance?' *Outside Online*, 8 November. Available at: https://velo.outsideonline.com/gear/modern-bicycles-the-end-of-diy-mechanic/ (Accessed: 9 November 2022).

9 What we mean is that for a mesological, everyday use, Newton's laws continue to be useful for mechanics and kinematics without the need to add more numerical precision.

References

Aristotle (2008) *Ética a Nicómaco* A. C. Caeiro (trans.) Lisboa: Quetzal Editores.

Baker, L. (2009) 'The Metaphysics of Malfunction', *Techné*, 13(2), pp. 82–92.

Boon, M. (2006) 'How science is applied in technology', *International Studies in the Philosophy of Science*, 20(1), pp. 27–48, https://doi.org/10.1080/02698590600640992

Broadie, S. (1991) *Ethics with Aristotle*. Oxford: Oxford University Press.

Bucciarelli, L. (1994) *Designing Engineers*. Cambridge: The MIT Press.

Coeckelbergh, M. (2019) 'Skillful coping with and through technologies', *AI & Society*, 34, pp. 269–287, https://doi.org/10.1007/s00146-018-0810-3

Dreyfus, H. L. (1997) *What Computers Still Can't do. A Critique of Artificial Reason*. Cambridge: The MIT Press.

——— (2002) 'Samuel Todes's account of non-conceptual perceptual knowledge and its relation to thought', *Ratio*, 15(4), pp. 392–409.

Feenberg, A. (1999) *Questioning Technology*. London: Routledge.

Gerrike, K., C. Eckert and M. Stacey (2017) 'What do we need to say about a design method?' In: 21th International Conference on Engineering Design (ICED 2017), 21–25 August 2017, Vancouver, Canada.

Harper, D. (1987) *Working Knowledge. Skill and Community in Small Shop*. Chicago and London: The University of Chicago Press.

Henke, C. and Sims, B. (2020) *Repairing Infrastructures. The Maintenance of Materiality and Power*. Massachusetts: The MIT Press.

Houkes, W. (2006) 'Knowledge of artifact functions', *Studies in History and Philosophy of Science* Part A, 37(1), pp. 102–113.

Houkes, W. and Vermaas, P.E. (2010) 'Malfunctioning' in *Technical Functions. Philosophy of Engineering and Technology*, vol 1. Dordrecht: Springer, https://doi.org/10.1007/978-90-481-3900-2_5

Husserl, E. (2008) *A crise das Ciências europeias e a fenomenologia transcendental: uma introdução à Filosofia fenomenológica*. Ferrer, D (trans). Lisboa: CFUL.

Ihde, D. (1990) *Technology and the Lifeworld: From Garden to Earth*. Indiana: Indiana University Press.

——— (2008) 'The Designer Fallacy and Technological Imagination' in *Philosophy and Design. From Engineering to Architecture*. P. Vermaas, P. Kroes, A. Light, S. Moore, (eds.) Dordrecht: Springer, https://doi.org/10.1007/978-1-4020-6591-0_4

Illich, I. (2009) *Tools for Conviviality*. New York: Marion Boyars.

Kant, V. and Kerr, E. (2019) 'Taking Stock of Engineering Epistemology: Multidisciplinary Perspectives', *Philosophy & Technology*, 32, pp. 685–726.

Latour, B. (1992) 'Where Are the Missing Masses? The Sociology of a Few Mundane Artifacts' in W. Bijker and J. Law (eds). *Shaping Technology/ Building Society. Studies in Sociotechnical Change*. Cambridge: The MIT Press, pp. 225–258.

MacIntyre, A. (1999) *Dependent Rational Animals - Why Human Beings Need the Virtues*. Chicago: Carus Publishing.

Mitcham, C. (1994) *Thinking Through Technology. The path between Engineering and Philosophy*. Chicago: University of Chicago Press.

Nye, D. (2006) *Technology Matters*. Cambridge: MIT Press Book.

Oroza, E. (2009) *RIKIMBILI. Une étude sur la désobéissance technologique et quelques formes de réinvention*. PU Saint-Etienne.

Petroski, H. (1996) *Invention by design: how engineers get from thought to thing*. Cambridge: Harvard University Press.

Polanyi, M. (1962) *Personal Knowledge*. London: Routledge.

Preston, B. (2003) 'Of Marigold Beer: A Reply to Vermaas and Houkes', *The British Journal for the Philosophy of Science*, 54(4), pp. 601–612.

——— (2009) 'Philosophical Theories of Artifact Function' in A. Meijers, (ed.) *Handbook of the Philosophy of Science, Philosophy of Technology and Engineering Sciences*, North-Holland, pp. 213–233, https://doi.org/10.1016/B978-0-444-51667-1.50013-6

Ryle, G. (1949) *The concept of mind*. Chicago: Chicago University Press.

Saramago, J. (1998) *The History of the Siege of Lisbon: A Novel*. Mariner Books.

Schön, D (1992) 'Designing as reflective conversation with the materials of a design situation', *Knowledge-Based Systems*, 5(1), pp. 3–14.

Smith, C. (2010) *Design for the other 90%*. Cooper-Hewitt.

Spelman, E. (2002) *Repair. The Impulse to Restore in a Fragile World*. Beacon Press. Boston.

Uexküll, J. (1988) *Dos animais e dos homens: digressões pelos seus mundos próprios* A. Candeias (trans.) Lisboa: Livros do Brasil.

Vallor, S. (2015) 'Moral Deskilling and Upskilling in a New Machine Age: Reflections on the Ambiguous Future of Character', *Philosophy & Technology*, 28, pp. 107–124.

Vermaas, P. and Houkes, W. (2003) 'Ascribing Functions to Technical Artifacts: A Challenge to Etiological Accounts of Functions', *The British Journal for the Philosophy of Science*, 54(2), pp. 261–289.

Verbeek, P.-P. (2006) 'Tecnopólis: a vida pública dos artefactos tecnológicos, *Análise Social*, XLI(181), pp. 1105–1125.

Vincenti, W. G. (1990) *What engineers know and how they know it: analytical studies from aeronautical history*. Baltimore: Johns Hopkins University Press.

Winner, L. (1977) *Autonomous Technology. Technics-out-of-Control as a Theme in Political Thought*. Cambridge: MIT Press.

Young, M.T. (2018) 'Intuition and Ineffability: Tacit Knowledge and Engineering Design' in A. Fritzsche and S. Oks, (eds.) *The Future of Engineering. Philosophy of Engineering and Technology*, vol 31. Cham: Springer.

5 A Standpoint Epistemology of Repair?

Cristina Bernabéu and Jesús Vega-Encabo

5.1 Introduction

In 'Rethinking Repair', Steven J. Jackson proposes an exercise of what he calls 'broken world thinking' (2014, 221). This notion reveals the fragility of the world we inhabit, which is shaped by a complex set of relationships between natural beings, social structures, technical infrastructures and everyday objects. Phenomena such as erosion, breakdown, decay or dissolution come to the fore, as opposed to the idealistic insistence on innovation and design. However, rather than passively dwelling on the fate and memory of ruin and debris, this exercise is grounded in a more optimistic notion. It recognises the potential of fragility to rethink how we collectively achieve order in the face of erosion and decay. To put it in more cosmic terms, it is as if we restore order and stability over and over again, conjuring up cosmic forces of destruction with our rituals.[1]

But fortunately, the discovery of fragility does not push us into this eternal confrontation between the forces of the cosmos. On the contrary, it leads us back to a forgotten realm, where the silent activities of maintenance, repair and care of those environments take place and where life and the world in Hannah Arendt's sense are reproduced (Arendt, 1958). Jackson integrates all of these activities into a broad concept of 'repair', that includes 'the subtle acts of care by which order and meaning in complex socio-technical systems are maintained and transformed, human value is preserved and extended, and the complicated work of fitting to the varied circumstances of organizations, systems, and lives is accomplished' (2014, 222). The fulcrum of a world that is maintained in its 'natural' order, full of meaning for those who inhabit it, seems to be found in the activities of repair.

Jackson also reminds us that behind the neglect of repair and maintenance activities lie issues related to the distribution of knowledge and power. We have become increasingly aware of how the invisibility of some activities in society often reveals profound inequalities, particularly in terms of the value we attach to the potential epistemic claims on the part

DOI: 10.4324/9781003316213-6

of those involved in those activities. Thus, Jackson suggests and briefly outlines the premises of a potential 'standpoint epistemology of repair'. This epistemology would first and foremost prioritise breakdown, allowing us to adopt a perspective that reveals what remains invisible in our everyday existence. In addition, it should endow repair itself with the potential to unearth those possibilities that have been crushed by the dominance of the forms of fixed objects and artefacts or the stability of socio-technical systems (Jackson 2014, 230–31).

Repair and maintenance activities are present in almost all of our interactions with the material world in which we live. Those who engage in such activities participate in shaping our technological worlds. The question we will explore in the following pages is the extent to which repairers can constitute a standpoint on which to ground certain epistemic claims about significant aspects of our socio-technical reality. Is a standpoint epistemology of repair possible? The answer to this question cannot be definitive, and we will point out some of the difficulties we face in attempting to delineate the collective of fixers from which a standpoint might emerge, along with the supposed epistemic advantages derived from their social location and the collective process of mutual recognition. The core of the argument rests on a broad characterisation of the fixer, which we introduce in Section 5.2: we are all repairers insofar as we participate in the (re)production of artefactual environments. If we identify fixers with a group of technicians specialised in repair, we are just narrowing a category that encompasses multiple agents and multiple aspects. It is then worth considering how a standpoint can emerge from such a diffused and heterogeneous figure. We will show that there is no expeditious way for establishing a specific epistemic privilege of the fixer because there is no way for the collective to be unified by their common needs and interests formed in a situation of oppression or subordination as standpoint epistemologies often require. Nevertheless, in the course of this chapter, we will highlight several aspects that make it still attractive to think about the role of the fixer in terms of critical epistemological conceptions. First, we will highlight the invisibility of repair and maintenance activities in our technological culture, as well as their hierarchical subordination to more visible figures. This hierarchy makes it difficult to recognise their knowledge claims (Section 5.3). Second, we will consider how our theories of the technological world can reintegrate the fixer into conceptions that align more closely with the technological process and even our everyday experience with artefacts (Section 5.4). Third, we will suggest that the category of the fixer is associated with specific forms of experience and practices in the (re)production of socio-technical environments, in line with the basic tenets of standpoint epistemologies (Sections 5.5 and 5.6). Fourth, we will recognise that repair activities can become sites of social critique with transformative potential, as Jackson suggests. However, repair as critique does not necessarily rely on a mechanism of reversing locations to gain access to a less distorted view of socio-technical reality, as standpoint epistemologies suggest. Instead, it stems from an awareness of the fragility and vulnerability of the material world. Anyone who engages critically as a repairer and maintainer, embodying an attitude of care and respect for things, can contribute to this consciousness. It is a form of consciousness that does not refer specifically to the social and epistemic conditions of subordination or oppression, as required by standpoint epistemologies in their more classical versions, but rather refers to the material world we inhabit (Section 5.7).

5.2 Repair

First, what should we understand by 'repair'? An appealingly *broad notion* of what repair is emerges from Jackson's previous quote. In our everyday environments, many artefacts and objects suddenly stop working, and we tend to attribute this failure to a misalignment of structural elements that affect their intended functionality (given what they are *supposed* to do). In most cases, repair consists of putting 'things' back in place, if possible. However, in many cases, the damage is irreversible and becomes an inexorable and definitive waste of components and materials, and it is advisable to discard the artefact. Repair can be done by users themselves, who adjust or replace components, or by technicians who *know* the mechanisms involved, the *expected* functioning of the device and the appropriate conditions for its use. In this sense, repair operates within a framework of shared expectations about the artefacts, objects and technical systems that surround us. It is not crucial for our purposes to delve into how these expectations are established, whether by designers and manufacturers or through open and diverse trajectories of use and exchange. Repair is an activity intimately connected to how we navigate our collective expectations and encompass various dimensions of our social and cultural life.

Take one of our most common household appliances, the refrigerator. We have established numerous cultural expectations around it, not only regarding its efficient operation but also regarding how it should be maintained, repaired or replaced. When one or more of these expectations are violated, we sense that something is wrong with the appliance and that intervention is necessary. However, we are often mistaken about the significance of this breach of expectation. Recently, one of the authors of this chapter experienced such a breach of expectation when their refrigerator stopped cooling during a summer heat wave. It appeared to be broken or at least not functioning as expected. However, as the technician promptly pointed out, restoring expectations did not require much effort, certainly not replacing the appliance. We had misinterpreted what was wrong with it: the appliance was certainly not in the best condition, but there were ways to restore its efficiency with more careful use. Repairing consisted of readjusting the actions and expectations we had for the device (Henke 2000). Therefore, it is essential to understand what we mean by *shared* expectations: not every artefact or technical system entails a fixed and non-negotiable set of expectations. On the contrary, they depend not only on the demands of various actors, including designers, manufacturers, users and especially repairers but also on the intervention capacities available to us. These expectations are shared in the sense that the reproduction of our artefactual and technical worlds necessitates – even if only locally and temporarily – adjustments among some of these expectations. In any repair context, negotiation occurs regarding the interpretation and significance of different expectations.

Once we understand repair from the perspective of the expectations, we build around objects (artefacts, systems and infrastructures) and their use (of which the failure of a particular component is only one aspect), there are many activities that can be considered part of 'repair' in this broad sense. In fact, fixing a broken object is only one possible *intervention* that seeks to maintain the expectations we share about the elements of our material culture and the socio-technical environments we build and inhabit. There are numerous routine interventions we make *to keep things in a good disposition*. We avoid neglecting them and give them the necessary attention: we clean them, perform routine checks, replace components and so on.[2] A similar experience can be observed with cars. Even if the technical checks are left to specialised technicians, we become aware of how frequently we need to intervene to keep the car in order, which is inseparable from our cultural expectations of cars. In most cases, this means an understanding (however minimally) of their own materiality, that is, of how their material condition (their mundane existence) embodies their cultural significance. Intervening in them also means arranging them in an orderly manner, avoiding certain mismatches with other objects and artefacts that are necessary for our practices of use.

What we might call maintenance and repair practices emerge from the flows of sustained activity in the cultural environments in which we live. In this continuous flow (what phenomenologists understand as being-in-the-world), there are constant breakdowns and small or large misalignments that we inadvertently *repair*. Only in some cases will it take a special effort to restore existing expectations in the flow of activity with things and thus rebuild the bonds that bind us to artefacts and things.

What else is there to say about these expectations we build around our material culture? As is the case in our cultural practices in general, we are dealing with a conglomerate of predictive and normative aspects.[3] We expect objects to behave in a regular way, according to certain purposes and modes of operation. But we also generate expectations about how other agents are supposed to use, manipulate and intervene in them. Many of these expectations have to do with what different actors should do or should have done. When there is a need for intervention, either to maintain regular behaviour or to reinforce compliance with certain norms, we contribute to repair in the sense of restoring broken expectations. We *rely on* this being borne out in our activities without the need to make specific inquiries. We intervene, or require others to intervene, when predictive or normative expectations break down. No doubt, certain standards have been affected: in our case, those related to the adaptation of the internal temperature regulator to the environmental conditions and to the needs (which could also be adapted) of intensive use during a heat wave, given the age and condition of the appliance itself.

The examples could be multiplied. They all capture aspects such as the following: (1) the interventions that I can or should make as a user/repairer are in principle culturally regulated; there are predictive and, above all, normative expectations about the nature, type, intensity and frequency of the interventions I am allowed or obliged to make (some of these are legal or contractual constraints); (2) so are the circumstances in which it is necessary to consult or engage a specialised technician, which points to a social division of repair/maintenance labour and thus highlights the figure of the repair technician or fixer, as we will call it in the rest of this chapter[4]; (3) furthermore, it should not be forgotten that these expectations (and how they are culturally regulated) change over time, but also with the social and economic situations of the actors and in the ways in which 'technical' knowledge is distributed in society.

This view of repair contrasts sharply with a purely functional approach. It does not focus only on specific interventions by a specialised technician who solves the problem caused by the failure. This is what we might call a *narrow view* of repair. In contrast, our multivariate model allows and requires a variety of interventions to maintain the good disposition of artefacts, systems and infrastructures. In maintenance and repair activities, it is not easy to isolate what could be considered the purely technical aspect of breakdown. This is crucial for the purposes of this chapter because it clearly shows that knowledge is necessarily distributed among different actors and cannot be attributed to a single, specific group of authorised and recognised experts. For this reason, we consider the fixer to be an ambiguous and diffuse figure. To reiterate, our approach does not focus only on technicians who repair devices or infrastructures but rather insists on how repair permeates all stages and activities involved in the (re)production of our material and artefactual environments. It is important to keep in mind that living in these environments means that we are engaged – to a greater or lesser extent, and with a more or less distinctive awareness of doing so – in tasks of maintenance and repair. By 'repairing' in this broad sense, we as actors contribute to reproducing both the conditions of existence/persistence of artefacts and the bonds we create through them with other actors. Repair, in this broad sense, encompasses a continuum of activities ranging from seemingly minor interventions in our everyday environments (such as cleaning or rearranging) to more elaborate restoration exercises that can improve the functioning of devices. Accordingly, repair, as we have said, consists of collaborating through our interventions in the persistence and reproduction of our environments.[5]

5.3 Knowledge Subordination

We have outlined a broad view of repair in which each of us is responsible, with a greater or lesser degree of involvement, for specific repair interventions that keep our expectations about artefacts, technical systems and infrastructures in place. It is also important to note that these differences in involvement are also reflected in the distribution of resources and epistemic competencies among the actors involved in the (re)production of artefactual environments. Even during the design phase, we are actively shaping the conditions for repair and influencing cultural expectations regarding what should be maintained and how maintenance and repair should be carried out. This implies that we assume the availability of knowledge and skills to be displayed when needed in order to keep things working. We sustain our technical and artefactual worlds with small gestures that bring specific knowledge and varying degrees of expertise to bear.

It is well known that societies exhibit a cognitive division of technical labour, which implies a differential distribution of cognitive resources and skills. But it is equally significant that the mechanisms of knowledge distribution are invariably accompanied by marks of visibility, recognition and acknowledgement. Some knowledge claims are more visible and more readily accepted than others. Let us take a simplified model of technical actors that distinguishes between what we will call makers, including designers and producers (product developers, manufacturers, etc.), and users or, more generally those people who inhabit, and take advantage of, technical environments. To these traditional figures (maker and user), we have added the fixer.[6] Now we think that one of the most deeply rooted assumptions regarding artefactual and technical knowledge is that makers (designers and manufacturers) are the bearers of the knowledge that is deemed necessary (even essential) to create and maintain these environments.[7] They also regulate the value of other possible knowledge claims. Consider the case of users who insist that this or that artefact *must* work in this or that way. Suppose they succeed in making it work. Are they entitled to claim knowledge about what the artefact does and how it does it? In a sense, they are not, as their knowledge about the artefact is measured against the expectations set by the maker. A similar argument can be made for fixers and their knowledge claims. The knowledge of fixers, developed in local contexts of repair, is subordinate to the normative requirements that define the expectations for keeping things in good order. Although we will return to this issue later in this chapter, in defining the possible epistemic privilege of the maker, it is important to remember that this epistemic subordination merely translates a model of agency into epistemic relations. Several philosophers and anthropologists have recently criticised (Ingold 2011; Preston 2013) ideational and hierarchical models of agency, according to which all production emanates from and conforms to an idea that hierarchically takes control of it.

We have seen that one of the main characteristics of artefactual environments is how their creation and maintenance involve multiple agents that deploy specific kinds of knowledge and skills. From ordinary users to expert technicians, they all have to acquire a particular experience with the objects and technical systems they install, use, maintain and repair. It is within these environments that artefacts gain special significance as they 'coexist' and 'interact' with numerous other artefacts and systems. The sustainability of artefactual and technical environments relies on complex *interdependent* relationships between objects and actors. Together with our experience and knowledge about the functioning, development and life of instruments, artefacts and technical systems, we build our expectations and manifest our dependencies. Indeed, we rely on artefacts for the most part of the activities we carry out and, in doing so, we exhibit a capacity to form certain *predictive* expectations about their behaviour. In addition, we need to do something else: we need to weave trustworthy networks that help shape our expectations about, for instance, the fact that they have been designed according to certain standards, that they will be properly maintained and repaired, that the skills and capabilities required for design, repair and use will be developed and so on. These are normative expectations about the behaviour of other agents, their knowledge and skills and the reliability of these artefacts and systems.

As it should be, all of these expectations that underlie our dependency networks, both predictive and normative, have a distinct *epistemic* dimension. Dependency networks only work if we collectively recognise the value and importance of certain forms of knowledge. This recognition is particularly crucial in the context of repair activities, where we rely on the specific knowledge and expertise of those who restore our expectations. However, it is often the case that this knowledge remains invisible and unrecognised. The local and distributed nature of this knowledge contributes to its lack of collective recognition.[8] As a result, the potential contributions of these repair experts to our understanding of technical environments are not fully realised.

It is an essential fact that much of this experience with artefacts and technical systems is considered expendable or incidental even in the reproduction of artefactual environments. It is hard to see how this experience could claim epistemic recognition and constitute an *authoritative* voice concerning the persistence and maintenance of our artefactual environments and the socio-technical systems that are sufficiently stable to respond to our needs and purposes. At least in our cultural imaginaries, and particularly in our theoretical views (as if they were ideological constructions),[9] the knowledge and experience of those who contribute to the maintenance and repair of our artefactual world play a secondary role, hardly capable of explaining the complexity of technical environments and their innovative and transformative potential. This image is at odds with how the reproduction of life in everyday environments depends not only on the incorporation of technical innovations but also on a complex and sometimes invisible social and technical division of labour in maintenance and repair contexts. Given this, it becomes relevant to consider a standpoint epistemology of repair that would succeed in inverting certain power relations in order to *authorise* subordinated forms of experience and knowledge within the cognitive division of socio-technical labour.

5.4 First Steps Towards a Theoretically Inspired Standpoint Epistemology of Repair

Jackson (2014) introduces the idea of a possible standpoint epistemology of repair as follows:

> The question is this: can repair sites and repair actors claim special insight or knowledge, by virtue of their positioning vis-à-vis the worlds of technology they engage? Can breakdown, maintenance, and repair confer a special epistemic advantage in our thinking about technology? Can the fixer know and see different things – indeed, different worlds – than the better-known figures of "designer" or "user"? Following on the claims of Hegelian, Marxian, and feminist theorists, can we identify anything like a standpoint epistemology of repair?
>
> (p. 229)

This series of rather rhetorical questions paints a fluctuating picture of what Jackson might be aiming at with his proposal for developing a standpoint epistemology of repair. We identify at least four possible issues here:

a) The first asks whether fixers can make particular knowledge claims about the technological world.

b) The second one subtly shifts the focus and asks whether our thinking and reflection about technology might gain some epistemic advantage by adopting the perspective of maintenance and repair. For now, let's understand the epistemic advantage in very neutral terms, such as obtaining more accurate or reliable representations, acquiring more knowledge or gaining a better understanding.
c) The third simply contrasts the epistemic positions of designers and users with those of fixers by pointing out the obvious fact that they know (or see) different things.
d) The fourth delves more deeply into traditional (Hegelian, Marxian and feminist) motivations for a standpoint epistemology, suggesting a potential inversion thesis: those who have been forgotten and 'unjustly oppressed' in the practices of socio-technical (re)production (the fixers) could claim an epistemic advantage over the makers (namely the designers) on certain issues.

It is easy to identify double ambiguity in this motivation for a standpoint epistemology of repair: the first is relative to those whose knowledge is at stake; the second concerns the nature and ground of the epistemic advantage. We will address the latter in the next sections, so let's focus on the first ambiguity now.

For "who is in a position" to claim knowledge in the face of the complexity of our technological world? The first question seems to point to the actors involved in the real practices of repair and maintenance, whose epistemic 'rights' have not been sufficiently recognised. The second question pertains to theorists who have the ability to navigate between the perspectives of different actors, such as designers, users and fixers. These theorists can shift their focus to a series of aspects that have been forgotten or remained invisible in our theories of technology.

True enough. Shifting the focus to the perspective of the ubiquitous phenomenon of repair and maintenance seems to confer the theorist a certain epistemic, even special, advantage. But this does not mean that it is the fixer (and the collective of actors involved in repair and maintenance activities) who is entitled to claim an epistemic privilege. Theorists could, in a sense, take ownership of their experience and knowledge by changing their approach and recognising that there are aspects of this experience of breakdown and decay that call for more appropriate attention to repair activities.

One might even suggest that what is being pursued is what we might call *a purely theoretical standpoint epistemology of repair*. Those theoretical approaches that seek to adopt the fixer's own perspective gain an epistemic advantage in terms of understanding how artefactual environments are (re) produced over those that continue to privilege *solely* (and *mainly*) the figures of the designer or the user.

So formulated, this approach only vaguely reflects the tenets of a standpoint epistemology. However, it helps to refresh some interesting points characteristic of these epistemological views. It can be said that this approach responds optimistically to the question of whether and how the theorist can *access* the point of view of those who are actively engaged in repair activities. Indeed, it does not assume that it is necessary to be located in the perspective of the oppressed or subordinated for this move to be epistemically fruitful.[10] If an epistemically advantaged standpoint is not given, but must be achieved, as we will show in section V.5, perhaps it is the theorist who carefully attends to the needs and interests of the fixers that might have access to an advantaged standpoint. As some feminist standpoint epistemologies have pointed out, it is not obvious that it is women as such who have epistemic privilege, but those groups with social and political consciousness who, through their theorising, respond objectively to the interests of the collective. It is thus the objective facts related to the role of repair activities in our technical and artefactual reality that ultimately animate more faithful and inclusive theories.

To some extent, this move toward a theoretically inspired standpoint of repair has been accomplished by more recent literature on maintenance and repair.[11] Nevertheless, how to access the specific point of view of the fixer is a question which is rarely addressed. New aspects of our technological world are revealed when we turn our attention to those objects and infrastructures that are broken or malfunctioning: their role in a complex network of other objects and actors becomes visible (Martínez and Laviolette 2019). The disruption of the continuity of our activity with objects when a breakdown occurs produces an experience of (thematic) awareness of what has remained in the background. Nevertheless, it is not enough to refer to this experience to bring technologies and infrastructures out of invisibility, let alone repair and maintenance activities. Mark Young, in a recent article, insists on the importance of analysing *how we conceptualise the phenomenon of breakdown*. It needs to be understood not as a simple anomaly or as a deviation from a form-norm that repair could (and should) bring back to its 'ideal' as fixed in design and manufacture (Young 2021). This understanding views artefacts and technology as permanent and stable (in their formal and functional sense), perpetuating a certain form of commodity fetishism that makes everything involved in making our artefactual worlds possible, such as the endless process of maintenance and repair, disappear from view.

Analytical clarity and explanatory power must be gained in order to get 'a deeper understanding of the aspects of everyday life which often reside in the background' (Young 2021, 116). But what connects – if necessary – a different conceptualisation of breakdown and decay in the artefactual world with the particular ways in which the fixers conceptualise their activity and become aware of their activity? One way to bridge this gap is by examining the growing ethnographic literature on maintenance and repair activities (Orr 1996; Henke 2000, Graham and Thrift 2007; Denis and Pontille 2015; Strebel et al. 2019; Denis 2020; Henke and Sims 2020), because it directly engages with the forms of knowledge and value that shape the fixer's point of view. There we 'discover' that breakdown is not only a functional mismatch connected to the form and matter of objects and infrastructures but above all a recurring aspect of our existence with artefacts, of the ongoing relationship between things and people, not a punctual failure. It is necessary to 'fix' people *and* things in normal contexts of use and function, and not to 'think' of breakdown and repair as (abnormal) events. In a way, this corrects a 'misrepresentation' of the technological process and of everyday experience with artefacts (as well as of the nature and role of infrastructures) that depend on a 'dominant' point of view in which the 'maker' ideally (and normatively) determines the mode of relations and experience with the technological realm.

However, the scope of this shift in perspective remains limited because it neglects to account for how it responds to specific concepts, experiences, values or practices of the particular standpoint of the fixer. Most importantly, it fails to explain how the scope and nature of a possible *epistemic advantage (or privilege)* can be grounded in the practice of the participants in repair activities. Is it possible to formulate a standpoint epistemology of repair that focuses on the knowledge claims of the fixers and how they are grounded? To address this question, it will be helpful to revisit some of the familiar claims of standpoint epistemologies, which we will explore in the next section.

5.5 Standpoint Epistemologies: Main Tenets

The very possibility of a standpoint epistemology rests on its ability to justify the epistemic privilege it claims for the dispossessed. It assumes that disadvantaged groups, by virtue of their oppression, have an epistemically privileged view of the social world, at least compared to the view of dominant groups, who have a more limited and perhaps distorted understanding of social life. However, this assumption goes beyond the existence of different perspectives on a given issue and does not simply emphasise the epistemic advantages derived from the fact of occupying a certain position. This is an uncontroversial thesis as it is a 'natural' condition for the existence of epistemic positions in general or a result of accumulated (social) experience (and the acquisition and deployment of expertise). Standpoint epistemologies hold – controversially – that certain knowledge claims, at least on a certain set of issues and after a process of consciousness-raising, are authorised by the fact that certain groups occupy certain positions of subordination.

The idea that epistemic privilege is conferred by occupying a subordinate position within the social structure has Marxist roots and is based, in particular, on Georgy Lukács's (1971) explanation of the epistemic grounds of the proletariat's consciousness-raising. However, feminist epistemology is one of the fields in which it has been most extensively developed in recent decades (Smith 1974; Collins 1990; Hartsock 1983; Harding 1991). It is not our purpose here to revisit the myriad of debates that standpoint epistemologies have generated. Therefore, we will summarise the main tenets of any standpoint epistemology in three characteristic claims: (i) the situated knowledge thesis, (ii) the epistemic privilege thesis and (iii) the inversion thesis.[12]

(i) *The situated knowledge thesis.* The clearest formulation of this first thesis appeals to an understanding of the nature of knowers. In contrast to traditional approaches to epistemology, which assume a transhistorical, universal and disembodied knowing subject, it is argued that we are all situated knowers, shaped by our social experience and location (Haraway 1988). In short, it states that '[k]nowledge claims are always socially situated' (Harding 1993, 54), according to social categories such as gender, ethnicity, class and so on. Knowers (and their knowledge) are shaped by their social and historical locations and differences in social factors create significant epistemic differences; they affect what one can know and also the epistemic claims to which one is entitled (Alcoff 2007).[13]

(ii) *The epistemic privilege thesis.* Standpoint epistemologies go beyond the descriptive claim that location shapes the epistemic capacities of the knowers. They argue, controversially, that some epistemic benefits derive from attaining a particular standpoint. Social location confers certain epistemic advantages on some groups, always relative to the perspectives of other groups. How to articulate these advantages is a matter of controversy within the field of standpoint epistemologies. To the extent that they manage to occupy such a standpoint, epistemic subjects are able to correct distorted views, at least of their own social situation, and to open paths to the discovery of new truths. They don't assume that all perspectives are equally valid in epistemic terms, but at the same time – otherwise, they would risk undermining the situated knowledge thesis – they deny the possibility of formulating an objective standard against which different perspectives can be measured.[14]

(iii) *The inversion thesis*. As Tanesini (2018) shows us, this thesis embodies the most characteristic emphasis of standpoint epistemologies. It is based on the idea that the standpoint of the socially oppressed is epistemically advantageous, compared to the perspectives of the oppressors. Those who occupy a socially (and in a sense, epistemically) subordinate role reverse their position with the socially (and in a sense also epistemically) privileged. Contrary to what common sense suggests (i.e., that those with lower incomes, fewer resources and lower levels of education are less likely to understand the social world), it is argued that 'the perspective of the downtrodden is epistemically privileged' (Tanesini 2018, 337). That is, the economically dispossessed and the politically oppressed, are able to see things that the privileged usually ignore (Wylie 2003). In this sense, some authors have spoken of a 'bifurcated consciousness' (Collins 1990); a consciousness that knows both the dominant group's vision, because it is the dominant one, and the oppressed group's vision, because it is the oppressed one. In this sense, the standpoint of the minority groups would have a higher probability of acquiring a more accurate view of the social world, since it would be able to describe it in a less distorted way; at least in the sense that it could be more faithful to the social relations which are produced by material conditions. For this reason, the inversion thesis can only be understood as being part of the political commitment of those who claim to be located in their particular standpoint.

To avoid misunderstandings, it is useful to add some comments to these three theses:

First, as several authors have pointed out (e.g., Wylie 2003; Tanesini 2018; Ashton 2020; Fricker 1999), the mere fact of being oppressed does not automatically make one's perspective a standpoint. Being oppressed is not a sufficient condition for epistemic privilege. This is crucial because it is what makes it an achievement, not something that is automatically acquired. Rather, the transition from a mere perspective to a standpoint can only be achieved through a process of critical reflection, that is, a 'conscious reflection on how power structures and resulting social locations influence knowledge production' (Intermann 2010: 785). This situates standpoint epistemologies as part of critical theory (Anderson 2020; Intermann 2010; Wylie 2003; Crary 2018). One acquires a certain epistemic advantage in the very process of becoming aware of one's social location, and this is the result of a collective process: knowledge is elaborated from the perspective of those who share collectively recognised experiences, needs or interests.

The second remark is not problem-free. How to make theses (i) and (ii) compatible has been the subject of intense debate within feminist epistemologies. For our purposes, it is sufficient to recall that claims to a less-distorted and even objective view of reality and a richer understanding of the social world do not imply a notion of objectivity as disinterestedness and as lacking any anchor in a particular social location. The debate has raised some troublesome points regarding broader epistemological positions on justification and knowledge, such as objectivism, contextualism or relativism. Whichever way out of these philosophical difficulties one is willing to accept, it must be agreed that adopting a particular standpoint must ensure epistemic improvement without abandoning the idea of non-neutrality.[15]

The third remark is widely accepted and has to do with the risk of essentialising the standpoint. If the experience of oppression due to belonging to a certain social group is a part of what explains epistemic privilege, there is a risk of thinking that all members of that group have *the same* experience. Indeed, when we assert that women share a common experience, we run the risk of assuming that all women suffer gender oppression equally. Alessandra Tanesini offers a good solution to this risk. In her pragmatist approach, the knowledge that women may have about their social situation is determined by shared interests and utility (which enable them to gain a better understanding of their situation), rather than by membership or subjective identification.[16] In this understanding, the standpoint is formulated through the identification of common interests rather than through an *a priori* demarcation of the group.

This clarifies the ambiguities of Jackson's proposal for a possible standpoint epistemology of repair. On the one hand, it cannot be reduced to a mere theoretical view of the main actors of the technological world (even if it reflects their interests and needs), but must emerge from the experiences and practices of the repairers themselves and of the collectives involved in the (re)production of the artefactual world; on the other hand, as can easily be seen from the brief reconstruction we have made of some of the central tenets of standpoint epistemologies, the nature and grounds of epistemic privilege relate to how relations of subordination (social and epistemic) are inverted in a collective process of identification or, at least, of adoption of shared (social and political) interests. In this sense, it is relevant to ask whether or not a standpoint epistemology of repair is possible.

5.6 A Standpoint Epistemology of Repair?

Can the fixer then achieve a standpoint? The answer to this question is neither clear nor straightforward. However, let us examine each of the three central tenets of standpoint epistemologies to explore some favourable considerations. Justifying the situated knowledge thesis should not pose much of a problem. As knowers, fixers seem to be located in a situation determined by their material and social conditions. Within this circumstance, they articulate their specific experience of the artefactual and socio-technical world. Therefore, there are no hindering circumstances that prevent this position from becoming an epistemic locus from which to explore its conditions concerning the technological world and, more specifically, its (invisible) contribution to its (re)production. Likewise, it does not seem problematic to accept that the category of fixer itself is a significant situating factor, as is the case with the category of gender concerning the (re)production of the social conditions of existence. We believe we have provided sufficient grounds to consider the category of fixers as delineating particular forms of experience in dealing with our shared expectations in socio-technical and artefactual environments. Potentially, these can serve as a starting point for constituting other forms of knowledge and for opening up new trajectories of socio-technical transformation.

The two normative theses (epistemic privilege and inversion) obviously pose more serious challenges, as the possibility of the fixer's perspective emerging as a genuine standpoint rests upon them. First, how can the fixer become a site of epistemic privilege? An answer to this question is necessary in order to satisfactorily explain how epistemic advantages are acquired. Second, how might we account for such privilege in terms of its subordinate location with respect to the epistemic practices of the technical world? Both aspects must necessarily be intertwined: this depends on the epistemic location that emerges as a standpoint. This is crucial because the standpoint is a *collective* achievement that should, for example, correct the conditions of invisibility of maintenance and repair practices or bring to light assumptions that distort our accounts of how our material and socio-technical environments are (re)produced. In this sense, the process that leads to achieving a standpoint is also a process of forming a knower who is able to make certain epistemic claims and who demands recognition, for they have become masters of their own voice, so to speak. Because the standpoint is rooted in the conditions of their lived experience as fixers, they are in a position to make visible aspects that would be denied to the dominant perspective. Therein, lies the potential epistemic advantage.

But this is one of the first problematic aspects: in what sense are relations of subordination so pronounced in the field of the distribution of knowledge and socio-technical skills that a reversal of roles is necessary to bring this transformative potential to light? Is there something in the disregard and invisibility of repair and maintenance activities, on which numerous theorists have insisted as we have seen, that makes it possible to speak of social *subordination* in the case of the reproduction of the technical conditions of reality?

Although the answer to these questions is not obvious, at least one point is clear: a form of subordination is reflected in hierarchical models that characterise our collective imagination and are projected into recognisable philosophical positions. Creators and inventors stand out as emblematic figures in the production of the technical world. In the background, the rest of the figures always appear, from the manufacturer to the user, but the fixer remains completely forgotten. This implies that each of these figures (which are, in principle, roles) occupies different (social and socio-technical) locations from which they can make their knowledge claims. A hierarchical space is created. In it, the makers dominate over the rest of the 'figures': they do so not only by virtue of the hierarchies that articulate social and political (and economic) forms of recognition but also in metaphysical terms. It is they who determine the metaphysical fate of the objects they create, who 'define' their nature.[17] All other figures must conform to the norm established by the makers' interventions in the world. But more importantly, this explains a certain epistemic privilege of the makers (the special authority on what an artefact really is): they possess practical knowledge as makers, as opposed to the mixture of observational, inferential and testimonial knowledge, certainly incomplete, of the users and of other figures who participate in the (re)production of the socio-technical world. Of course, users can either intervene and reshape the intended uses of an artefact or idiosyncratically adapt artefacts to their uses, but they can hardly claim a hierarchical inversion, since the prevailing paradigm is a unidirectional communication model in which messages go from makers to users. Moreover, it is not easy to see how the inversion could affect epistemic privilege since it is ostensibly based on a specific form of authority that only the maker can exhibit.[18]

It is an obvious fact that repair and maintenance practices have been overlooked in our understanding of how we collectively shape our artefactual and technical worlds. In this sense, they share with other collectives a certain subordination to 'nobler' tasks, as well as to those social groups that are more decisive when it comes to shaping socio-technical systems. The question, then, is whether the particular location of fixers can become a legitimate framework for epistemic claims about what the technical world is and how it is created and maintained. However, if we examine the most relevant claims of standpoint theory, such a framework should not only be legitimate but also superior to other locations, such as those of designers or users. It is important to clarify that, within this framework, the location of the fixers should be understood as superior in the sense that they are able to access more significant truths about the technical world and thus can gain a better understanding of it by denouncing the distorted visions that 'dominant groups' have naturalised and protected from criticism.

Accordingly, we can point out two major difficulties in elaborating a standpoint epistemology of repair. The first is to identify the collective that embodies the figure of the fixer. Who are the fixers? The activities of maintenance and repair, as we have seen, permeate the entire socio-technical system; moreover, they are part of the essential activities that sustain our material culture. In what sense, then, can we identify a group or a collective of fixers? In what sense is it a collective that can make specific epistemic claims? It is not clear whether 'fixers' constitute a collective with distinct social group identities or established social roles that share the same set of cognitive and practical skills. Furthermore, it is uncertain whether there exists a fixer's standpoint either in terms of identification with a social group or in terms of having 'shared objective interests' that would motivate them to adopt that standpoint in a way that contributes to their interests and goals. It is worth remembering once again that repair, in our sense, implies a wide variety of tasks, ranging from infrastructure maintenance, cleaning, care of objects, minor repairs, replacements, etc. Additionally, the 'cultural' diversity in our understanding of repair should be considered: there are different historical–cultural trajectories in how we understand and engage in repair, and not all of them have 'subordinated' repair to nobler projects of endless innovation and change.

Second, even if it were possible to identify a group with its own experience and knowledge, we lack an explanation for the potential epistemic privilege they would seek. No one denies that they might be able to focus on specific knowledge related to their activities (practical knowledge, tacit knowledge, etc.) and even forms of (thematic) understanding of the technical world in which they intervene. One might think that it is precisely this fact that is often forgotten, that it represents unacknowledged knowledge and unheeded claims to be part of the game of mutual recognition. But this does not give us the rationale for the superiority of their own perspective, the sense in which it would be less distorted. And it does not even allow us to assume that fixers as a collective possess the resources to elaborate a particular epistemic standpoint. We cannot adopt a standpoint epistemology in which mere positioning confers epistemic privilege. If they are capable of articulating their experience with their specific cognitive resources and by doing so reach an enhanced understanding, it must be an achievement that requires the collective itself to engage in the development of tools that strengthen their epistemic claims and transform the practices involved in the (re)production of artefactual worlds. They must have the ability to intervene in the transformation of these practices themselves.

5.7 Repair as Critique[19]

Section 5.6 highlighted two challenges in applying standpoint epistemologies to repair: first, it is difficult to define fixers as an epistemic subject based on shared interests or needs; second, it seems impossible to apply an inversion thesis to collectives involved in repair and maintenance compared to other collectives engaged in technical (re)production. Indeed, as long as the criteria used by standpoint epistemologies to identify and delimit oppressed social groups (workers, women, people of colour, etc.) susceptible to developing a standpoint remain unchanged, these difficulties appear insurmountable. This does not mean that the insights provided by critical epistemologies like standpoint theory do not help recognise the transformative role that can arise from considering the experiences of repair and maintenance. In fact, limits and perspectives illuminate each other in an analysis of the paths that collectives take to arrive at a standpoint.

Within traditional standpoint epistemologies, there are usually two paths that serve to demarcate social collectives that are capable of constituting genuine standpoints. We will refer to them as the identity path and the material path. Once again, drawing a comparison with gender will help us illustrate the challenges. On the identity path, women are considered as such when they subjectively identify with the norms of femininity. In the case of the material path, women are defined as such based on the position of subordination they occupy in the sexual division of labour (reproductive sphere).

If the fixer were to follow the path of identity, becoming a fixer so as to be part of the collective would consist of self-consciously assuming such a role, making it a defining aspect of one's own identity. At first glance, this path does not appear to justify the claim that people involved in maintenance and repair activities occupy a particular standpoint. Since anyone who subjectively identifies with such a figure can become a repairer, this path results in a category of potential knowers that is too diluted and diffuse. This category includes those who engage in Do-It-Yourself (DIY) practices, which Wolf and McQuitty define as a 'specific form of prosumption (...) in which individuals use raw and semi-raw materials and component parts to produce, transform, or reconstruct material possessions, including those extracted from the natural environment' (2011: 154). As is well known, DIY practices apply to all areas of our lives, from hacking with Ikea furniture to DIY urbanism to the transformation of automobiles. In the context of maintenance and repair, there are some well-known reference sites, such as iFixit (Fixing the World 2023), a website that offers repair guides for various devices and forum support for users. However, it is far from clear whether fixers are framed by a set of norms with which they could identify in order to participate in the collective. Norms of femininity, for example, are shaped by structural conditions that dictate social and cultural expectations for women. Nothing similar can be seen in the various movements that are involved in the 'visibilisation' of repair and maintenance tasks or even the production and manufacture of products. It should be noted, however, that DIY practices have recently adopted a more collective approach, as can be seen in the shift towards DIT (Do It Together) strategies, the proliferation of repair cafés (Repaircafé 2022) or the growing trend of the libraries of things (Library of Things n.d.).

The material path seems more promising. Reproductive and care work is, within this path, the material and social place that collectively defines women or that makes them an oppressed subject, as opposed to the collective of men, who perform the productive tasks. It is the tasks of care and reproduction that can lead some women to adopt a specific point of view about themselves as a collective, as well as about the hierarchical structuring of social reality. The epistemic privilege of those who become feminists lies in the fact that they are in a position to unmask the fantasy of patriarchal reality, which is built on a fantasy of autonomy, individuality and productivity. It reveals the fact that reproductive labour, despite being treated as dispensable, is an essential activity of the social world.

Similarly, in the material and cognitive divisions of the technical world, the fixers should be demarcated by their subordinate position. However, the question arises: subordinate in relation to whom? We have previously insisted on the hierarchical model we project in our images and ideas of technical (re)production, which seems to subordinate all the figures and actors to the makers. But this, as such, does not delineate the collectives according to the social, economic or material conditions of work. The economic and material conditions under which groups in the Global North perform their technical maintenance tasks may differ significantly from those in the Global South, where individuals engage in demanding repair and maintenance work under poorly paid labour contracts. Nevertheless, it could be argued that just as women are identified by the assigned tasks in the sexual division of labour, the collective of fixers is identified by their particular care and treatment of artefacts and technical environments – their 'reproductive' tasks, which may endow them with a particular vision for unmasking the fantasy of domination and technical control of environments. This does not mean that the collective of fixers as such acquires a privileged standpoint on the technological world by virtue of being in conditions of subordination, but it does mean that the experiences of fixers and users reveal specific ways of understanding the conditions of reproduction of the technical world. The awareness stemming from this knowledge can undoubtedly contribute to correcting distorted perceptions of the nature of technology itself.

It is necessary to mention something relevant here: the hierarchical structure of the technical model is not entirely similar to the hierarchical structure that defines the social world. Mainly because the positions, roles or figures that define it (designers, users and fixers) are transversal. Thus, the superiority that designers have over users and fixers in this model is always related to a specific aspect that only a careful theoretical work of contextualisation can adequately identify. Thus, IT staff fixes artefacts in obedience to what the designs of the device establish as functionally appropriate. However, they do not do so in the same sense as maintainers in the Global South countries, such as workers in India, do so, nor in the same sense as iPhone employees or users who repair their devices at home following tutorials on YouTube. To be clear, the oppression or subordination of fixers in the tech world is intersectional. As far as this question is concerned, suffice it to say that the intersectional nature of their subordination prevents the inversion thesis from being completely fulfilled or at least prevents it from being fulfilled in the same way as it is fulfilled in the standpoint model. Its intersectional and delocalised character means that this inversion can only be claimed in relation to those specific aspects that can be analysed from the specific contexts in which the work of fixers is inscribed.[20]

Should we then abandon the promises of such critical epistemologies in the case of repair? Perhaps it would be more feasible to approach repair and maintenance activities through the lens of the ethics of care. Yes, there is an urgency to take care of the world we inhabit together, so the ethics of care is essential to understanding many of the normative cues at play in repair and maintenance. However, this ethical commitment is inseparable from an epistemology that emphasises the transformative potential that derives from a process of becoming aware of the meaning of these activities, the experiences gained and their social, cultural, and, of course, technical insertion and impact (de la Bellacasa and Puig 2011; de Wilde 2021; Martin et al. 2015; Liao and Huebner 2021). The experience of fixers, as well as their skills and the possibility of bringing to light our expectations about artefactual environments, should inform our knowledge about what needs to be transformed if we are to take care of the world (Houston et al. 2016). Although fixers as a group may lack a clearly defined epistemically privileged standpoint, there is a growing social demand, particularly from certain collectives, for paying attention to the conditions in which repair and maintenance activities occur. Movements advocating for the rights of repairers have highlighted the need to unify demands and recognise shared interests. Consider, for example, the way some communities are drawing our attention to the need to create 'a more caring and well-maintained world' (themaintainers.org) by defending certain values, under-recognised embodied experiences of maintenance and disrupting ties to material things.

The current social and political reality has led certain groups of citizens and fixers to claim themselves as a potentially transformative subject linked to their perspective as fixers and maintainers. In a way, they embody a certain social awareness of the cultural and technical centrality of these activities and advocate for their rights as a collective. This suggests that the social movement for the Right to Repair (2023) (which has been gaining strength) is not only focused on individual rights but also encompasses a broader, collective understanding of the technical and social worlds. At its core, this movement seeks to create a better world through political engagement.

Let us take stock. Fixers do not fit neatly into the framework of classic oppressed collectives that have been the focus of standpoint epistemologies. Their position does not arise from the same material conditions or group identification as traditionally marginalised groups. It is a collective in which consciousness-raising does not pass through the first-person experience of oppression. In this sense, it has nothing to do with the consciousness-raising characteristic of identity politics. However, it is still possible to preserve the critical and transformative potential of adopting the stance of repair and maintenance.

In her analysis of the standpoint as critical consciousness, Crary (2018) suggests that what is important to engage in the development of a radical political perspective is the acquisition of a certain sensitivity to harm. However, we need to remember that the harm here pertains more to the material world itself, with its fragility and vulnerability, rather than to specific groups or collectives and their material or identity conditions. This form of consciousness-raising emphasises our responsibility to care for and protect the material world we inhabit, rather than focusing solely on self-identity. From a philosophical perspective, this approach aligns more naturally with proposals of an ecological nature, such as Lorraine Code's epistemology (2006) or interactionist approaches like José Medina's epistemology of resistance (2013). The former involves a contextualist localisation of epistemic practices, while the latter relies on forms of epistemic friction between collectives. Both frameworks allow us to understand, without relinquishing the situated knowledge thesis, how to address the contextually shaped claims to recognition of experiences and practices related to the (re)production of our artefactual world, particularly regarding our ways of inhabiting it. What we are witnessing, therefore, is a shift in orientation, where consciousness-raising no longer stems from a dialectic of opposition and inversion, but from an awareness of the fragility of existence and the constant negotiation that is required to inhabit our environment and engage with the material world. By embracing this perspective, we can acknowledge the significance of repair and maintenance as practices that contribute to our understanding of, and our responsibility, towards the world we inhabit.

5.8 Conclusion

Is a standpoint epistemology of repair possible? The question remains open. We have argued that several aspects point to the impracticality of using standpoint theories to ground clear epistemic advantages to fixers over other socio-technical actors. We have introduced a broad notion of repair and maintenance that we believe more adequately describes the way we collectively sustain our artefactual environments. We all play this role in one way or another. This is not to say that there are no groups that engage in specific repair activities, with their own knowledge and skills. What is difficult is to delineate the collective that specifically adopts a standpoint from which to claim some epistemic advantage over the intimate workings of our technological world or its social or political conditions. Thus, it may not be easy to see how to substantiate the theses of epistemic privilege and inversion that would define the normative framework of standpoint epistemologies in the case of repair.

On a more optimistic note, however, shifting the focus to the complex and heterogeneous experiences of ourselves as fixers does indeed contribute to highlighting its critical potential. Three brief observations will suffice to remind us of its epistemic and transformative nature. First, the fixer's perspective helps us to demystify certain fantasies of domination and control over technical environments; the fixer reminds us that the effective conditions of habitability of our technological world are always underpinned by constant attention and care. Second, without necessarily emerging as a specific standpoint, there are collectives that contribute to the awareness of these conditions, which in turn has increased the demand to transform the social and cultural conditions in which we develop our repair activities. Third, the shared experience of repair and maintenance – at its various levels – gives rise to the articulation of a collective awareness of the fragility of things. In a sense, the authority of the fixer's voice functions less as an inversion of epistemic positions with other socio-technical actors than as a way of respecting the voice of our complex material and technical worlds. Perhaps attending to the intricate and heterogeneous activity of repair will help us better understand the ties that bind us to things and the sometimes illusory expectations we project onto the artefactual environments we inhabit.[21]

Notes

1 This is why Jackson can claim that repair is the fulcrum of two worlds that confront (Jackson 2014, 222).

2 Steven Vogel in a chapter included in this volume ('Maintenance and the Humanness of Infrastructure') insists on this idea of maintenance as an ongoing activity, but also in the continuity between the activities of building and maintenance (Vogel, 2024).

3 Technological systems are also subject to stochastic processes that do not generate precise predictive expectations, but that are part now of how we need to take care of our artificial worlds. Theoretically, we can build, so to say, stochastic expectations about how these systems and environment could behave and 'evolve'.

4 In fact, in our highly differentiated technological environments, at least for some devices and certainly for complex technical systems and infrastructures, the expectation by default is that others – specialised technicians – will do the repair and maintenance.

5 This insistence in the conditions of persistence (and reproduction) of the environments should not be viewed from a purely static dimension. We assume that our artefactual worlds change constantly and with them our expectations. In repair and maintenance, we care about their *stability*, which does not have to be mistaken with the mere preservation of unchanged environments. An earlier chapter in the volume explores this idea in further detail (Young, 2024).

6 The figures are idealised; they reflect roles and not specific actors, although it is obvious that one can later identify each role more or less closely with specific groups. This is true of both designers and fixers, as it is of any other possible technical roles.

7 In ancient philosophy, Plato was critical of this idea and considered that it is users who are entitled to judge how artefacts should be produced by artisans (see *Republic* 601c–602a and *Cratylus* 390b–d). But since at least the Middle Ages, when Maimon introduced the idea of maker's knowledge in his *Guide for the Perplexed*, a particular image of the epistemic privilege of makers in relation to their products has been consolidating. See also Nicolas of Cusa in his dialogues *De Idiota*, or Vico in his *Scienza Nuova*. For recent reconstructions of the tradition of maker's knowledge, see Pérez-Ramos (1988) or Floridi (2011). See also comments on Section 5.6. We thank the editors of this volume for the suggestion to note this broad historical change in the consideration of the figure of the maker.

8 This has been an insistence in STS studies since the 1980s. The localised and distributed dimension of knowledge related to repair and maintenance is much more evident, as well as its invisibility. For more on the (in)visibility of repair practices, see Young and Coeckelbergh (2024).

9 This can be detected both in philosophical discourses on technology and in the socio-technical imaginaries we project in our cultural works. Technical worlds seem to be suspended in the air, as if they were worlds of ideas. From its Platonic matrix, thinking about technical production only recognises the role of makers and users. Or since when in our collective imagination about technology do repairers or maintainers take centre stage? Consider, for example, the imaginary projection of our digital future: where and how are the work or skills of those who maintain and repair the endless series of software failures, the network crashes, the machines, the huge data factories, represented? Needless to say, that this does not correspond to the very experience of technical workers.

10 Social location is crucial in delineating the epistemic advantage but only to the extent that we gain access to the standpoint through critical reflection on the position of a group within a framework of unequal distribution of knowledge. So, it cannot be denied that we may gain access to the relevant standpoint through theorising, insofar as there are objective facts that are publicly accessible. See the comments by E. Anderson 2020 on this possibility.

11 This is certainly true of Jackson's pioneering paper, but also of a recent contribution by Young (2021).
12 These designations respond to a reconstruction recently proposed by A. Tanesini (2018). But theses defined in similar terms can be found in Bowell (2023) (Internet Encyclopedia of Philosophy) or Ashton (2020). The core ideas are related to how social location has epistemic consequences captured in the theses of situated knowledge and the claims of epistemic privilege or advantage. More controversial, but as we will see not less essential, is the inversion thesis.
13 The core insight of standpoint theory, that is, the idea that the epistemic perspective (and thus the justification of beliefs) depends on the experience that subjects have in relation to the specific social position they occupy (whether privileged or oppressive) is built on the situated knowledge thesis (Ashton 2020).
14 The very fact that different experiences, needs and interests shape different epistemic practices adds a potential epistemic improvement to the extent that it promotes inquiring from a perspective in which bias and distortion are at least counterbalanced. Rather, the controversial issue lies in how such a non-biased and non-distorted understanding can be achieved if it is anchored to a particular standpoint. That is, it is not enough to argue that the standpoint adds a new perspective on a subject: it must seek to reinforce the epistemic virtues of good research.
15 For a discussion of the different positions, see Harding (2004), Rolin (2006) and Ashton (2020).
16 'Putative knowledge is thus said to be socially situated in the sense that, given their shared interests, some knowledge claims are more useful than others for collectives and their members' (Tanesini 2018, 336). In our opinion, her proposal is more faithful to the distinctive political character of standpoint epistemology (which has to do with gaining a better understanding of the social world in order to be able to transform it), rooted in feminism as an emancipatory movement.
17 The model has been current in western metaphysics at least since the work of Nicholas of Cusa. It is recognisable in those views on the metaphysics of artefacts that delineate the artefactual kinds on the basis of the intentions of the makers, who in turn have privileged knowledge (in a very strong sense implying the impossibility of error and ignorance) about their nature. See, for example, Thomasson (2007).
18 What we mean by this is that given the way in which privilege is understood, based on a form of knowledge inaccessible as such to the other figures, a genuine inversion between the figures of the maker and the user is not possible (obviously, neither is it possible for the fixer). Hence, the viability of a standpoint epistemology of repair that emerges from the effective conditions of experience and knowledge of the fixer would depend on a correct understanding of the thesis of epistemic privilege.
19 This section has benefitted greatly from very insightful comments from the editors of this volume and from the reviewers of the chapter.
20 Note that the requirement to analyse oppression through the filter of intersectionality is not exclusive to the technical model. It is also the case of contemporary feminist theories, including standpoint epistemologies, which seek to account for the fact that gender oppression affects us in complex, non-univocal ways and through diverse axes of oppression that intersect with each other, giving shape to the specific oppression in each case or context.

21 We would like to thank Diego Lawler and Fernando Broncano for many discussions on the nature of the artefacts and the reproduction of the technical world. This chapter has been enriched by the observations and comments of the reviewers and, particularly, of the editors of this book. The first version of this chapter was presented at a seminar on Maintenance and Philosophy of Technology organised by Mark T. Young, which was followed by a very fruitful discussion that helped us to polish our ideas.

References

Alcoff, L.M. (2007) 'Epistemologies of Ignorance: Three Types.', in S. Sullivan and N. Tuana (eds), *Race and Epistemologies of Ignorance*, pp. 39–58. New York, NY: State University of New York Press.

Anderson, Elizabeth (2020) 'Feminist Epistemology and Philosophy of Science', *The Stanford Encyclopedia of Philosophy* (Spring 2020 Edition), Edward N. Zalta (ed.), https://plato.stanford.edu/archives/spr2020/entries/feminism-epistemology/

Arendt, H. (1958). *The human condition*. Chicago University Press.

Ashton, Natalie Alana (2020) 'Relativising Epistemic Advantage', in Martin K (ed.) *The Routledge Handbook of Philosophy of Relativism*. New York: Routledge.

Bowell, T. 'Feminist Standpoint Theory', *Internet Encyclopedia of Philosophy*, online: https://iep.utm.edu/fem-stan/, accessed April 11 2023.

Code, L. (2006) *Ecological Thinking: The Politics of Epistemic Location*. Oxford: Oxford University Press.

Collins, P. (1990) *Black Feminist Thought: Knowledge, Consciousness, and the Politics of Empowerment*. London: Routledge.

Crary, A. (2018) 'The methodological is political: What's the matter with 'analytic feminism'. *Radical Philosophy*, 202, pp. 47–60.

de la Bellacasa and Maria Puig (2011) 'Matters of care in technoscience: Assembling neglected things'. *Social Studies of Science*, 41(1): 85–106. https://doi.org/10.1177/0306312710380301

de Wilde, M. (2021) 'A Heat Pump Needs a Bit of Care': On Maintainability and Repairing Gender–Technology Relations. *Science, Technology, & Human Values*, 46(6): 1261–1285. https://doi.org/10.1177/0162243920978301

Denis, D.J. (2020) 'Why do maintenance and repair matter', in Anders B. et al (eds.) *The Routledge Companion to Actor-Network Theory*. New York: Routledge, pp. 283–293. https://doi.org/10.4324/9781315111667-31

Denis, J. and Pontille, D. (2015) 'Material Ordering and the Care of Things', *Science, Technology, & Human Values*, 40(3), pp. 338–367. https://doi.org/10.1177/0162243914553129

Fixing the world, one gizmo at a time. | iFixit News (2023) Available at: https://www.ifixit.com/

Floridi, L. (2011) 'A Defence of Constructionism: Philosophy as Conceptual Engineering', *Metaphilosophy*, 42(3): 282–304.

Fricker, M. (1999) 'Epistemic Oppression and Epistemic Privilege', *Canadian Journal of Philosophy* 29(Suppl.), pp. 191–210.
Graham, S. and Thrift, N.J. (2007) 'Out of order: understanding repair and maintenance', *Theory*, pp. 1–25. https://doi.org/10.1177/0263276407075954
Haraway, D. (1988) 'Situated Knowledges: The Science Question in Feminism and the Privilege of Partial Perspective', *Feminist Studies*, 14(3), pp. 575–599. https://doi.org/10.2307/3178066
Harding, S.G. (1991) *Whose science? Whose knowledge?* Ithaca, NY: Cornell University Press.
Harding, S. G. (1993) 'Rethinking Standpoint Epistemology: What is Strong Objectivity?', in L. Alcoff and E. Potter (eds.), *Feminist Epistemologies*, New York; London: Routledge, pp. 49–82.
Harding, S. G. ed. (2004) *The Feminist Standpoint Theory Reader*. New York and London: Routledge.
Hartsock, N. (1983) 'The Feminist Standpoint: Developing the Ground for a Specifically Feminist Historical Materialism', in Sandra H. and Merrill H. (eds.) *Discovering Reality: Feminist Perspectives on Epistemology, Metaphysics, Methodology, and the Philosophy of Science*. Dordrecht: D. Reidel, pp. 283–310.
Henke, C.R. (2000) 'The mechanics of workplace order: Towards a sociology of repair', *Berkeley Journal of Sociology*, 44, pp. 55–81.
Henke, C.R. and Sims, B. (2020) *Repairing Infrastructures*. Cambridge, Massachusetts, London, England: The MIT Press.
Houston, L. et al. (2016) 'Values in Repair'. CHI '16: Proceedings of the 2016 CHI Conference on Human Factors in Computing Systems May 2016, pp. 1403–1414. https://doi.org/10.1145/2858036.2858470
Ingold, T. (2011) *Being alive: essays on movement, knowledge, and description*. Abingdon, Oxon; New York, NY: Routledge.
Intermann, K. (2010) '25 Years of Feminist Empiricism and Standpoint Theory: Where Are We Now?', *Hypatia*, 25(4), pp. 778–796. https://doi.org/10.1111/j.1527-2001.2010.01138.x
Jackson, S.J. (2014) 'Rethinking Repair', in Taleton, J. et al. (eds.) *Media Technologies: Essays on Communication, Materiality, and Society*. The MIT Press. https://doi.org/10.7551/mitpress/9780262525374.003.0011
Liao, S. and Huebner, B. (2021) 'Oppressive Things'. *Philosophy and Phenomenological Research*, 103(1): 92–113. https://doi.org/10.1111/phpr.12701
Library of Things. (n.d.) Available at: https://www.libraryofthings.co.uk/
Lukács, G. (1971) *History and Class Consciousness; Studies in Marxist Dialectics*. Translated by Rodney Livingstone. Cambridge, MA: MIT Press.
Martin, A., Myers, N. and Viseu, A. (2015) 'The politics of care in technoscience'. *Social Studies of Science*, 45(5): 625–641. https://doi.org/10.1177/0306312715602073
Martínez, F. and Laviolette, P. (2019) *Repair, brokenness, breakthrough*. 1st ed. New York: Berghahn Books.
Medina, J. (2013) *The Epistemology of Resistance: Gender and Racial Oppression, Epistemic, Injustice, and Resistant Imaginations*. Nueva York: Oxford University Press.

Orr, J. E. (1996) *Talking about machines: Ethnography of a modern job*. New York: ILR Press.
Pérez-Ramos, A. (1988) *Francis Bacon's Idea of Science and the Maker's Knowledge Tradition*, Oxford: Clarendon.
Preston, B. (2013) *A Philosophy of Material Culture*. London: Routledge.
Repaircafe (2022) *About – Repaircafe*. Available at: https://www.repaircafe.org/en/about/
Right to Repair Europe (2023) *Home – Right to Repair Europe*. Available at: https://repair.eu/
Rolin, K. (2006) 'The bias paradox in feminist standpoint epistemology', *Episteme* 3(1–2), pp. 125–136.
Smith, D.E. (1974) 'Women's Perspective as a Radical Critique of Sociology', *Sociological Inquiry*, 44(1), pp. 7–13. https://doi.org/10.1111/j.1475-682x
Strebel, I., Bovet, A. and Sormani, P. (2019) *Repair work ethnographies*. Singapore: Palgrave Macmillan.
Tanesini, A. (2018) 'Standpoint Then and Now', in Miranda F. et al. (eds.) *The Routledge Handbook of Social Epistemology*. 1st ed. New York: Routledge, pp. 335–343. https://doi.org/10.4324/9781315717937-33
Thomasson, A.L. (2007) 'Artifacts and human concepts', in Eric Margolis & Stephen Laurence (eds.), *Creations of the Mind: Theories of Artifacts and Their Representation*. Oxford University Press. pp. 52–73.
Vogel, S. (2024) 'Maintenance and the Humanness of Infrastructure', in Mark T. Young and Mark Coeckelbergh (eds) *Maintenance and Philosophy of Technology. Keeping Things Going*. New York, NY; London: Routledge, pp. 1–36.
Wolf, M. and McQuitty S. (2011) 'Understanding the Do-it-yourself Consumer: DIY Motivations and Outcomes', *AMS Review* 1, no. 3–4, pp. 154–170.
Wylie, A. (2003) 'Why Standpoint Matters', in Robert F. and Sandra H. (eds) *Science and Other Cultures: Issues in Philosophies of Science and Technology*. New York, NY; London: Routledge, pp. 26–48.
Young, Mark T. (2021) 'Now You See It (Now You Don't): Users, Maintainers and the Invisibility of Infrastructure', in Michael N. et al. (eds.) *Technology and the City*. Dordrecht: Springer, pp. 101–119.
Young, Mark T. (2024) 'Technology in Process: Maintenance and the Metaphysics of Artifacts', in Mark T. Young and Mark Coeckelbergh (eds) *Maintenance and Philosophy of Technology. Keeping Things Going*. New York, NY; London: Routledge, pp. 1–36.
Young, Mark T. & Coeckelbergh, M. (2024) 'Keeping Things Going: Maintenance and Philosophy of Technology', in Mark T. Young and Mark Coeckelbergh (eds) *Maintenance and Philosophy of Technology. Keeping Things Going*. New York, NY; London: Routledge, pp. 1–36.

6 Sustainability as Planetary Maintenance

Jochem Zwier

6.1 Introduction

The following investigation asks in which sense sustainability is to be understood in terms of maintenance and what the limits and implications of such an understanding are. While it may be clear that sustainability and maintenance are closely related since sustainability seeks to maintain habitats and habitability, it is less clear what this relation ultimately entails. Sustainability remains a contested concept that has spawned a growing discourse about which techno-scientific interventions are necessary and desirable, which ethical and political implications are justified and how sustainable solutions are to be governed (Salas-Zapata and Ortiz-Muñoz, 2019; Marshall and Toffel, 2005). For instance, questions have been raised about whether and why sustaining food supplies trumps sustaining the character of the land or ways of farming (Mariola, 2005). Although the importance of such debates can hardly be overstated in light of the looming climate catastrophe, it is worth noting that as a byproduct of the increasingly resounding call for more effective action and concrete sustainable initiatives, the ontological implications of the concept of sustainability are often left by the wayside. This is unsurprising since it is usually sufficiently clear that whatever sustainability or ecological maintenance ultimately means, sustainable and renewable developments appear to be superior to fossil-fuelled developments. Be that as it may, this chapter departs from the supposition that a philosophical inquiry into sustainability is nonetheless worthwhile since it helps understand and critically examine the horizon towards which sustainable developments are oriented. Accordingly, the purpose here is to try and clarify the metaphysical and ontological implications of the concept of sustainability, which can be done by explicating its relation to maintenance.

To this end, Section 6.2 attempts to clarify the metaphysical and ontological roots of the concept of maintenance. It aims to demonstrate that these roots lead to Aristotle. As a result, the concept of maintenance turns

DOI: 10.4324/9781003316213-7

out to correspond to a hylomorphic ontology and correlated technical–antagonistic apprehension of reality. The purpose of the section is to explicate this correspondence and show how the concept of maintenance lies at the heart of technics. This prepares the question regarding how it informs and maps onto the concept of sustainability.

Section 6.3 develops this question by discussing the Anthropocene. It demonstrates how the technical–antagonistic aspect of maintenance returns in sustainability, now conceptualised as the striving to maintain the habitability of an increasingly unruly Earth. The question then becomes whether the hylomorphic ontology of maintenance returns similarly. Section 6.3 develops a negative answer to this question by contrasting the verticality and transcendence of Aristotle's hylomorphism with the horizontality and nihilism of sustainability. This problematises the concept of sustainability by showing how the question regarding what or who is to be sustained necessarily remains unanswered. This helps articulate the limits of understanding sustainability as technical maintenance, as it demonstrates that while maintenance and sustainability both involve a kind of movement, the movement of sustainability must be negatively characterised as a fleeing movement: while it may be clear what to move away from (unsustainable fossil-fuelled civilisation), a positive determination is kept in abeyance.

Section 6.4 argues that as long as sustainability mirrors the schema of technical maintenance, it must either ignore its nihilistic horizon or affirm its fleeing movement. In raising doubts regarding the sustainability of fleeing itself, the work of Simondon is discussed to explore the possibilities of planetary maintenance beyond hylomorphism. It concludes with an evaluation of these possibilities and their implications.

6.2 Maintenance as Technics, Technics as Maintenance

This section inquires about the metaphysical characteristics of the concept of maintenance. A glance at the concept suggests that it necessarily concerns individuals and antagonism. An individual being and associated functioning must be maintained to antagonistically 'strife against' (*anti-agon*)[1] its degradation and corruption. One gives the woodwork of a house the occasional fresh coat of paint to keep it from weathering and withering away, thus maintaining habitability. One changes the oils and filters of a car to prevent a clogged engine, thus keeping it operational. One cultivates the garden to subdue the proliferation of weeds and encourage the proliferation of crops, thus keeping a garden rather than an overgrowth. Such examples all involve some individual (house, car and garden) and functions that are prone to wear and tear. Accordingly, to maintain is to counteract degeneration and corruption, thereby literally 'keeping in hand' *(manu-tenere)* the original identity and ideal functioning of things. Such 'keeping in hand' is characterised by a double keeping: a keeping-under[2] (the elements, the ravages of time) and a keeping-up (things: identity and functioning).

However elementary, these observations bear witness to nothing less than an apprehension of reality or an ontology: there are individuals, there is their potential decay and there is the care for their upkeep. Rather than simply given, such an apprehension grows from philosophical roots, which are in this case to be found in the fertile ground of Aristotle's metaphysics. Two of his metaphysical considerations particularly resonate with the above outline of maintenance: the first is hylomorphism, according to which things or individuals are considered in terms of form *(morphe)* and matter *(hyle)* and the second concerns the relation between nature *(physis)* and technics *(technè).*

In Aristotle's hylomorphism, any fabricated individual[3] is constituted because a technician (*technitès*) brings matter in form. Following the canonical example, the smith knows how to manipulate the silver matter in such a way that a chalice is formed to serve its sacrificial purpose. The name for the smith's knowledge is technics (*technè*) (Aristotle, *Physics* 2.1.193a 31–2). Given the theme of this volume, it is worth observing that this can be literally understood in terms of maintenance, meaning that a conception of maintenance lies at the heart of *technè*. Just as carpenters do not by themselves create wood, smiths do not create the silver matter, since both originate in nature as trees and ores. They are given to technics rather than being its product.[4] With reference to the etymology of maintenance (*manu-tenere*), smiths take and keep this given matter into their hands to mould it. Such 'handling' is done by observing the outlook *(eidos)* of the fully formed and functional individual *(telos)*, where observing both means 'seeing' and 'complying with': the smith sees and complies with the outlook of the form of the chalice and correspondingly handles matter to bring it in the form of the completed chalice (cf. Glazebrook 2000, p. 104). Accordingly, while belonging 'to bringing-forth, to poiesis' (Heidegger 1977, p. 13), technics is less a matter of pure creation or absolute imposition than of 'taking in hand', that is, maintenance and manipulation of what is given. By implication, when defined as the *archè tès kinèseoos* (Visser, 2014, pp. 68–73), this cannot mean that the technician is *archè* in the sense of sole principle and origin. Rather, reading the genitive *tès* the other way around, *archè tès kinéseos* means to take command[5] *of* a movement, much akin to how in commandeering a ship, one takes possession, one handles a given. As Heidegger (1977, p. 8) explains: 'The silversmith considers carefully and gathers together' rather than creating from scratch.

And does this not point to both the heart of the smithy and the archetype of all technics, namely fire? Mythologically speaking, it is by making up for his forgetful brother's mistake (cf. Stiegler, 1998) that Prometheus offers fire and technics to otherwise ill-equipped humankind, placing it into their hands to maintain and manipulate. As pyrotechnicians, humans do not create fire from scratch, but handle and manipulate a natural given to their ends, be it in the form of torches, ovens or eventually internal combustion engines. As the latter example suggests, this understanding of technics as maintenance does not only concern so-called (and arguably mislabelled) primitive technics like torches and fireplaces (cf. Clark and Yusoff, 2014) but equally holds for modern technics like electronics, where rather than creating it ex nihilo, engineers handle, manipulate and modulate a naturally given movement (of electrons, or, put archaically, lightning) to achieve their ends. Such manipulation demonstrates how a conception of maintenance understood as 'taking-in-hand' lies at the heart of technics itself.

Considering technics in this way further points to the antagonistic aspect of maintenance. If not maintained properly, the fire dies or overgrows into a wild conflagration. The technics of fire accordingly consist of maintaining a rather precarious equilibrium, which Aristotle's thought generally articulates as the middle. This middle is most often associated with Aristotle's ethics, where it marks the virtuous position between two extremes, for example, courage between cowardice and recklessness, (Aristotle, *Nicomachean Ethics*, 1115b 17–9). Yet, the same figure returns in technics as maintenance. To maintain the middle is to domesticate the movements of nature *(physis*, which is always understood in terms of movement or *kinesis*, cf. Glazebrook 2000, p. 103*)*[6] by way of striving against two extremes that it harbours, namely corruption or going into itself on the one hand (*phftoras*, the fire dying), and unrestrained going outward on the other (*genesis*, wildfire). Such middle maintenance can be observed in modern electronics as well, where circuits have an equilibrium 'operational window' between inactivity (*phftoras*, e.g., a dead battery) and hyperactivity (*genesis*, a short circuit).

Interpreted along these lines, technics handles and maintains naturally generated movements (fire; lightning, silver and wood) and temporarily stabilises them into a form. Maintaining stabilised informed matter such as a particular wooden bed is antagonistic since it implies striving against the very natural movements that gave rise to the bed, but which also risks upsetting its equilibrium or middle. To clarify, consider how Aristotle famously states that 'if you planted a bed and the rotting wood acquired the power of sending up a shoot, it would not be a bed that would come up, but *wood*' (Aristotle, *Physics II*,1, 193a). Aristotle's example expresses the precarious equilibrium of the wooden bed: if not maintained, for instance, by just leaving it outside, exposed to the natural elements, it de-forms and recedes into the movement of physis again, either by sprouting (*genesis*) or rotting away (*phftorás*). Aristotle famously says that technics 'completes what nature cannot bring to a finish' (Aristotle, *Physics II*,1, 199a) to express how the bed as informed matter owes its completion to nature as well as to technics. Yet, this always comes with the potential of its reverse, where nature consumes what technics completed: the bed is consumed by worms, the iron by rust, etc. This antagonism concerns, in Arendt's words, the 'unending fight against the processes of growth and decay through which nature forever invades the human artifice, threatening the durability of the world and its fitness for human use' (1958, p. 100).

This then lays out the metaphysical and ontological roots of the concept of maintenance, which corresponds to a hylomorphic ontology and correlated technical–antagonistic apprehension of reality. Maintenance understood as 'taking-in-hand' or 'handling' lies at the heart of technics itself. If sustainability at the very least resonates with maintenance inasmuch as sustainability generally concerns the maintenance of habitats and habitation, the ensuing question is whether and how the mentioned metaphysical characteristics of maintenance inform or disinform the concept of sustainability.

6.3 Sustainability as Maintenance

Sustainability closely resembles the above interpretation of maintenance. To sustain is to maintain, here not so much a specific bed or chalice, but the environment, an ecosystem, and one's *oikos* or habitat. As such, sustainability echoes the antagonism found in maintenance. As Horatius famously noted (*Epist. I*, 10,24) that 'you may drive nature out with a pitchfork, but she will keep coming back', the concept of sustainability necessarily expresses this strive against the return of nature.

After all, it now seems abundantly clear that we either find ways of becoming sustainable or degenerate into the physis whence we came. The discourse of the Anthropocene expresses this antagonism particularly well. On the one hand, it stages a highly technologised humanity, the *anthropoi*, as the dominant Earth-shaping force. On the other hand, it stages the physis, now considered beyond local natural environments, ecosystems and habitats as the planetary physis named Earth, and articulates how it has awoken from its slumber (Hamilton 2017, p. 45). While this slumber, which might be referred to as the Holocene, may have given the impression of a stable physis that sustained ever-increasing extraction of (fossil) resources, the Anthropocene orchestrates a new agon between technologised geo-humanity and the Earthly, geo-physis (cf. Zwier and de Boer 2023).

Among others, this strife can be witnessed in extreme weather events, global warming, desertification, droughts, crop failures, dwindling biodiversity, etc. The general response to such extremes and loss of a Holocenic stable equilibrium or middle appears to consist in the maintenance of both protagonists of the Anthropocene: the *anthropoi* and Earth qua *physis*. If, as suggested, maintenance involves the domestication and handling of physis, it is now up to geo-engineers and sustainable developers to tame the unruly Earth system and abate its threat of becoming uninhabitable (cf. Baskin 2015; Hamilton 2013). Conversely, it appears up to technology designers, educators, ethicists and policymakers to help (re)domesticate the *anthropoi* who have lost their way in excessive fossil-fuelled consumerism and who must become responsible and responsive to the Earth and its new climatic regime.

This general response serves to indicate how the antagonistic metaphysics inscribed in technical maintenance returns in sustainability, as emphatically expressed in the Anthropocene's strive between geo-technè and geo-physis. However, examination of the other aforementioned metaphysical dimension of maintenance, namely hylomorphism, will reveal a significant difference.

6.3.1 Hylomorphism and Verticality

In the Aristotelean conceptualisation, what is to be maintained or kept in equilibrium is guided by the outlook of the form. By complying with the observed form of the bed, the technician seizes the material movements of nature (trees growing and decaying) to maintain them in the form of a bed. Technical things thus 'stand upright' over against the 'growth and decay through which nature forever invades the human artifice' that Arendt mentioned (1958, p. 100). The vertical character of such 'standing upright', which also resounds in maintenance as 'upkeep' is critical to hylomorphism and deserves further attention.

To clarify this verticality: consider how the bed that rots away and thus re-enters the movements of nature (as worm food, then as bird food, etc.) only concerns the particular artefact. The general form remains untouched. It is for this reason that Aristotle ties the notion of form to that of *entelecheia*, which Heidegger renders as *en-telos-echein*, 'something's holding (or maintaining) itself-in-its-completion-(or limit)' (Heidegger 2000, p. 63; cf. Glazebrook 2000, p. 99–100). The form denotes a completion, an optimum, the ideal thing in *optima forma*. In a wordplay that references his own name, Aristotle also refers to it as the 'best goal' or 'optimal goal', the *telos ariston* (Aristotle, *Nichomachean Ethics*, 1099b.17). Put simply, it is only by observation of its complete and optimal form that one can recognise a broken bed and construct a new one (cf. Section 6.2) – how else could one recognise a thing as being broken or deficient? With regards to the verticality of 'standing upright' and 'upkeep', this means the following: if technics domesticate the material movements of nature to maintain them in form (e.g., of a bed), the complete and optimal form that guides this process itself *stands out from* these movements of corruption and generation: it stands out and is beyond the growth and decay of physis, and is therefore metaphysical.[7]

Notwithstanding their other disputes, Aristotle remains beholden to his teacher Plato on this point. Plato's Cratylus demonstrates a similar metaphysical verticality of form, where in the discussion of a broken shuttle of a loom, the craftsman tasked with making a new shuttle observes not the broken artefact, but 'looks up to' its ideal, general and incorruptible form (Plato, *Cratylus*, 1997a, 389b; cf. Aydin 2021, p. 19). In Platonic doctrine, the particular, *material*, and therefore transitory shuttle participates in this form. Similarly, in Socrates' famous metaphor of the three beds in *The Republic* X, the carpenter imitates the ideal form of a bed (and the painter imitates that imitation) (Plato, *Republic*, 1997b, 598). Further, Plato's speculation on a *demi*urge (literally a technician or artifex) as a world-maker does not create ex nihilo but observes and is guided by outstanding, metaphysical ideas as blueprints (Plato, *Timaeus*, 1997c, 28a6).[8]

What follows is that the verticality of the hylomorphic schema is best understood as an interrupted verticality that transcends to the metaphysical. The outlook (*eidos*) or *idea* concerns a higher kind of being, because, unlike particular shuttles, beds and chalices, Plato's ideas and Aristotle's forms are themselves ever-present, untouched by vicissitude and self-sustaining, and thus require no maintenance. These general forms are observed (in the double sense of seen and complied with) in the sublunary, material and particular dealings of mortals.[9]

This may then give rise to the following question: how does the hylomorphic verticality found in technics as maintenance relate to sustainability? Does sustainability similarly look up to and observe a transcendent *telos ariston* or *optima forma*? What follows attempts to demonstrate that the answer to this question must be negative. The hypothesis to be developed is that rather than vertical-metaphysical, the figure of thought at stake in sustainability is horizontal-nihilistic.

6.3.2 Sustainability and Horizontality

To clarify this point, consider how sustainability expresses a kind of movement. This does not primarily mean the scientific, technological and political-economic progressive movement required to achieve sustainability. Rather, in the more fundamental terms of the hylomorphic scheme, sustainability expresses movement insofar as it seeks to maintain, domesticate or handle the movement of unruly matter (*hyle*) such as CO_2 back into formation (*morphe*). Although CO_2 may be the most telling example of a material movement of physis in disequilibrium (consider its resonance with the previous example of wildfire as extreme physical genesis), one can also think of animals as matter to be maintained (sustainability as maintenance of biodiversity), of ecosystems (sustainability as maintenance of the environment) or even the Earth (sustainability as maintenance of the dynamics of the Earth system). All these examples concern a natural material movement in disequilibrium, either on the side of growth (CO_2) or on the side of decay (biodiversity). Sustainability is the attempt to maintain and to handle this movement, similar to how the upkeep of a bed or shed handles the natural movement of decay.

However, where the hylomorphic scheme in technical maintenance is guided by a positively determined metaphysical form such as the complete and optimal bed or chalice, movement in sustainability is primarily determined negatively: away from pollution, no more externalities and no longer depleting fossil resources (and, on the associated political spectrum: no more inequality, no poverty, no exclusion, etc.).[10] In short, a movement away from a fossil-fuelled Holocenic mode of existence that has become untenable in the Anthropocene. But if the ideal of sustainability seeks to depart from the fossil-fuelled era, which destination does this departure seek? Or, put in the terms used in the above interpretation, if sustainability is a kind of maintenance, and if technical maintenance is guided by an optimal form that can be observed, what would be the corresponding *optima forma* guiding sustainability?

As it appears, the negatively determined ideal of sustainability hardly resembles the positively determined Platonic *idea* and Aristotelean *eidos*. It is here that sustainability meets nihilism, which in Nietzsche's formulation means '[t]hat the highest values devaluate themselves. The aim is lacking; "why?" finds no answer' (1986, p. 9). Plato's ideas and Aristotle's forms can be understood as such 'highest values', but they have lost their fecundity. Leaving aside the genealogy of this loss, the aforementioned antagonism between geo-technics and geo-physis in the Anthropocene makes it manifest. The Aristotelean conception ultimately implied cosmic stability with an enduring presence of optimal, outstanding forms. This was paradigmatically expressed in the unchanging and perfect circular movement of the superlunary sphere (moved by a still more perfect unmoved mover) but similarly engages the optimal forms of technics such as the ideal, unchanging and stable form of the chalice, bed or Plato's shuttle. By contrast, today's geophysics, geology and Earth systems science attest to a volatile cosmos, in general, and Earth, in particular. As a result, it becomes impossible to speak of the Earth's optimal form. If still meaningful at all, one must face the metaphysical *nihil* of this optimum and admit its contingency. Although the Earth can indeed be said to have gone through an optimum, where relative stability and climatic conditions allowed for civilisation to flourish, this optimum now appears as a happenstance exception rather than the rule, the rule being an unruly planet where contingency rather than perfect cosmic stability reigns (cf. Clark 2011; Hamilton 2017).

Responding to being deprived of an ultimate positive idea or eidos, sustainability formulates its ideals negatively: no more externalities, no more pollution, zero waste, etc. If there are 'sustainable development goals', the goals in question primarily indicate reparations and corrections of wasteful ways that now appear erroneous.

A symptom of this situation is that all ultimate positive determinations now appear deeply suspect: is there an optimal form of the relation between technics and nature? Offering a formulation of this optimum would be a hazardous undertaking. At best, it would be unmasked as some romantic and harmonious projection. At worst, it would be ridiculed as some green fantasy. Furthermore, political suspicions are bound to arise: for whom exactly would it be an optimum? For humans? Which ones? For Global Northmen who wish to prolong their security, rule and comfort? In the words of Stacy Alaimo (2012, p. 562):

> Scholars in the humanities, or, more aptly, the posthumanities, may well ask what it is that sustainability seeks to sustain and for whom. Questions of social justice, global capitalist rapacity, and unequal relations between the global North and the global South are invaluable for developing models of sustainability that do more than try to maintain the current, brutally unjust status quo.

Timothy Morton (2018, p. 77) similarly asks:

> What exactly are we sustaining, if not the one-size-fits-all agricultural temporality pipe that has sucked all lifeforms into it like a vacuum cleaner, pretty much, over its 12,500-year run?

Irrespective of whether such suspicions are justified, the relative ease by which they now emerge[11] serves to indicate, in Nietzsche's terms, the devaluation of the ideal optimum, which thereby no longer serves as a beacon to orient the movement of sustainability.

In this qualified sense, sustainability can be said to be nihilistic. Rather than a pejorative moral judgement, this names a formal characteristic, or rather an in-formal characteristic, since the outlook of an optimal form is absent. This does not necessarily mean that sustainability lacks any ideals whatsoever but articulates how the status and topology of the ideal have changed. To clarify, consider the potential objection that in the final analysis, sustainability is not only negatively determined (no more externalities, etc.)

but also positively oriented towards an optimal outlook or ultimate idea. This idea could then be formulated as something like 'enduring sufficient quality of life for everyone'. Yet on the one hand, such a formulation paints such a vague vista that in comparison to the idea of eidos of Aristotle's optimal bed, it rather appears as an idyl. On the other hand, as remarked previously, a closer examination of such an ideal demonstrates its deeply contested character: sufficient quality for whom? Which quality exactly? Who is excluded from 'everyone' (humans? animals? other life forms, etc.)?

Accordingly, the ideal no longer follows the verticality of a metaphysical optimum that speaks for itself from on high. Instead, the ideal of sustainability corresponds to a horizontal topology. The mentioned suspicions (sustainability for whom? which quality of which lives? etc.) demonstrate that rather than speaking from transcendent heights, the ideal of sustainability becomes contested on a horizontal terrain.[12] Accordingly, it becomes less a question of looking up to an ideal than of constructing one from the ground up, a process inescapably accompanied by political contestations of who and what is included and excluded. The observation of optimal forms makes place for their generation. In Latour's words, the transcendent notion of 'Nature' makes place for 'world' as 'that which opens to the multiplicity of existents, on the one hand, and to the multiplicity of ways they have of existing, on the other' (Latour 2017, p. 35). The political contestation regarding sustainability ultimately comes down to 'what existents have been *chosen* [to exist], and what forms of existence have been *preferred?*' (ibid. p. 37; cf. Zwier and de Boer 2023, pp. 76–81). The question concerning what exactly is to be constructed thereby necessarily recurs. However, since the blueprints from Plato's Timaeus have gone missing, their negatives must suffice: whatever is to be constructed, it must be less polluting, less exclusivist, less anthropocentric, etc. The ideal guiding sustainability thus appears negatively determined as 'less suboptimal', while a positive determination recedes like a horizon when approaching it.

Yet, if sustainability faces a receding horizon, if what ultimately is to be sustained remains elusive, this destitution hardly constitutes a hindrance to its movement. This is unsurprising when considering that not every movement seeks a destination. In fleeing a crime scene or the site of a disaster, one is not concerned with arriving anywhere in particular. For sustainability, the designated crime scene and the site of disaster is the globally warming wasteland of Holocenic fossil-fuelled civilisation. Rather than requiring a clear destination, the necessity of fleeing the scene suffices to instigate its movement. Today's virtually ubiquitous recital of sustainability as 'sustainable development', 'sustainable entrepreneurship' and 'sustainable growth' (e.g., Ploum et al. 2018) attests to this imperative of movement, which neither has nor requires any ideal goal or optimal form.

To recapitulate, the initial observation was that sustainability mirrors technical maintenance, given how it concerns a metaphysical antagonism between (geo)technè and (geo)physis. The question was whether the hylomorphic metaphysics that characterises technical maintenance correspond similarly. The answer is, on the one hand, sustainability follows hylomorphism insofar as it anxiously seeks to maintain and contain matter (*hyle*) such as CO_2 in formation (*morphe*). On the other hand, due to its nihilistic character, sustainability parts with hylomorphism and is no longer guided by a transcendent optimal form. Since sustainability moves away from something determinate (unsustainable fossil-fuelled civilisation) towards nothing determinate, its movement is best characterised as flight. Maintenance in the sense of moving matter (*hyle* such as CO_2) away from disequilibrium remains central, but the *morphe* found in the hylomorphic scheme no longer guides such movement. As a result, not some optimal form, but the fleeing movement itself is to be maintained.

6.4 Technical Maintenance of Actuality or Potentiality?

The above diagnosis of sustainability's nihilism and flight of course does not imply that it should be altogether abandoned. It could be argued that the widely advocated idea of 'becoming more sustainable' can simply do without a positively determined optimum since determinations of the sub-optimum from which to flee are sufficiently clear. Habitability simply trumps the alternative. Be that as it may, this raises the question of whether such flight is itself sustainable or is itself to be sustained. Does moving towards sustainable technical maintenance of the Earth ultimately imply perennially fleeing from sub-optima? It appears that since it mirrors technical maintenance, sustainability must either ignore this question and prioritise actual evasive maneuvers or affirm a sustained fleeing movement. Perhaps, both options offer a sufficient response to the situation of the Anthropocene. Still, and this is the wager of this final section, it may be worthwhile to further explore the limits of this situation. Although such an exploration cannot be expected to formulate a full-fledged alternative and some post-nihilistic conceptualisation of sustainability, it can shed light on the presuppositions that undergird current conceptualisations to open alternative avenues. One way in which this can be done is by once more returning to the technical character of sustainable planetary maintenance.

The technical maintenance mirrored by sustainability corresponds to the cybernetic figure of the homeostat in its specific guise as a thermostat. The now popular image of the Earth-as-thermostat shows this: relatively modest technologies like solar panels and the more ambitious technologies like solar radiation management are alike in aiming to dial back the runaway thermostat of global warming (Zwier and Blok, 2019, pp. 622–7).

Such dialling back corresponds to the negative feedback in cybernetic homeostats, where the system's output (classical example: observed steam engine speed) feeds back into the system's input (the steam inlet, by way of a centrifugal governor), to minimise the difference between the intended and observed behaviour and maintain an operational equilibrium condition. This equilibrium echoes the aforementioned Aristotelean equilibrium as optimum (*telos ariston)* between *genesis* (e.g., an overdriven steam engine) and *corruption* (e.g., a halting engine). Dialling back the Earth-as-thermostat similarly feeds back observed outputs such as '5°C too high' to inputs such as 'produced emissions'. Yet, because of sustainability's nihilistic character, the difference is that the optimum setting of the thermostat remains unresolved at worst and a perennial political contestation at best (cf. Zwier and de Boer, 2023).

If the thermostat thus portrays the technical character of sustainability, this raises the question of whether technics primarily consists of maintaining homeostasis and equilibrium. Aristotle and sustainability answer positively: technics maintains the material movement of *physis* in equilibrium. Conversely, Gilbert Simondon answers negatively. His work is therefore instructive for exploring the limits of sustainability.

6.4.1 *Simondon versus Hylomorphism*

Simondon criticises the hylomorphic scheme for being an abstracted, deadlocked and ultimately misguided way of metaphysically considering technics and ontology in general. It is abstract because it corresponds to a specifically aristocratic and detached way of perceiving reality. Seeing the technical process in terms of pre-given matter to be 'handled' (i.e., maintained) into pre-established forms 'corresponds to the knowledge of someone who remains outside the workshop and considers nothing but what enters and exits it' (Simondon 2020, p. 30). For Simondon, such a perspective not only concerns workshops, artisans and technics but also fundamentally characterises metaphysical thinking and the tradition of philosophy (cf. Combes 2012, p. 1). This tradition privileges static *individual beings* (cf. Section 6.2) such as 'final optimal form' and 'pre-given matter', and in so doing wrongfully overlooks the process of their *individuation.* Conversely, in attempting to 'grasp ontogenesis in the whole unfolding of its reality', Simondon's project consists of seeking 'to know the individual through individuation rather than individuation through the individual' (2020, p. 30).

Thinking ontogenetically requires a transfiguration of the hylomorphic scheme. Among many other examples (cf. Voss 2018, p. 110), Simondon concretely develops such thinking in his analysis of brick moulding, where he replaces stability with the concept of metastability. Rejecting the hylomorphic differentiation between pure matter and form, Simondon (2020, p. 22) explains:

> To give a form to the clay is not to impose the parallelepiped form onto raw clay: it is to pack the prepared clay into a fabricated mold. If we start from the two ends of the technological chain, the parallelepiped and the clay in the quarry, then we can experience the impression of realizing in the technical operation an encounter between two realities of heterogeneous domains and of instituting a mediation through communication

A brick cannot be made from wet sand, which would simply disintegrate upon drying. Hence, rather than passive matter waiting to be actively in-formed, the clay matter is positively determined as a potential for deformation (Simondon, 2020, p. 25; cf. Combes 2012, p. 5). On a molecular level, clay is already in-formed in the swampy soil in such a way that the *other* material form of the mold:

> limits and stabilizes rather than imposing a form: it provides the goal of deformation and achieves it by interrupting it according to a definite contour [...] the mold plays the role of a fixed set of modeling hands, acting like halted kneading hands.
>
> (Simondon 2020, p. 25)

The individuation of a brick establishes *communication* between hitherto heterogeneous domains of becoming. In this sense, it is similar to how the individuation or becoming of plants first established communication between a cosmic order (light energy) and an intramolecular order (mineral salts, oxygen, etc., cf. Combes 2012, p. 4). Similar to how a plant is the communicative medium connecting these orders, brick moulding establishes communication between the clay and the brick mould. Since it is itself technically prepared, 'the pure form [of the mold] already contains actions, and the raw material is the capacity of becoming; the actions contained in the form encounter the becoming of the matter and modulate it' (ibid., p. 25). For Simondon (ibid., p. 41), this marks the insufficiency of the hylomorphic schema, which:

> does not account for implicit forms, since it distinguishes between the pure form (which is called form) and the implicit form, which is conflated with other features of matter under the name quality.

The reference to 'kneading hands' and 'modulation' recalls the interpretation of technical maintenance as taking in hand or 'handling' of *physis*

(Section 6.2). When Simondon criticises the hylomorphic schema of the metaphysical tradition, it accordingly remains questionable whether this critique can be fully directed at Aristotle himself. After all, as the example of the bed made clear, the wooden matter out of which the bed is made is, in Simondon's terms, 'implicitly formed'.[13] Still, Simondon's ontogenetic consideration of communication between two chains of becoming (the clay with potential for deformation and the mould as a potential limit to deformation) marks a significant departure from hylomorphism. When these chains enter into communication in the technical operation of brick moulding, something new enters into reality. This is to say that in contra hylomorphism, there is no pre-given, transcendent and complete optima forma that serves as the ultimate orientation point (*telos ariston)* for a technician. Instead, the form is an outcome or 'endpoint' (Combes 2012, p. 5) of the technical operation: 'The technical operation integrates implicit forms rather than imposing a totally new and foreign form on a matter that would remain passive vis-a-vis this form' (Simondon 2020, p. 42).

Instead of unchanging, stable beings like Aristotelean forms or Platonic ideals, Simondon privileges the metastable: 'The original state of being is a state that goes beyond coherence with itself, that exceeds its proper limits: original being is not stable, it is metastable' (Ibid. p. 369).[14] As Yuk Hui (2016, p. 84) explains:

> 'metastable' describes a transitional state through which the individual as product is given to us, but further individuation can take place when the individual is stimulated by external information or by energetic excitation from within.

Simondon's paradigmatic example is crystallisation. It cannot be understood as the imposition of form, but only as communication. When a supersaturated solution enters into communication with heat energy, it starts to crystallise, which in turn triggers more crystallisations. The condition for this occurrence is that the amorphous solution is neither stable nor unstable, but metastable: the solution is charged with potential and when a threshold is passed, the amorphous state transitions to a crystalline structure, without 'the contribution of some foreign form' (Simondon 2020, p. xxiii). Ascribing the process of becoming found in crystallisation to reality as such, and referring to the potential charge of a supersaturated solution, Simondon (2020, p. 352) points out that

> This charge of the undetermined can be called nature (...) a veritable reality charged with potentials actually existing as potentials, i.e. as an energy of a metastable system.

If hylomorphic ontology ultimately understands reality as self-maintaining *actuality* in terms of transcendent forms (see Section 6.3.1 and footnote 9), Simondon's ontogenesis understands reality as self-maintaining *potentiality*, always maintaining 'a reserve of becoming' (2020, p. 358) that gives rise to new communications and individuations.

6.4.2 *Metastability or Elusive Homeostasis*

Simondon's conceptualisation of technics and metastability raises the following question: if, as argued (Section 6.3.2), sustainability qua planetary technical maintenance is characterised by flight, what does Simondon's revision of technics and hylomorphism imply for this characterisation?

The philosophy of individuation does not counter nihilistic devaluation by formulating a new optimal form to guide sustainable planetary maintenance. It does not indicate the optimum setting to which the Earth's runaway thermostat should be dialled back. The possibility of such static anchoring points is principally refused. Instead, Simondon sidesteps this metaphysical trajectory. Rather than maintenance geared towards some *actuality*, which is to say, recalling Heidegger's characterisation of metaphysics, some 'enduringly present' optimal form (cf. Section 6.3.1), Simondon's thought considers the question of sustainability as one of maintaining *potentiality*. Not the homeostat, thermostat or any other stable state, but becoming is to be taken in-hand. Such taking in-hand or maintenance would then not consist of *the conservation* of form against physical corruption and generation, but in establishing novel *communications*.

Within the scope of sustainability and the theme of the Anthropocene, this primarily indicates communication between technologised humanity and Earth. Whereas sustainability here sees antagonism between the geo-technè of humanity and the geo-physis (Section 6.3.1), Simondon's thought allows for seeing novel communications and individuations. This not only means that an otherwise statically determined humanity now wields newly individuated technologies such as solar panels and geo-engineering facilities but also indicates the ongoing individuation of humanity and the Earth. A significant aspect of this individuation is that humanity no longer merely lives *on* Earth but must live *with* the Earth. This can be regarded as a novel communication, similar to how the living individuation of plants first established communication between a cosmic order and an intramolecular order.

While investigating the implications of this novel communication remains an ongoing task (cf. Hamilton 2017; Zwier and de Boer 2023), it is clear that it cannot be constrained to an Anthropocentric perspective according to which the situation is imposed by humanity. Just like the clay and the mould are considered as two chains of becoming, technologised humanity and the Earth likewise find themselves communicating anew. Regarding technics, specifically, Simondon holds that its evolution is not imposed by human forms, but it individuates by itself as a quasi-living being (cf. Lindberg 2022, p. 54). His interpretation of the Guimbal turbine shows that although human anticipation and invention essentially belong to the process, the turbine shapes its own conditions of operation and thus '*renders itself possible*' by constituting a 'techno-geographic milieu' (Simondon 2017, pp. 57–8). The turbine only works when inserted into the water, only when it establishes novel communications that condition the water to function both as the energy source and the cooling mechanism. Whereas Heidegger still prioritised Aristotle's silversmith as 'carefully gathering together' the material movements of nature into an observed form (cf. Section 6.2), for Simondon such gathering or self-conditioning is non-anthropocentrically interpreted as the outcome of several individuating chains: the technical individuations internal to the turbine which contain 'an entire electrical factory' (2017, p. 58); the water becoming milieu for its functioning and the social individuations resulting from its generated energy, etc.

Viewed from this perspective, the question comes to concern the new techno-geographic milieu of the Anthropocene. Rather than sustaining some ineffable optimal form, the task then consists of understanding what living *with* the Earth entails. Simondon's ontogenetic reworking of hylomorphism, technics and ontology opens avenues for investigating this question beyond the limits of sustainability understood as planetary technical maintenance. Although such an investigation remains to be carried out further, two presumably informative considerations can be offered in closing.

First, a critical dialogue with Simondon appears in order. It could be argued that the Earth becoming uninhabitable of itself nothing else than a physical individuation on the level of the Earth System, where a situation crystallises that no longer happens to support the individuations that are associated with civilisation. This raises the critical question of how Simondonean thinking can accommodate this kind of 'negative individuation' or what Bernard Stiegler refers to as 'disindividuation' (2006).[15] As noted, Simondon abandons metaphysical dualistic *oppositions* of form versus matter and (human) technics versus nature, to instead privilege *compositions* of physical, living, psychical and technical chains or 'regimes of individuation' (Simondon 2020, p. 12). Yet, this seems to imply a certain unidirectional or even upward development moving from the physical (e.g., crystals) to the living (e.g., organisms), to the technical and to the psychical (human) and social. What, however, if the physical individuation of the planetary habitat that renders it uninhabitable, thus comes to undermine rather than support 'higher level' individuations like psychical and social individuations? If such a question is on point, it could be argued that in stressing individuation as becoming or Aristotelian *genesis*, its companion term of *corruption* now comes to haunt Simondon's philosophy of potentiality and becoming.[16] In any case, Simondon's attempt to move beyond a rigid opposition between matter and form, as well as between nature and technics, allows for addressing such critical ecological questions anew (cf. Lindberg 2022, p. 62).

Second, it may be clear that abandoning maintenance as conservation of an actuality in favour of maintenance as communication with a potentiality comes at a cost: it becomes experimental rather than cybernetically controlled (cf. Bardin 2021, p. 38). Experimental does thereby not necessarily refer to a strictly controlled scientific experiment but involves receptivity and generosity to what remains beyond control and what remains potentially new. This calls for a different comportment: rather than either seeking to conserve some optimal form or flee from the sub-optimum (cf. Section 6.3.2), Simondon's thinking invites a progressive comportment that welcomes new individuations. Instead of the homeostatic control according to some existing yet elusive optimum, it opens towards mediation with what remains hidden, with what is always *'richer than its self-coherence'* (Simondon 2020, p. 369). It is here that Simondon (2020, p. 411) equates the philosopher and the technician:

> Thales, Anaximander, and Anaximenes are technicians above all else … the veritable technician is the one who is a mediator between the community and the hidden or inaccessible object.

Concerning the abovementioned question of whether sustainable planetary maintenance is finally characterised as flight, it remains to be decided whether this experimental cost is too high. If the currently accepted view understands sustainability as the maintenance of either some actual planetary optimum, or at least a habitable minimum, then rethinking maintenance in terms of potentiality may appear too naively optimistic about individuations and its 'potential becomings', too audacious or even reckless in light of what needs actual attention. On the other hand, the alternative appears to be the affirmation of the sustained fleeing movement of sustainability. What is more, even when (rather favourably) assuming that the optimal thermostat setting is sufficiently clear and uncontested, past attempts to dial it in can be evaluated by their fruits. It then appears that today's rather blatant failures to reset the thermostat point to more than a few rotten apples (cf. Malm, 2018). As the 2023 IPCC Synthesis Report states:

> In 2018, IPCC highlighted the unprecedented scale of the challenge required to keep warming to 1.5°C. Five years later, that challenge has become even greater due to a continued increase in greenhouse gas

> emissions. The pace and scale of what has been done so far, and current plans, are insufficient to tackle climate change.
>
> (IPCC 2023)

Seen in this light, even if Simondon's privileging of potentially novel communications and progressive comportment appear preposterous in the light of looming catastrophe, the failures of sustainability's planetary technical maintenance might be read as a justification to nonetheless venture in this direction.

6.5 Conclusion

This chapter asked in which sense sustainability is to be understood as maintenance and what the limits and implications of such understanding are. It argued that sustainability follows the metaphysical schema of technical maintenance, but runs up against a limit of form. Whereas technical maintenance knows the optimal form of what to maintain (a bed, a house and a bicycle), sustainability only knows the suboptimal form that should not be maintained. As a result, it engages in a fleeing movement. Although understandable and even justifiable in light of the present climatic situation, the question remains whether such flight determines the limit of what planetary maintenance means. By way of its critique of hylomorphism and emphasis on individuation and potentiality, Simondon's work opens an alternative avenue. Exploring this avenue suggests that sustainability and planetary maintenance need not mirror the schema of technical maintenance, but might investigate newly established communication between technics, humanity and the Earth. If Simondon is thereby read as a critique of maintenance and its adherence to hylomorphism, a critique cannot be understood as rejection or refusal, but as delimitation. Accordingly, while technical maintenance of individuals such as chalices, beds or even planets is necessary, Simondon suggests that it is not sufficient. It neither exhausts the potential of technics nor how it is maintained or 'taken in hand' nor how it enters into individuating communication with the Earth.

Whether one opts for sustainability's flight or Simondonean experiments, it appears that the conditions for considering these options are given for now. Perhaps, the response to this gift must be to cultivate it further, to ask which potential Earth is to be maintained, for which individuals or individuations and to which potential ends.

Notes

1 Throughout this chapter, etymological references are used, not to prove or fix their final meaning, but simply to suggest that the genealogy of the words we use is not inconsequential and can inform and inspire reflection on what we – often implicitly and unreflectively – mean by them.

2 The Dutch word for maintenance is 'onderhoud', which literally translates as 'under-keep', thus expressing this characterisation of maintenance.
3 Fabricated individuals (e.g., a chair) differ from natural or physical individuals (e.g., a tree) insofar as the latter carry their movement towards from within themselves (the tree becoming full-grown), whereas fabricated individuals involve an external movement in which the technician or craftsperson plays a role (a tree never grows into a chair). Given the present focus on maintenance and sustainability, this chapter focuses on fabricated individuals and leaves the question of natural individuals (and whether they ultimately derive from a technical paradigm) open. See Heidegger (1998, pp. 202–4); Glazebrook (2000, p. 99).
4 The required intermediary technical steps of turning trees into lumber or ores into silver do not contradict their physical givenness.
5 Next to 'origin' and 'principle', *archè* also means 'command', as Agamben (2019) makes clear in 'What is a command?'
6 The word 'nature' similarly expresses movement, as it derives from 'nāscī' or 'being born, being generated' or simply 'growing'.
7 Meta means after, beyond, above or transcendent, which is how metaphysics is traditionally understood: after and above the physical.
8 In modern philosophical thought on technology, the clearest echoes of this vertical metaphysical thinking can be found in the work of Friedrich Dessauer (cf. Dessauer, 1956; cf. Mitcham, 1994, pp. 29–33).
9 The metaphysics of interrupted verticality thereby resonates with Heidegger's definitions of metaphysics as 'enduring presence' (*Beständige Anwesenheit*) and 'onto-theology' since observed forms or ideas are enduring and constitute the highest *(theos)* instance of being (cf. Heidegger 2002, pp. 42–75).
10 This can be observed in the 'sustainable development goals' such as 'no poverty', 'zero hunger' and 'reduced inequalities', combined with industrial development with 'no externalities' like pollution, use of fossil resources, etc.
11 The very phenomenon of this 'relative ease' may of course well be interpreted as a laborious and hard-fought achievement by critical scholars, activists, etc.
12 In Morton's work, 'ontology [...] is a vital and contested political terrain' (2013, p. 20). One could find the devaluation of the highest goals anticipated in Descartes' mechanistic universe, where true goals are restricted to the domain of human morality, yet only as 'conjectures', whereas real optima forma are inaccessible: 'for it does not appear to me that I can without temerity seek to investigate the [inscrutable] ends of God [les fins impenetrables de Dieu]' (Descartes, 4th meditation (1911, p. 20).
13 Simondon acknowledges this in his discussion of Aristotle's bed: 'Here, the technical operation accommodates the living form and partially diverts the latter for its own benefit' (2020, p. 42).
14 According to Anne Sauvagnargues, Simondon's philosophical extension of the thermodynamic notion of metastability is crucial for his account of reality: 'Metastable being, in disequilibrium, involves this state of asymmetrical disequilibrium which accounts for tension and the production of the new' (2013, p. 58).
15 Where it could be argued, as Stiegler does, that the becoming uninhabitable of the Earth primarily points to the bankruptcy of psychical and social-collective (dis)individuation since the 'physical individuation' of ecological collapse is rooted in unfettered consumerism (cf. Stiegler et al. 2021).
16 In 'The Limits of Human Progress', Simondon touches on this problem but is mostly concerned with the 'saturation' of potentials, rather than corruption (cf. 2010).

References

Agamben, G. (2019) *Creation and anarchy: The work of art and the religion of capitalism*. Stanford University Press.

Aydin, C. (2021) *Extimate technology: Self-formation in a technological world*. Taylor & Francis.

Baskin, J. (2015) Paradigm dressed as epoch: The ideology of the Anthropocene. *Environmental Values*, *24*, 9–29.

Bardin, A. (2021) Simondon Contra New Materialism: Political anthropology reloaded. *Theory, Culture, and Society*, 38(5), 25–44.

Clark, N. (2011) *Inhuman nature. Sociable life on a dynamic planet*. Los Angeles: Sage.

Clark, N., & Yusoff, K. (2014) Combustion and society: A fire-centred history of energy use. *Theory, Culture & Society*, *31*(5), 203–226.

Combes, M. (2012) *Gilbert Simondon and the Philosophy of the Transindividual* (T. Lamarre, Trans.). Cambridge: The MIT Press.

Descartes, R. (1911) *The Philosophical Works of Descartes* (E. S. Haldane, Trans.). Cambridge: Cambridge University Press.

Dessauer, F. (1956) *Streit um die Technik*. Frankfurt: J. Knecht.

Glazebrook, T. (2000) From "physis" [Greek] to Nature, "techne" [Greek] to Technology: Heidegger on Aristotle, Galileo, and Newton. *The Southern Journal of Philosophy*, *38*(1), 95–118.

Hamilton, C. (2013) *Earthmasters: The Dawn of the Age of Climate Engineering*. New Haven: Yale University Press.

Hamilton, C. (2017) *Defiant Earth: The Fate of Humans in the Anthropocene*. Cambridge: Polity Press.

Heidegger, M. (1977) *The question concerning technology and other essays* (W. Lovitt, Trans.). New York: Harper and Row.

Heidegger, M. (1998) *On the essence and concept of physis* (T. Sheehan, Trans.) In W. NcNeill (Ed.) *Pathmarks* (pp. 183–232) Cambridge: Cambridge University Press.

Heidegger, M. (2000) *Introduction to Metaphysics* (G. Fried & R. Polt, Trans.). New Haven & London: Yale University Press.

Heidegger, M. (2002) *Identity and difference*. (J. Stambaugh, Trans.) Chicago: University of Chicago Press.

Hui, Y. (2016) The parallax of individuation. *Angelaki. Journal of the Theoretical Humanities*, *21*(4), 77–89.

IPCC (2023) *Urgent climate action can secure a liveable future for all*. Press Release 20 March 2023. IPCC_AR6_SYR_PressRelease_en.pdf [accessed: 20 March 2023].

Latour, B. (2017) *Facing Gaia: Eight Lectures on the New Climatic Regime*. (C. Porter, Trans.) London: Polity Press.

Lindberg, S. E. (2022) Simondon écologiste? In J. H. Barthélémy & L. Duhem (eds.), *Écologie et technologie: redéfinier le progrès après Simondon* (pp. 49–62). Paris: Éditions matériologiques.

Malm, A. (2018) *The progress of this storm: Nature and society in a warming world*. Verso Books.

Marshall, J. D., & Toffel, M. W. (2005) Framing the elusive concept of sustainability: A sustainability hierarchy. *Environmental Science & Technology*, *39*(3), 673–682.

Mariola, M.J. (2005) Losing Ground: Farmland preservation, economic utilitarianism, and the erosion of the agrarian ideal. *Agriculture and Human Values 22*, 209–223.

Mitcham, C. (1994) *Thinking Through Technology: The Path Between Engineering and Philosophy*. Chicago: The University of Chicago Press.

Morton, T. (2013) *Hyperobjects. Philosophy and Ecology After the End of the World*. Minneapolis and London: University of Minnesota Press.

Morton, T. (2018) *Being Ecological*. Cambridge, MA: MIT Press.

Nietzsche, F. (1986) *The Will to Power* (W. Kaufmann & R. J. Hollingdale, Trans.) (W. Kaufmann, ed.). New York: Random House.

Plato (1997a) 'Cratylus', In J. M. Cooper (ed.). *Plato: Complete Works*. Indianapolis: Hackett.

Plato (1997b) 'The Republic', In J. M. Cooper (ed.). *Plato: Complete Works*. Indianapolis: Hackett.

Plato (1997c) 'Timaeus', In J. M. Cooper (ed.). *Plato: Complete Works*. Indianapolis: Hackett.

Ploum, L., Blok, V. Lans, T., & Omta, O. (2018) Toward a validated competence framework for sustainable entrepreneurship. *Organization & Environment*, *31*(2), 113–132.

Salas-Zapata, W. A., & Ortiz-Muñoz, S. M. (2019) Analysis of meanings of the concept of sustainability. *Sustainable Development*, 27(1), 153–161.

Sauvagnargues, A. (2013) 'Crystals and Membranes: Individuation and Temporality', In A. de Boever et al. (Eds.) *Gilbert Simondon: Being and Technology*. Edinburgh: Edinburgh University Press. pp. 57–70.

Simondon, G. (2010) The limits of human progress: A critical study. (S. Cubitt, Trans.) *Cultural Politics*, 6(2), 229–236.

Simondon, G. (2017) *On the Mode of Existence of Technical Objects*. (C. Malaspina & J. Rogove, Trans.). Minneapolis: Univocal Publishing.

Simondon, G. (2020) *Individuation in Light of Notions of Form and Information*. (T. Adkins, Trans.). Minneapolis: University of Minnesota Press.

Stiegler, B. (1998) *Technics and time: The fault of Epimetheus*. (R. Beardsworth & G. Collins, Trans.). Stanford University Press.

Stiegler, B. (2006) The Disaffected Individual in the Process of Psychic and Collective Disindividuation. (P. Crogan & D. Ross, Trans.). Available at: https://arsindustrialis.org/disaffected-individual-process-psychic-and-collective-disindividuation (Accessed: March 2023).

Stiegler, B. & The Internation Collective (Eds.) (2021) *Bifurcate: There is No Alternative*. (D. Ross, Trans.). Open Humanities Press. Available at: http://www.openhumanitiespress.org/books/titles/bifurcate/ (Accessed: March 2023).

Visser, G. (2014) *Heideggers vraag naar de techniek. Een commentaar*. Nijmegen: Van Tilt.

Voss, D. (2018) Simondon on the notion of the problem. A genetic schema of indivudation. *Angelaki. Journal of the Theoretical Humanities*, *23*(2), 94–112.

Zwier, J., & Blok, V. (2019) Seeing through the fumes: Technology and asymmetry in the Anthropocene. *Human Studies*, *42*(4), 621–646.

Zwier, J., & de Boer, B. (2023) Earth becomes world? Scientific objects, nonmodern worlds, and the metaphysics of the Anthropocene. *Environmental Humanities*, *15*(1), 64–86.

7 Towards a Realist Metaphysics of Software Maintenance

Keith Begley

7.1 Introduction

Software would seem to be an unusual object of maintenance because it is not subject to degradation in the way that physical components are and so the maintenance of software has a different character to the maintenance of hardware. This has been recognised by software engineers themselves. For example, consider the following passage on software maintenance from a software engineering textbook.

> [...] *whiledo* constructs do not wear out after 10,000 loops, and semicolons do not fall off the ends of statements. Unlike hardware, software does not degrade or require periodic maintenance. Thus, software systems are different from hardware, and we cannot use hardware analogies successfully the way we can for other aspects of software engineering.
>
> (Pfleeger & Atlee 2010, pp. 535–6)

The 'difference' that is at issue here, put ontologically, is a difference in ontological status between abstract and concrete objects. Similar observations have been made by others. For example, Gruner has asked the question 'Something that is not material cannot physically decay or wear out; in what sense is it then possible to speak of "software maintenance?"' (2011, p. 296). There is a presupposition here that software is abstract in some way and also that maintenance is merely preservative or conservative. However, if software were abstract, in the sense of something non-concrete and unchanging, then it seems that software maintenance would not be possible at all, at least not in a way that would have some effect upon an abstract object. Realism about software need not commit us to such an otiose position. As we shall see, we can have our abstract software and maintain it too, just not in any way that has an effect upon an abstract object.

DOI: 10.4324/9781003316213-8

Some things that computer scientists and users often call 'software', that is, its *copies*, are in part physical and are involved in interactions with hardware. Many philosophers and computer scientists have been concerned with drawing attention to software's dual nature, its being both symbolic *and* physical, abstract *and* concrete (e.g., Moor, 1978; Colburn, 2000; Turner, 2018; Coeckelbergh, 2022), and some propose that it is an 'abstract artifact' that is temporal but non-spatial while its copies are concreta (Irmak, 2012). In Section 7.2, I will delineate and assess some of these views.

There are strong connections to be found between this topic and recent investigations in the philosophy of linguistics, particularly the ontology of words, where software has often been mentioned in comparison. My approach draws upon a recent parallel project to reconstruct Katz's metaphysics of words (Begley, 2023). In Sections 7.3 and 7.4, I will argue, *contra* Irmak (2012), that software can be best understood through a realist type-token distinction that was put forward by Katz (1998). This approach employs both non-spatiotemporal abstract types and tokens of those types. I show that employing and further developing Katz's realist approach can help us to make better sense of computational artefacts and their maintenance than the alternatives outlined. In Section 7.3, I provide an outline of Katz's ontology of *composite objects*, before, in Section 7.4, applying it to software and computational artefacts in general.

In Section 7.5, I discuss how we should make sense of the notion of software maintenance in light of the realist view of software developed. I also address a problem that has been raised by Young (2021) for theories of artefacts that employ intentions, such as the one involved with the realist view of software and software maintenance for which I shall argue. I will show that software maintenance on this view is not solely conservative but also adaptive and enhancive, constituting a source of counterexamples and so is not susceptible to the problem.

7.2 The Ontological Dualities of Computational Entities

It has long been noted that computers and computational objects appear to exhibit a dual or 'mixed' nature. This has been put in a variety of ways by authors over the years. In this section, we will consider some of these ways, coming to focus in particular on the use of the abstract-concrete distinction. An early example that is often mentioned in the literature comes from an article by James H. Moor.

> Computers can be understood on two levels. Computers can be analysed on the physical level just as any physical system can. If the computer is electronic, then the laws of physics concerning electronics can

> explain the computer activity. But, it is the understanding on the symbolic level which makes computers calculating devices, for it is under this kind of interpretation that various structures or processes of the computer are understood as symbols. Similarly, a computer program can be understood on two levels. Physically, computer programs may be a series of punched cards, configurations on magnetic tape, or in any number of other forms. Symbolically, computer programs are understood as instructions to a computer.
>
> (Moor, 1978, p. 213)

Here Moor makes a distinction between a physical and a symbolic level, both of computers and their programs. The physical level and its laws provide the explanation for how the physical components of the computer end up in certain physical states,[1] with, for example, quantities of charge arranged thus and so. However, this is not sufficient for computation, because computation is a kind of abstract operation that has been characterized in various ways, for example, as involving abstract states and transitions between those states on input or, equivalently, as general recursive functions, etc. (Turing, 1937 [1936]; Church, 1936). That is, nowhere in such abstract descriptions need quantities of charge, etc., be mentioned.

Colburn (2000) takes the title of an undergraduate textbook in computer science, *Concrete Abstractions* (Hailperin, Kaiser, & Knight, 1999), as his starting point for reflection, and others have followed him (e.g., Irmak, 2012; Turner, 2018). However, the authors of the textbook do not discuss ontology beyond the first few paragraphs of the preface in which they attempt a brief initial justification for calling computer science 'the discipline of concrete abstractions'.[2] Colburn examines the view that software is a 'concrete abstraction' more closely.

> While computer scientists often enthusiastically embrace this duality, a metaphysician will view it as a puzzle to be explained. How can something, namely a computer program, be at once concrete and abstract?
>
> (Colburn, 2000, p. 205)

Colburn approaches the puzzle by employing a framework that is ordinarily used to classify responses to the mind-body problem, that is, as being either monistic or dualistic. Nevertheless, he recognised that the approach results in 'not an entirely appropriate language, for describing programs as concrete abstractions' (2000, p. 205). The approach turns out to be rather underproductive, as in the end, he entertains notions of pre-established harmony. As Turner points out, this is merely a metaphor for a kind of harmony that is better explained as being due to correct implementation

(2018, p. 51). Furthermore, as we shall see presently, there are much more promising lines of metaphysical theorising in neighbouring areas such as the philosophy of linguistics, where we find very similar issues being discussed regarding the apparently mixed or dual ontological status of words and other linguistic entities.

The 'puzzle' recognised by Colburn has been recognised by others. For example, Turner raises a similar question.

> [...] everywhere we turn in computer science we find the same pattern: a fusion of an abstract thing with a concrete one. Does this herald the introduction of a new kind of ontological entity? How are we to categorize and conceptualize the entities of computer science?
>
> (Turner, 2018, p. 15)

Despite this initial characterization in terms of the abstract-concrete distinction, Turner instead appropriates a conception of the dual nature of what Kroes (2012) calls *technical artefacts*. Kroes' approach takes the duality involved in this case to be between structural and functional properties, with the structural side being satisfied by the physical objects involved and the functional side being satisfied on the basis of intention (Turner, 2018, p. 25).[3] Where I agree with Turner is that there must be intention involved in the creation of a computational artefact. However, the functional side is put forward in terms of a design specification only for what a physical system should do, and so the required symbolic or abstract level that was mentioned by Moor and Colburn appears to be deprecated. In Sections 7.3 and 7.4, we will give further consideration to what is intended in the intention, including what the ontological status of that thing intended is.

This is an apt point to mention a way in which the conception of software maintenance is related to the conception of software as an artefact constituted through an intentional act. Young (2021) is rightly concerned to show that the usual conception of maintenance as preservation or conservation fails to recognise the full role that maintenance plays in the production of artefacts and their efficacy. He aims to show that the widespread adoption of this conservative view is due to the acceptance of a conception of the production of artefacts that involves intentions. He first points out that 'among philosophers [...] artifacts are nearly always understood as manifestations of the intentions of their creators' (2021, p. 358). He then argues that such an understanding 'often encourages us to privilege the moment construction finishes as the point in an artifact's history where the intentions of the designer are most fully realized' (p. 359). The claim here is that this privileging of the moment of full realisation of the intentions of a design leads us to think of maintenance merely as a means of conserving this full realisation. Additional considerations would also appear to apply to realist or essentialist views in particular.

> Because the essence of an artifact, the form, is understood to be atemporal, time and change are cast as external influences against which a technology must be protected, giving rise to a conception of maintenance as conservation.
>
> (Young, 2021, p. 400)

These arguments would appear *prima facie* to present a problem for any view that maintains both that its artefacts are realisations of intentions and that maintenance is not merely conservative, but also, for example, adaptive and enhancive. So, it would appear to be a potential problem for the realist view of software and software maintenance that I will put forward in the following sections, based on Katz's ontology for composite objects. This is because Katz too provides a role for intentions as aspects of creative relations (or patterns of relations) that token artefacts of a certain abstract type. I will return to address Young's problem again in Section 7.5, after developing the realist approach employing intentions in Sections 7.3 and 7.4.

Returning to our discussion of the dual nature of computational artefacts, we should note that Coeckelbergh (2022) has similarly recognised the 'hybrid' nature of computational objects, with particular regard given to the role of language.

> Language is often literally connected to technology, or is even part of what the technology is. Consider some of the technologies most of us use daily, such as computers and mobile phones (or less common technologies such as robots). These are material artifacts, to be sure, but they run on software, which is based on a software language, on code. Both the material aspects and the linguistic aspects constitute the technology. For the technology to work, it is crucial that they are connected. Code by itself cannot do anything in the world; the hardware (and other software) is needed. Vice versa, the hardware needs the language of software to do things. When it comes to contemporary information and communication technologies (ICTs), it is clear that language and materiality are entangled. The agency and ontology of such technologies cannot be adequately described without taking language into account. Even at the so-called technical level, these technological artifacts already have a hybrid nature: they are material and linguistic at the same time.
>
> (Coeckelbergh, 2022, p. 343)

Whether or not the hardware-software distinction is an ontologically salient one remains to be seen. That is, *both* hardware and software would appear to have dual natures. Further, both can be emulated by another computer, a piece of software or a piece of hardware. I will return to this point again briefly in Section 7.4.

Coeckelbergh focuses in particular on the mediating roles that language plays in technological practices and argues that the insights of a particular brand of *use*-based philosophy of language have been neglected in this regard in the philosophy of technology. As is well-known,[4] and is emphasised by Coeckelbergh, there was an 'Empirical Turn' in the philosophy of technology roughly 40 years ago, that is, a limitation of considerations to those regarding material artefacts and our embodied engagements with them. However, it is clear that the questions regarding the dual nature of software and other entities in computer science have not abated and that the abstracta have stubbornly refused to be turned away.

In my view, which is especially concerned with the *objects* under consideration, the insights of another neighbouring subfield have also been neglected, that of the philosophy of linguistics, where metaphysical debates regarding the status of linguistic entities have been ongoing since at least the publication of Jerrold J. Katz's *Language and Other Abstract Objects* (1981).[5] More recently, work has been done in the philosophy of linguistics, and on the philosophy of words in particular, which treats of issues similar to those that we come across when considering the dual nature of objects in computer science.[6] There should be no surprise in this, as the domains overlap in many ways. I will discuss Katz's approach further in the next section.

One author who has treated of both areas in parallel ways is Irmak (2012, 2019).[7] Irmak claims that most philosophical accounts of software have failed 'to capture the nature of software as an artifact' (2012, p. 56),[8] mentioning Moor and Colburn in particular. Further, he says that 'software is a kind of abstract artifact which fails to fit the classical Platonic picture of abstract objects [....] because it is not an eternal and mind-independent entity but is created by human beings with certain intentions' (2012, p. 56). A realist, such as Wetzel (2006) or Katz (1998), will at least minimally agree that artefacts have temporal properties and so could not be non-spatiotemporal abstracta. However, Irmak's ontology recognises both these created temporal abstracta that he calls 'software' *and* the concrete physical objects that he calls its 'copies'. On Irmak's account, it is the temporal abstracta that he calls 'abstract artefacts', while presumably also accepting that the concrete copies are also artefacts of some other variety.

In contrast, Irmak thinks that algorithms are indeed 'Platonic' abstracta, that is, eternal and mind-independent, and says that he assumes this without an argument for the position (2012, p. 59). He says that there is a type-token relation between an algorithm and a particular instance of it, but he is largely unconcerned with the natures of types and tokens and follows others in taking types to be abstract universals and tokens to be concrete. Nevertheless, he does leave open the possibility that there may be some abstract tokens (2012, p. 59). He also mentions that a similar distinction holds for the text or code of software that is written in a particular programming language (p. 60).[9]

Irmak argues that software is created, and so is not an abstract object in the 'Platonic' sense, and, further, that it can survive 'changes in its algorithm or text' (p. 68). So, in the absence of a theory of changing types or universals, he concludes that it is not a type or universal. Thus, he thinks that 'the relation between computer programs and their physical copies cannot be understood in terms of the type/token distinction' (p. 68). Why he thinks that the distinction nevertheless applies to the code of these changing programs is not addressed (cf. p. 60). He argues that software can be destroyed if all of its copies, executions, and memories of it are destroyed. Thus, he thinks that software is abstract because it lacks spatial location, but nevertheless has temporal location (p. 70).

7.3 Katz's Realist Ontology of Composite Objects

In this section, I will argue against Irmak's view, which was presented in Section 7.2, and in favour of Katz's ontology of composite objects before applying it to software in Section 7.4. Perhaps the greatest tool available to philosophers is the principle of non-contradiction (PNC), first put forward explicitly in terms of negation by Aristotle (*Metaphysics*, Gamma 3–4, $1005^{b}19$–$1006^{b}34$). However, as employed in metaphysics, and not merely as a logical principle, it is often not appreciated that PNC is not simply a prohibition on the positing of contradictory objects or facts, which it certainly is, it is also a heuristic by which parsimonious ontological divisions can be made. That is, contradictions must be contradictions with reference to one and the same respect of a unitary thing.[10]

For example, take Irmak's notion of software as not being identifiable with a concrete object, but also being a created entity under certain intentions, that is, having a temporal beginning. On the standard distinction between abstract and concrete, he claims that this is a contradiction. Irmak then effectively interprets PNC as being a reason to revise the distinction to allow for abstract objects that are temporal (Irmak, 2012, p. 56). However, this is not the only option available and should not be the first port of call for theory modification in view of the fact that the distinction is so fundamental in metaphysics, and so revisions to it can have global theoretical effects extending beyond the local problem under consideration.[11] *A fortiori* rejecting PNC is not considered to be a further option here either. A potential third option is instead to interpret PNC as being, for example, a reason to analyse merely the object in question with a finer grain and so to dissolve the contradiction by way of some further analysis. On this view, it is correct to predicate being temporal of a piece of software relative to a certain aspect or component of it or, rather, its copies, and also to predicate being atemporal of a copy relative to a different aspect or component of it. This is analogous to my standing straddling the threshold of a room and being said to be both inside and not inside the room. It would be proper in such a case to distinguish different parts of me and apply the predicates to me but only relative to those parts (Politis, 2004, p. 145).

In his work *Realistic Rationalism* (1998, hereafter *RR*),[12] Katz introduces a category of objects that he calls *composite objects*, which are *heterogeneous* with respect to the abstract-concrete distinction by having both abstract and concrete constituents. Katz's theory of composite objects effectively provides us with a finer-grained analysis of tokens, including artefacts, which is in accord with the latter kind of theorising described above, and for which there is no need to revise the usual distinction between abstract and concrete. So, in addition to what he called objects that are *homogeneous* with respect to the distinction, that is, abstract and concrete objects, Katz argues that there are objects that are *heterogeneous* with respect to the distinction, those of a third category, namely, composite objects. Thereby, the distinction remains unrevised, but these three categories are nevertheless disjoint and jointly exhaustive (Katz, 1998, p. 145).

A composite object is formed by virtue of a *creative relation*, or a pattern of such relations, holding among its *component* objects. That is, components are the relata of creative relations (1998, p. 141). Katz says that 'In the case of artifacts, the intention of the producer or the intention of the user can be an aspect of the creative relation' (1998, p. 141). Nevertheless, the intention must be of the right sort. Katz uses the example of a dog swallowing biscuits because they taste pleasant, but which does not create a new object having the dog's stomach and the biscuits as components (p. 141). In the case of tokens, the intention must be to form a token of a specific type.

Further, some relations, such as 'number of, identity, between, greater than, and inside of' (1998, p. 141), are *uncreative* because they do not involve 'the notion of some further object of which the objects satisfying the relation are components' (p. 154). Katz uses the example of 17 cherries falling from a basket, which would merely instantiate the number 17 and not create a further object with the 17 cherries and the abstract object, the number 17, as its components (p. 154).

There are three kinds of composite objects: *simple*, *complex*, and *compound*. *Simple* composite objects have only homogeneous components, that is, only abstract and concrete objects (and, of course, at least one of each). *Complex* composite objects have both homogeneous and heterogeneous objects as components. *Compound* composite objects have only heterogeneous objects as components, that is, at least two composite objects (Katz, 1998, p. 142). It is important to note here that the latter two kinds of composite objects can be hierarchically structured objects having many layers because these composite objects can have other composite objects as components. This makes them very useful for explaining the kind of linguistic structures that we find in natural and artificial languages, and, indeed, as I shall show, in software.

The ontology of composite objects suggests the following general principle for specifying identity conditions: 'Composite objects are identical when they have the same components and the same creative relation holds of them in the same way', that is, 'corresponding objects saturate the same places of the relation' (1998, pp. 145–6; cf. p. 167). It also suggests a general principle for specifying existence conditions: 'The condition for the existence of a composite object is the sum of the conditions for the existence of their components plus the condition that the relevant creative relation obtain among them' (1998, p. 146; cf. p. 167).

Composite objects can be of different denominations depending on their components and the relations involved. For example, impure sets are composite objects formed by a *containment* relation holding between the null set, which is an abstract object, and the other objects and sets that are contained by the null set (1998, p. 142).[13] On Katz's account, *tokens* are composite objects formed by a *type-token* relation and (in at least the simple case) have a *type* as an abstract component. This makes them a significantly different kind of entity to those posited by other accounts of the type-token distinction, which are often taken merely to be physical objects. Tokens on this view are categorized in virtue of their type components or constituents. That is, on the basis of whether the type or types involved are mathematical, logical, linguistic, or musical, etc., or some combination of kinds of type (1998, p. 146). As we will see in Section 7.4, we may also add computational objects to this list. In Katz's view, types are not universals and tokening is distinct from instantiation (1998, pp. 154–5). In the case of instantiation, an object instantiates a property by virtue of its relation to a universal. In the case of tokening, a composite object *co-possesses* many of its properties by virtue of having the components that it does. Tokens have certain properties of their abstract components in a relative way, that is, relative to those abstract components, whereas the abstract components themselves have them in an absolute way. The reason that tokens appear to have inconsistent sets of properties, and why they may seem to be both abstract and concrete, such as in the case of the dual or 'hybrid' nature of software discussed earlier, is that these properties are possessed relative to the appropriate component. That is, it is as much of a genuine contradiction as saying of me that I am standing both inside and not inside the room when I am straddling the threshold, which is not a contradiction at all under the correct analysis (Politis, 2004, p. 145).

7.4 Types, Tokens and Software

In this section, I will apply Katz's ontology to software and its copies. Most of the remarks that Katz made about computation in his writing tended to be included as parts of technical discussions of linguistic theory. However, in a footnote of an article co-authored with Postal from 1991, in which the ontological status of language is a central focus, they compare linguistics to computer science. In their view, the core of theoretical linguistics is a formal science of natural languages (NLs), but there are also other areas of linguistics.

> [...] historical linguistics, sociolinguistics, etc., are genuinely empirical because they concern the relation between NLs and the people who know them. [...]
>
> Note that computer science similarly has a dual nature: part of it studies physical objects (e.g., computer hardware) and part abstract objects (e.g., algorithms).
>
> (Katz & Postal, 1991, p. 538: n. 11)

At this stage, Katz had not yet developed the ontology of composite objects, which had its first and only airing in *RR* (1998). There he also uses that ontology to explain and expand further upon the distinctions between grammar and other areas of linguistics (cf. Katz, 1998, p. 165). Later in their 1991 article, while delineating two possible readings of what they take to be an equivocation by Chomsky regarding the nature of I-languages,[14] Katz and Postal make an analogy between I-languages as abstract, mathematical objects and software.

> One term of the equivocation is to define an I-grammar as 'a characterization of a function in the mathematical sense'. The other term claims that I-languages are physical objects existing in space-time and entering into causal relations. On the former term of the equivocation, I-languages are analogs of computer *programs*; they are software. On the latter, they are analogs of physical states of computers which instantiate programs. They are then hardware. But they can't be *both*.
>
> (Katz & Postal, 1991, p. 544)

Ultimately, their argument is that conceptualism, which is the position that claims that NLs are I-languages, 'must limit itself to the hardware interpretation of both grammars and sentences. For the software interpretation is tantamount to realism' (p. 549), which is the position that NLs are neither E- nor I-languages but something like P-languages.[15] It is quite striking here that software, computer programs, are used as paradigmatic examples of an abstract object. This analogy was never pursued further in later work. However, I will now apply the later ontology of composite objects to this underdeveloped position and attempt to extend it in line with how Katz later develops the metaphysics of linguistic entities in *RR*. The aim will be to have the main aspects of the theory in place, so that we may then attend to the notion of software maintenance further on that basis. Further details and developments extraneous to this aim will be left for another time.

Notwithstanding his taking software to be an abstract object, Katz would have recognised that software can have copies that are executed by computers. In his ontology, these copies would be tokens of a certain type. Irmak denied that this was possible because he claimed that types could not accommodate change. This is for the most part accurate in that, for Katz, types are unchanging abstract entities. However, Irmak's general ontological position is not novel. Indeed, Katz responded to a very similar position that was put forward by Hale (1987), which claimed that things like games and languages, although abstract, have temporal location because they are created (cf. Katz, 1998, pp. 132–3). Katz also responds to a very similar charge that was made by Hale against the notion that languages are abstract objects. Hale's concern was that realism requires us 'to regard what we are pleased to call the history of the English language as really a record of a succession of replacements of one language by another' (Hale, 1987, p. 49; cf. Katz, 1998, p. 137). Katz responded that in the final analysis of what is surely a matter for linguistics, the individuation of languages will be explained in terms of individuating sets of sets of sentence types. For example, in the case of English, it is such a set that includes relations between idiolects, dialects, and languages. The history of English is a history of its speakers' changing competence in a succession of sets of sentences (1998, pp. 137–8; cf. Katz, 1981, p. 9).

Similarly, we can think of, for example, the named items of software that Irmak mentions, such as Windows 7 (no longer maintained), as a set of sets of interrelated program types that collectively individuate that software. In the same way that it would be up to the linguist to determine the exact set in the case of a natural language, it is up to the software engineer or other specialists to determine the exact set in the case of Windows 7. The essential message here is that whatever abstract means of explanation or whatever configuration of sets of types is needed is available to them and this avoids the problems raised by Hale and Irmak, respectively.[16] In the case of software, the proper arrangement will of course be complicated. It may resemble something like a branching Git tree,[17] which has many more actual versions and development lines than those that eventually become available to the public, some of which are simultaneous lines of development. Indeed, there are many more possible trees, potential lines of development and maintenance, which are never realised. Irmak, perhaps unwittingly, proposed to explain all this complexity by way of only a single changeable abstract object. Although it is clear that such an ontology would allow for a single line of change, it is unclear how, if at all, it could account for copies of previous versions of the software as still being copies of the software. The software, on Irmak's account, a temporal abstract object, has changed and so is no longer instantiated by the 'copy' in question. In light of the above, we should adopt the distinction between the software and the individual computer programs (or systems of multiple programs and their documentation, etc.) that make up the tree, each distinct from the other but each a node of the tree. Software then is properly identified as being a set of sets of program types (and other ancillary types) to an arbitrary depth.

Attending to the program types themselves, we can see that they are made up of algorithms, that is, generally, finite progressions of operation types.[18] As we saw earlier, although Irmak accepts that algorithms are non-spatiotemporal abstract objects, he neglects to provide an argument for the claim. Nevertheless, it is clear that they must be, because algorithms, and progressions of operation types more generally, are what are under discussion in, for example, formal program verification and analytic determinations of computational complexity, just as other kinds of abstract objects are under discussion in other kinds of mathematical proof (cf. Katz, 1998, p. 167). One cannot with certainty formally verify a sequence of physical events in this sense due to their dependence 'upon the causal properties of physical systems, whose presence or absence is only ascertainable by means of inductive procedures' (Fetzer, 1988, p. 1059). It follows that algorithms are not sequences of physical events, rather, they are abstract progressions. As we will see later, in Section 7.5, this also has the consequence that software maintenance must be in part a formal discipline that enhances the efficacy of software.

Program types are analogous to *discourse types* in Katz's theory of linguistic entities, which are progressions of sentence types.[19] Discourses are things such as novels, poems, speeches, etc., and compositions constructed by arranging sentences and expressions of a language. Such types are tokened by a compound composite object, the third basic kind of composite object mentioned earlier. A *succession relation* holds as the creative relation between the composite objects that are the components of the token. In the case of discourse tokens, these are often sentence tokens or other expressions arranged in a sequence that is in accord with the progression of the discourse type (Katz, 1998, p. 167). Analogously, in the case of programs, operation tokens are arranged in a sequence that is in accord with the progression of the operation types in a program type.

Program types may also be represented in the high-level code of a programming language, which has the property of being able to be compiled, interpreted or otherwise translated into low-level assembly code and assembled into machine code. The high-level code itself does not have (or need not have) this low-level code as a component. Rather, the components of the high-level code are the character set in use, including, for example, alpha-numeric characters and various special characters, which are themselves tokens of those character types. At this level, such objects are close to those of natural language representations, and may even contain some natural language, which is what makes them useful to programmers. The code at any level may also be displayed to a user via a printout on paper or a monitor, etc., which is effectively a translation by some pointillistic algorithm designed for this purpose. Each character is translated into a certain sequence of flashes of light or dots on the page, firing positions, etc., which together re-token the character in a new physical medium.

The operation tokens that are the components of token programs are complex composite objects that are basic or atomic manipulations of data. Their components are data objects, which are composite, and abstract operation types. These can be more or less complex. Data objects may take various forms depending on the use case. Usually, data objects are binary sequences that are compound composite objects. At the most basic level, each token binary value in such a sequence is a simple composite object. Its components are certain physical quantities, be they charges, pressures, voltage ranges, positions, punched holes, etc., and an abstract binary value type. The concrete physical states may happen to *instantiate* the items of the binary sequence, that is, be counted zero or one, but this is not what is operative, because they could just as well be used to *token* the binary value of the opposite type than they do. Rather, they token the values that they do in virtue of the way in which the computer system was set up, that is, as a product of the intentional acts of some company of designers and engineers. Similarly, a certain sequence of tokens of binary values may token a compound composite object, a data object that is a binary numeral of a certain type or a coded instruction.

Notice here that the operation token itself may be a complex composite object, while its coded instruction may be a long compound composite object, for example, written in machine code. So, the ontology accords with the well-founded intuition that these are distinct objects/events. The arrangement of the hardware of the computer too may be coded as further binary sequences, compound composite objects. After all, the states and transition functions of a computer are abstract objects that can themselves be encoded as a binary sequence. It should also be considered that computers can be emulated as part of a program running on another computer. Famous examples include Turing machines built in Conway's Game of Life (Rendell, 2016) and computers in game worlds, such as redstone computers in Minecraft. Such cases can also be explained using the ontology of composite objects. In these cases, the role otherwise played by the concrete physical layer is played by a layer of composite objects. For example, instead of simple composite objects tokening binary values, complex composite objects will play the same role. What matters is that tokens formed in each of these ways will be of the same type. We will leave further discussion of these more complex areas of investigation aside for present purposes,[20] except to say that ultimately the distinction between hardware and software is really one based on what we choose to encode and token where and in what medium. At a certain level of abstraction, the distinction breaks down. This is shown in the ontology by there being a close similarity between the composite objects involved in hardware and software, at the particular level of their involvement.

7.5 Software Maintenance

With this ontology for software in hand, we can proceed to ask what it can tell us about the metaphysics of software maintenance. On the realist account, that I have developed above by employing Katz's ontology, we can see that software is an abstract object and its copies are tokened by composite objects that have a hierarchical structure. Software maintenance then is a discipline that on the one hand has a formal side that attempts to discover and thereby token a range of valid program types appropriate and peculiar to the software, that is, in accord with a specification or specifications of its functions. The software that is being maintained in this sense is a set of program types that encompass the range of valid realisations of a specification or, more broadly, an intention for a computer to perform a certain computation or computations. The set and the program types remain unaffected by this, as they are abstracta. All of these are relations that are extrinsic to the types in question. On the other hand, software maintenance is a discipline that constrains any copies produced to the tokening of those program types. For example, corrupted copies must be repaired and, in the most usual case, lower versions must be updated to higher versions.

Let us now address Young's problem for this view, especially with regard to how the nature of software maintenance relates to the nature of software, or its copies, as artefacts that are constituted in part by way of an intention. Young (2021) has been rightly concerned to show that the usual conception of maintenance as preservation or conservation fails to recognise the full role that maintenance plays in the production of artefacts and their efficacy. It is perhaps understandable why, say, a civil engineer might have such a one-sided conception of the role of maintenance in the production of, for example, a bridge. Such an engineer produces the bridge to a certain specification and intention. At that point, the bridge is complete and, in their eyes, the primary maintenance role to be played is conservative and restorative against the usual vicissitudes of physical matter (cf. Young, 2021, p. 359). However, the case of software engineering is importantly different because the role of maintenance is difficult to ignore, and software engineers are under no illusions about its importance and character.

> This distinction between development and maintenance is increasingly irrelevant. Very few software systems are completely new systems, and it makes much more sense to see development and maintenance as a continuum. Rather than two separate processes, it is more realistic to think of software engineering as an evolutionary process [...] where software is continually changed over its lifetime in response to changing requirements and customer needs.
>
> (Sommerville, 2016, p. 60)

From the very outset of their basic training – I happen to have been through such training myself – software engineers are told that the initial software development for any project is merely the tip of the iceberg, representing only 20%, and that 80% of the effort will be spent doing maintenance (Pfleeger & Atlee, 2010, p. 540), usually involving code that they did not write themselves. Of this maintenance, approximately 24% will be the correction of errors, another 19% will be necessary for adaptation to changing environments, but 58% will be for the addition of new features and other enhancements (Sommerville, 2016, p. 272; cf. Pfleeger & Atlee, 2010, p. 546). Hence, they will often expend over twice as much effort doing maintenance that adds enhancements as they did on the initial development. Software engineers are only too aware of the reasons for this and are well ahead of the philosophers in this regard. One reason is that programmers generally prefer to work on new projects rather than on maintaining old ones. This can lead to the break up of teams and thereby the loss of experience and knowledge of the system to be maintained, which must then be acquired by any subsequent maintainers at some not insignificant cost of days or weeks of work. The distinction between development and maintenance also results in there being little incentive to write easily maintainable software, because it will most likely be another team that ends up maintaining it (Sommerville, 2016, pp. 90, 272–3). The general attitude to software maintenance is evinced by the following passage.

> [...] these problems stem from the fact that many organizations still consider software development and maintenance to be separate activities. Maintenance is seen as a second-class activity, and there is no incentive to spend money during development to reduce the costs of

> system change. The only long term solution to this problem is to think of systems as evolving throughout their lifetime through a continual development process. Maintenance should have as high a status as new software development.
>
> (Sommerville, 2016, p. 273)

As we saw earlier, Young instead places the blame for the misunderstanding of the role of maintenance generally on the ontological conception of artefacts as realisations of design intentions, supposedly due to a pernicious kind of Aristotelianism that sees change as antithetical to a completed artefact that fits its form and essence (2021, p. 359).[21] Many acts of maintenance of artefacts, conservative or otherwise, do nevertheless fulfil something of the initial intention. That is, an intention and what is intended need not extend merely to the initial development of an artefact. It is clear that an intention can also encompass an artefact's maintenance. Indeed, Young agrees that artefacts are often designed with the expectation of future and unanticipated changes (Young, 2021, pp. 361 ff.). Although I agree with Young that a rebalancing of our conception of maintenance may be warranted in order to take account of it in all its aspects, the cause of the imbalance has, I believe, been misidentified.

Young's underlying claim is that there is an entailment between the conception of the nature of artefacts as realisations of their producers' intentions and the conception of the maintenance of those artefacts as being solely conservative.[22] A corollary of this claim is that if Young is correct that maintenance is not merely conservative, which he surely is, then it follows that many, if not all, artefacts are not realisations of their producers' intentions. So, seen in this light, the claimed entailment constitutes a broader claim about the nature of artefacts and their production. This consequence could only be drawn if it were shown that such an involvement of intentions in a theory of artefacts is *inconsistent* with maintenance having a non-conservative aspect, which, I take it, has not been shown. This is, I think, important for the possibility of the realist view of software and software maintenance for which I have argued, because they involve the employment of intention in the formation of tokens of an abstract type.

Furthermore, I take software engineering and software maintenance to be sources of straightforward counterexamples to the claimed entailment. In almost all cases, the producers of software intend for it to be periodically updated and/or enhanced, and in order to implement changes to software, knowledge of its intended purpose and design is required (Sommerville, 2016, p. 90). The exact nature of future changes could outstrip the producer's knowledge of the software (on the Katzian view outlined, a complicated abstract object) of which they intend to produce a token. Sometimes, it is choreographed to some degree in advance what features, etc., will be added at certain stages, but often it is not. Nonetheless, there is *ab initio* an intention to maintain the software in an adaptive or enhancive way, or continually to develop it throughout the duration of its use, as Sommerville advocates. This intention can also be observed more widely in the practice of releasing alpha and beta versions, and sub-unitary versions (for example, version 0.1), which are in one sense complete and functioning pieces of software, but are also intended to be further developed and enhanced. That is, such realisations of an intention or intentions need not be an all or nothing linear affair. The thing intended is an abstract structure that can be more or less fully comprehended, adumbrated, and realised due to its complexity, that is, both single-program complexity and complexity of the tree of programs, etc., that make up the software according to the view. The force of such counterexamples is that the entailment between the intentional conception of an artefact's production and its maintenance being conceived of as conservative only holds in cases where the intentions are restricted to the realisation of a 'fully completed' artefact, so to speak, and where it is not in the nature of the artefact to be periodically enhanced, etc. Perhaps, the entailment holds for an ordinary table or a bridge, so conceived, but it does not hold for most software,[23] which, as we are all too aware, and expect, is continually patched, updated, and upgraded. It is also questionable whether the entailment could hold even for such artefacts as tables and bridges. The same question should be asked of these cases, regarding whether the producers of these artefacts intend for them to be maintained (whether or not they understand that this can often be adaptive or enhancive and not merely conservative). Perhaps, they even take intended maintenance into account when, for example, choosing their materials and other design features and constraints. So, it seems likely that they often do so intend. In any event, a single counterexample is all that is needed to show that a generalised version of the claimed entailment is false.

This is not to say that there are no entailments of worth between the nature of software, or the production of its copies, and the nature of its maintenance. I have already pointed to examples of such entailments. The reason why an aspect of software maintenance, an aspect that is involved with increasing software's efficacy, is a formal discipline aimed at verification and determination of computational complexity is because software is an abstract object made up of sets of sets of program types. The non-abstract constituents of its copies are not amenable to such a formal discipline because they depend upon causal properties that are ascertainable only by inductive means. In that case, wider considerations regarding the environment and use case are taken into account. For example, whether the software's ability to correct errors due to interstellar radiation impacting a space probe is a pertinent factor or whether it will matter for the use case. All of this kind of adaptive maintenance is distinct from whether the algorithms involved in the software are formally verifiable, because it is about a distinct constituent of the composite object.

If Young's claim is not the reason for the imbalanced view of maintenance, then we should ask what is. It is more likely that something like the situation that Sommerville presents, as outlined earlier, is the correct diagnosis of the tendency to view maintenance as solely or primarily conservative. So, on this view it is some feature or features of our society, market economics and perhaps even our psychology, etc., which is the cause of the imbalanced view. This becomes all the more stark when the imbalanced view is set against the extent of adaptive and enhancive maintenance practices involved with something like software. The cause of taking maintenance to be solely conservative is not any particular philosophical understanding (or misunderstanding) of the nature of artefacts generally, and especially not that they are beholden to intentions and what is intended. The example of software maintenance, I think, shows that this could not be the case because the software engineers have precisely a design-oriented perspective on their artefacts, though they are pragmatic about what can be achieved given the complexity of their task (cf. Young, 2021, pp. 361 ff.). They discuss such designs and specifications at length and form elaborate practical theories about the best methods of formulating and achieving them (for example, Waterfall, Agile, etc.). It is also clear that they are not under any illusions regarding the important role of maintenance, including its non-conservative, adaptive and enhancive aspects, because this is what they spend the majority of their time doing.

7.6 Conclusion

In this chapter, we have seen that while many authors have taken computational objects like software to have dual or 'mixed' natures, this should be taken as an ontological puzzle to be solved. Some proposals to solve this puzzle have not been very fruitful or have skirted the main issue regarding the abstract-concrete distinction. Other proposals, such as Irmak's, have taken more drastic measures in reformulating the traditional distinction between abstract and concrete objects to include temporal abstracta. I argued that this was based on what is an unwarranted appeal to PNC, in view of the fact that we could instead opt for a finer-grained analysis that left the distinction unchanged. Such a finer-grained analysis is provided by Katz's ontology of composite objects. I argued that this ontology could be used in a way roughly analogous to how Katz uses it to explain linguistic entities in order to explain computational objects such as software.

This allowed us to see that the notion of a realist metaphysics of software maintenance is not as otiose as the initial worries might have had us believe. We saw instead that software can be understood as being a set of sets of related program types in accord with a specification or, more broadly, an intention. Software maintenance, on this view, involves both discovery of a valid range of program types in accord with this intention and also the use of programming and engineering to token the appropriate types that are in accord with this set. We also saw that software engineers have an understanding of maintenance as being continuous with development, and in which maintenance is not merely conservative and about fixing errors but is in fact mostly adaptive and enhancive. Software maintenance thereby constitutes a source of counterexamples to Young's claimed entailment between the conception of the nature of artefacts as realisations of their producers' intentions and the conception of the maintenance of those artefacts as being solely conservative. The reason for that imbalanced view of maintenance as being solely conservative is much more mundane, and primarily due to factors stemming from our society, economics, and psychology, rather than our understanding of the nature of artefacts and their production.

Acknowledgements

The author wishes to thank three anonymous reviewers and the editors for their comments on drafts of this chapter.

Notes

1 For the purposes of this discussion, I leave to one side the issue of laws and the nature of physical objects, etc.
2 The philosophical view that they express is ambiguous, perhaps between a conceptualism and a nominalism about abstracta. For example, they say that 'the term abstraction refers to an idea or general description, divorced from physical objects. [....] although these abstractions such as "word processors" and "documents" are merely convenient ways of describing patterns of electrical activity, they are also things that we can buy, sell, copy, and use' (1999, p. ix). This is at least an odd use of the word 'merely'. Colburn seems not to recognise this ambiguity in their position, for he is more clear that there is an apparent incompatibility at play in the notion of a concrete abstraction.
3 It is perhaps worth observing in passing that there is a potential problem with Turner's appropriation of Kroes' work because Kroes explicitly excluded software from his notion of a 'technical artefact': 'software programs fall outside the scope of this book. I consider software programs to be "incomplete" technical artefacts; only in combination with the appropriate hardware that executes software programs are they able to fulfil their technical function' (Kroes, 2012, p. 2: n. 4).
4 See especially: Achterhuis (2001); and Verbeek (2022), for an overview.
5 Katz himself raises the question earlier, regarding meaning, in his *Semantic Theory*, but leaves it without a 'final answer' (1972, pp. 16 & 38).

6 For an overview of some of the central questions in the philosophy of linguistics, see Stainton (2014). For a recent survey of the ontology of words, see Miller (2020). Also, see the recent *Synthese* topical collection on 'The Philosophy of Words': https://link.springer.com/collections/cgicihfcie
7 I respond to Irmak's ontology of words (2019), and in particular his criticism of Katz in that regard, in Begley (2023). So, I will not get into those arguments here, except where there is a direct connection.
8 Irmak says that 'an object is an artifact if and only if it is an intentional product of human activity' (Irmak, 2012, pp. 55–6). He claims that this is a widely accepted definition and references Hilpinen (2011) for further discussion. However, Hilpinen discusses a number of definitions and debates, and it is not clear that Irmak's definition is widely accepted. Further, Irmak's definition would seem to be an overproductive one. That an artefact is an intentional product of someone's activity is perhaps necessary, but is it sufficient? Counterexamples and difficult cases abound. For example, are smiles or incredulous stares artefacts?
9 Irmak quotes what he says is the text of a program for adding two values that is written in the language C++, but what follows is the text of a program written in assembly language (cf. Irmak, 2012 p. 60). Irmak here appears to misread Colburn (2000, p. 202), without citation.
10 For a thorough explanation of this principle in the context of Aristotle's *Metaphysics*, see Politis, 2004, pp. 122 ff.
11 I believe that Katz provides such an example: 'An ontology that says that all concrete objects are corporeal would preclude concrete objects that are mental in the Cartesian sense, and hence make the classical mind/body problem unstatable' (1998, p. 125). By making spatial properties the only mark of the concrete, Irmak's ontology is of this kind.
12 Much of my discussion here has benefitted from a parallel project to reconstruct Katz's metaphysics of words. For further in-depth discussion, see Begley (2023).
13 Katz is careful to distinguish the containment relation from membership and inclusion, as it is more general than those relations and should not be confused with them (1998, p. 142). See Begley (2023) for further discussion.
14 Chomsky introduced the distinction between *E-language, I-language*, and *P-language* in part to provide clear and concise labels for categories introduced in his earlier work. E-language or 'externalized language' is a construct 'understood independently of the properties of the mind/brain', including 'the notion of language as a collection (or system) of actions or behaviors of some sort' (1986, p. 20). I-language or 'internalized language' is 'some element of the mind of the person who knows the language, acquired by the learner, and used by the speaker-hearer' (p. 22). P-language or 'Platonic language' is the name he gives to 'something else additional [...] existing in a Platonic heaven alongside of arithmetic and (perhaps) set-theory' (p. 33), which he says is claimed by Katz and others (p. 49).
15 See footnote 14 above.
16 For a further discussion of Katz's response to Hale on the issue of natural language, see Begley (2023).
17 Git is an open-source distributed version control system that was invented by Linus Torvalds in 2005 for the development of the Linux Kernel. It is now widely used for software development, and on repository websites such as GitHub. However, as Irmak rightly points out, there are many other methods (Irmak, 2012, p. 63).

18 Katz tends to reserve 'progression' for reference to the abstract types, and 'sequence' for the tokens.
19 Northover et al. (2008, p. 93) make a broadly similar analogy between token performances or readings of a novel and token executions of computer programs.
20 For a full discussion, and with regard to virtual reality, see Begley (forthcoming).
21 Whether this is an accurate interpretation of Aristotle or not is a vexed issue in the literature. Aristotle might not have taken artefacts to be *oūsiai* (primary beings or 'substances') (cf. *Metaphysics*, Book Eta 3, 1043b22). In any event, it does not appear to be essential to the point at issue about intentions.
22 As we saw in Section 7.2, Young's argument as presented refers to our *understanding* of artefacts and how it 'encourages' us to think about maintenance (2021, p. 359), rather than referring to an entailment between the two conceptions. However, it will be more charitable to assume that such an underlying entailment is claimed, especially as Young has confirmed this to me (personal communication).
23 There are of course edge cases. For example, if I were to install a copy of Windows 7, or some other final version of now unsupported software, I would be creating a token the maintenance for which will be conservative, if any maintenance occurs at all.

Bibliography

Achterhuis, H. (ed.) 2001. *American Philosophy of Technology: The Empirical Turn*. Bloomington: Indiana University Press.

Aristotle 1952. *Metaphysics*. Trans. Richard Hope. New York: Columbia University Press.

Begley, K. 2023. 'Katz Got Your Tongue? The Metaphysics of Words', *Synthese*, 202(4): 107. https://doi.org/10.1007/s11229-023-04324-x

Begley, K. forthcoming. 'The Metaphysics of Communication in Virtual Reality'. In: T. Ahlers & K. Senkbeil (eds.), *Virtual Reality in den Geisteswissenschaften – Konzepte, Methoden und interkulturelle Anwendungen [Virtual Reality in the Humanities: Concepts, Methods and Intercultural Applications]*. Peter Lang.

Chomsky, N. 1986. *Knowledge of language: Its nature, origin and use*. New York: Praeger.

Church, A. 1936. 'An Unsolvable Problem of Elementary Number Theory', *American Journal of Mathematics*, 58: 345–363.

Coeckelbergh, M. 2022. 'Using Philosophy of Language in Philosophy of Technology', In: S. Vallor (ed.), *The Oxford Handbook of Philosophy of Technology*. Oxford: OUP, 341–357.

Colburn, T. R. 2000. *Philosophy and Computer Science*. London: M. E. Sharpe.

Fetzer, J.H. 1988. 'Program Verification: The Very Idea', *Communications of the ACM* 37(9): 1048–1063.

Gruner, S. 2011. 'Problems for a Philosophy of Software Engineering', Minds & Machines 21: 275–299.

Hailperin, M., Kaiser, B., Knight, K. 1999. *Concrtete Abstractions*. Pacific Grove: Brooks/Cole. https://www.gustavus.edu/+max/concrete-abstractions.html

Hale, B. 1987. *Abstract objects*. Oxford: Basil Blackwell.

Hilpinen, R. 2011. 'Artifact', *The Stanford Encyclopedia of Philosophy*. https://plato.stanford.edu/archives/win2011/entries/artifact/

Irmak, N. 2012. 'Software is an Abstract Artifact', *Grazer Philosophische Studien* 86: 55–72.

Irmak, N. 2019. 'An Ontology of Words', *Erkenntnis* 84: 1139–1158.

Katz, J.J. 1972. *Semantic Theory*. New York: Harper & Row.

Katz, J.J. 1981. *Language and Other Abstract Objects*. Totowa, NJ: Rowman & Littlefield.

Katz, J.J. 1998. *Realistic Rationalism*. Cambridge, MA: MIT Press.

Katz, J.J. and Postal, P.M. 1991. Realism vs. Conceptualism in Linguistics. *Linguistics and Philosophy*, 14(5): 515–554. https://www.jstor.org/stable/25001444

Kroes, P. 2012. *Technical Artefacts: Creations of Mind and Matter. A Philosophy of Engineering Design*. Springer.

Miller, J.T.M. 2020. 'The Ontology of Words: Realism, Nominalism, and Eliminativism', *Philosophy Compass*, 15(7): 1–13. https://doi.org/10.1111/phc3.12691

Moor, J.H. 1978. 'Three Myths of Computer Science', *The British Journal for the Philosophy of Science* 29(3): 213–222.

Northover, M., Kourie, D. G., Boake, A., Gruner, S., & Northover, A. 2008. 'Towards a Philosophy of Software Development: 40 Years after the Birth of Software Engineering', *Journal for General Philosophy of Science* 39: 85–113.

Pfleeger, S.L. & Atlee, J.M. 2010. *Software Engineering: Theory and Practice*. 4th Edition. Upper Saddle River, NJ: Prentice Hall.

Politis, V. 2004. *Aristotle and the Metaphysics*. London: Routledge.

Rendell, P. 2016. *Turing Machine Universality of the Game of Life*. Cham: Springer.

Sommerville, I. 2016. *Software Engineering*. 10th Edition. London: Pearson.

Stainton, R. J. 2014. 'Philosophy of Linguistics', *Oxford Handbooks Online*. https://doi.org/10.1093/oxfordhb/9780199935314.013.002

Turing, A.M. 1937. 'On Computable Numbers, with an Application to the *Entscheidungsproblem*', *Proceedings of the London Mathematical Society*, s2-42: 230–265.

Turner, R. 2018. *Computational Artifacts: Towards a Philosophy of Computer Science*. Berlin: Springer.

Verbeek, P.-P. 2022. 'The Empirical Turn, In: S. Vallor (ed.), *The Oxford Handbook of Philosophy of Technology*. Oxford: OUP, 35–54.

Wetzel, L. 2006. 'Types and Tokens'. *The Stanford Encyclopedia of Philosophy*. https://plato.stanford.edu/archives/fall2018/entries/types-tokens/

Young, M.T. 2021. 'Maintenance'. In: D.P. Michelfelder and N. Doorn (eds.), *The Routledge Handbook of the Philosophy of Engineering*. London: Routledge, 356–368.

8 Maintaining Perpetual Actuality in the Digital Age?

Simondon's Conception of Maintenance and the Networked Era

Johannes F.M. Schick

'Maintenance' encompasses the control and repair of technical objects, a prerequisite for this being that technical objects are constructed in such a way that they *can* and *must* be maintained.[1] This implies a distinction between living beings and technical objects; technical objects require maintenance since they lack the self-sustaining and self-repairing structure of organisms.[2] While organisms are 'concrete to begin with', technical objects are – as constructed beings – always perfectible (Simondon, 2017, p. 51). These ontological and temporal aspects of maintenance are key to Gilbert Simondon's conception of technical objects (2017, p. 255; 2009, p. 19). For Simondon, the ideal technical object is open and has a 'reticular structure' (i.e., networked), which allows it to be 'completed, improved, maintained in the state of perpetual actuality' (Schick, 2021a; Simondon, 2009, p. 24). The construction of technical objects should already imply that improved parts (e.g., a faster computer chip, a more efficient battery, etc.) can and should be replaced. Technical objects should thus be open for improvements provided by the infrastructural networks of production.

For Simondon, 'maintenance' implies a temporal aspect that relates technical objects with the present. Maintaining a technical object in this emphatic sense means to continue the engagement with the technical object so that it is kept – perpetually – up to date. The openness of technical objects therefore enables a relationship of human beings towards techniques in which technical objects are not something to be bought and discarded, but instigate participation (Simondon, 2014a, p. 364).

Yet, while it is still plausible to think about open objects on a mechanical level, evoking the image of bricoleurs that maintain machines and tinker with objects of all kinds in their workshops, such objects are much harder to come by in the digital age. 'Maintenance' in an etymological sense means to hold things in one's hands, which allows 'manipulation' and sustainment of a certain 'tension' in relation to the present moment. Can these two aspects of maintenance be more than an anachronism in the digital age?[3] On the one hand, it seems we are rarely without a smart device in our

DOI: 10.4324/9781003316213-9

hands, maintaining, establishing and creating relations with technical ensembles – and with other human beings. On the other hand, we are not manipulating the technical infrastructure directly, but rather using it to participate in different forms of publics that can be either 'calculated publics' (Gillespie, 2014, p. 188) (i.e., generated through algorithms) or 'data publics', where data is used for a different purpose than it was initially generated for (Gray et al., 2018).

In Chapter 8, I will attempt to reconstruct the concept of maintenance in Simondon's philosophy of technology and furthermore explore how this conception is relevant to the current technical phenomena of the digital age.

In the first part of this chapter, I will develop Simondon's conception of maintenance in terms of the process of 'concretization' and 'technicity' to address the temporal dimensions of technical objects. Simondon shows that technical objects evolve with the process of concretisation. Technical objects have a temporal condition and tend to become 'concrete' (Simondon, 2017, p. 25ff.), that is, a state where the elements of the technical object are in an intrinsic relation with each other, serve multiple functions and cannot be replaced by any other part. The so-called 'concrete technical objects' increasingly run independently (*they self-maintain*) and create their own milieu that allows their proper functioning. The actualisation of maintenance for the digital age exemplifies the difference between 'maintenance' and self-maintenance (*auto-entretien*): smart devices are 'concretized objects'. Their production relies on an infrastructure and their material elements are intricately linked to each other. Spare parts cannot be fabricated by the user, programming languages need to be learned in order to access multiple layers of the object and engagement with the technical object is initiated through the user interface by mediated bodily movements. 'Maintaining' smart tools thus cannot mean manipulation of each of the elements, but rather providing the conditions for their internal functioning. While specialised skills have always been necessary for maintenance, smart tools are far harder for the user to maintain since the interaction with the technical object occurs on the interface. The internal and external relations with users, other technical objects and networks are what technicity refers to. These relations have to be captured to maintain technical objects. I will thus introduce the conception of maintenance in the philosophy of Simondon as a condition to entertain a relationship with technical objects and to fathom their 'technicity', that is, the concrete technical aspects of an object's functioning and its embeddedness in socio-technical ensembles. Acts of maintenance do not merely refer to acts of construction, but are intrinsically linked to their use, as Mark Thomas Young points out (2022, p. 363f.).

Maintenance relies, however, not only on the skill of the user but also on the structure of the technical object. Both the user and technical object have to be in recursive relations with each other so that technical networks participate in cultural networks and vice versa. To explore this reciprocity, I will, in the second part of this chapter, introduce the analogy between (hi)story-telling and maintenance. This analogy draws on Simondon's pedagogical work and his call for 'poets of techniques' (Simondon, 2014b, p. 416). I will combine these two aspects of Simondon's philosophy to show that maintenance 'tells' the (hi)story of an object and adapts it to the actual situation. Furthermore, the notion of 'story-telling' provides the means to understand Simondon's call for a 'technical mentality' as an attitude towards technical objects and its relevance for the digital age.

In the concluding part, I will elaborate on whether Simondon's notion of maintenance can be actualised in an era where engagement is mediated in the aforementioned data-driven way. This may be a point of departure from which to understand and maintain a relation with the technicity of the networks we are living in and is in line with current efforts to 'reflect over the temporal nature of technologies' (Young, 2022, p. 363).

8.1 The Temporality and Evolution of Technical Objects

In this first part, I will turn to Simondon's description of the evolution of technical objects as a process of concretisation, that is, an evolution from an abstract technical object, where each element can be replaced by another regardless of material constitution providing this element fulfils its function, to a more concrete technical object, where each element is in an intimate, recursive relation with other elements. Each element of the 'concrete technical object' has a specific material structure and fulfils a specific function in relation to the object's other elements (Simondon, 2017, p. 27). This evolution thus concerns the interrelatedness of the technical object's components.

This process has implications for the phenomenon of maintenance. On the one hand, one could consider it an evolution of objects that increasingly maintain themselves (*auto-entretien*) (Simondon 2017, p. 50). Less human intervention is required to guarantee the functioning of these objects, while these enter multiple technical networks and rely on their infrastructure (Simondon 2017, p. 50). On the other hand, maintenance of 'concrete' technical objects requires an understanding of how their elements relate to each other in order to guarantee proper functioning.

The evolution of a technical object establishes new relations in two directions: towards its inner constitution and milieus, as well as with the natural world, the external milieus. These relations are modified over the course of the evolution of technical objects. They are the expressions of a process of organisation; as technical objects *tend* towards the self-organisation of an organism, a network with other natural and technical objects is established via adaption to the external and the internal milieu:

> This object needed a regulative external milieu in the beginning, the laboratory, workshop, or sometimes the factory; it gradually increases its concretization, it becomes capable of doing without the artificial milieu, because its internal coherence increases, its functional systematicity closes as it organizes itself. The concretized object is comparable to the spontaneously produced object; the object frees itself from the originally associated laboratory and dynamically incorporates the laboratory into itself through the play of its functions; what enables the self-maintenance (*auto-entretien*) of the objects conditions of functioning is its relation to other technical and natural objects, and it is this relation that becomes regulative; this object is no longer isolated; it associates itself with other objects, or suffices unto itself, whereas at first it was isolated and heteronomous.
>
> (Simondon, 2017, p. 50)

Simondon describes a process where human beings intervene less and less in the actual functioning of the technical object. This process can certainly be seen – from both the perspective of use and of construction – as progress.

The process of concretisation of the technical object has as its ideal the organisational form of the living being. The understanding of the technical object as a 'unit of coming-into-being' that is 'present at each stage of its becoming' (Simondon, 2017, p. 26) is oriented towards the actual developmental processes of technical objects: technical problem-solving is most effective when new elements do not have to be added from the outside, but new relations are invented within the technical objects, thus reducing the complexity and the elements through 'condensation and concretization' to a 'minimum that is at the same time an optimum' (Simondon, 2009, p. 19). The example of the engine illustrates the concretisation process very clearly:

> One could say that the contemporary engine is a concrete engine, whereas the old engine is an abstract engine. In the old engine each element intervenes at a certain moment in the cycle, and then is expected no longer to act upon the other elements; the pieces of the engine are like people who work together, each in their own turn, but who do not know one another.
>
> (Simondon, 2017, p. 27)[4]

The relation of the elements to each other is transformed in the 'concrete engine' and through this new relation, problems are solved. The function, for instance, of 'protecting engines from overheating to prevent deformation of the pistons inside the engine' led to the problem that as long as elements for cooling were added to the engine from the outside, it necessarily became larger, heavier and less efficient (ibid.). The solution to the problem was to use new materials to integrate cooling fins within the piston design, thus preventing piston deformation and providing cooling (Simondon, 2017, p. 27f.). Another example is the turbine of the GUIMBAL group: there, the interaction of different elements (a very small alternator is placed in a housing wrapped in oil, which in turn is placed precisely in the piping and in the turbine) allows for more effective cooling than would be possible with generators in the air (Simondon, 2009, p. 19).[5]

The concretisation process describes the genesis of the technical object as an individual. That is, as a being that possesses a genuine mode of existence, which consists of striving from an abstract to a concrete state. This process relates the elements of the technical object to each other. Within the technical object, natural elements (water, gas and fire) and technical elements (cylinder pistons, etc.) interact. This recursive causality creates a specific technical milieu that contains the potential for possible solutions to problems, becoming increasingly stabilised and interacting with natural elements. A technical individual therefore has an 'associated milieu' (Simondon, 2017, p. 59) that ensures its functioning.

While this milieu was – and still is – 'regulative' in craftsmanship and relies on the tool-wielding human being, in the post-industrial age the associated milieu consists of technical ensembles and the relationship of the human being with the technical object has been transformed. It is no longer necessary to maintain in the literal sense, that is to manually hold the technical object in its existence, but it is possible to concentrate on different aspects of their mode of existence: the focus can shift to the organisation of relations to other natural and technical objects. The ideal relation of human beings with technical objects would be as a coordinator in the midst of technical objects (Simondon, 2017, p. 17). This effort to be in the midst of technical objects goes against the mode of production in the factory, where one either remains on the level of construction or on the level of the manager, who deals with the integration of the technical object in the networks beyond the factory, that is, on the level of the network (Simondon, 2017, p. 78ff.).

Whilst the artisan may also be a 'coordinator' in providing the 'regulative milieu' for his/her tools, a contemporary user of a concretised object (a *smartphone*, for instance) maintains it by concentrating on a different operative level, since the object 'frees itself' from its original space of production and enters into a network where it can *self-maintain* (*auto-entretien*).

This means it creates its own 'milieu that conditions it, just as it is conditioned by it' (Simondon, 2017, p. 59). It 'mediates the relation between technical, fabricated elements and natural elements, at the heart of which the technical being functions' (ibid.) It is thus this transformation of milieus – both the regulative milieu, that necessitates the interference of the human being, and the associated milieu created with the operations of the technical object – that allows for *self-maintenance* (*auto-entretien*). The concept of *auto-entretien* indicates the different relationships human beings enter into with concrete technical objects. For concrete technical objects, it is no longer necessary to *hold* them literally *in hand* (*main-tenir*) to keep them functioning. Instead, maintenance is directed at the relations that allow for *self-maintenance* (*auto-entretien*). While the object was 'in the beginning isolated and heteronomous' (Simondon 2017, p. 50) and needed the support of the infrastructure provided by the laboratory, the integration of these conditions in its functioning and the association with other natural or technical objects generates its autonomy (ibid.).

Maintenance thus refers to an activity that provides and cares for the requisite conditions for *self-regulation* and *self-maintenance* (*auto-entretien*) of the technical object. It refers both to the internal relations of the technical object as well as the relations with the external milieu.

This has a couple of consequences: maintenance becomes increasingly inherent to technical objects and therefore specialised; technical objects generally seem to tend towards an organisation that allows for specific forms of participation. In order to understand the relationships within the technical object, that is, the conditions of *auto-entretien* and the relations with the world, specialised knowledge and skills are required. Simondon even calls for a 'relation of equality' with technical objects 'that is, a reciprocity of exchanges; a social relations of sorts' (Simondon 2017, 105). Human beings should therefore aspire to understand technical objects and their genesis.

I argue that maintenance plays a crucial role in this task. Maintenance relates the technical object spatially and temporally within the network of production and use. It implies a specific understanding of the 'technicity' of technical objects. Here, technicity does not simply refer to the inner workings or construction of a technical object. Rather, technicity refers to Simondon's theory of individuation. 'Technicity is one of the two fundamental phases of the mode of existence of the whole constituted by man and the world' (Simondon 2017, 173).[6] It refers not only to the technical constitution of the object but also shows that techniques and technical objects are essential to the being-in-the-world of human beings and the way in which they structure the world. Technicity refers therefore to the relations within the technical object as well as to the relations of the technical object in the wider world. In technical objects, it is the 'degree of the object's concretization'. It expresses the concrete relation of form and matter (Simondon, 2017, p. 72). Furthermore, technicity encompasses both theoretical knowledge and the practices performed with technical objects (Simondon, 2017, p. 171). As such, technicity is inscribed in technical objects. Tools and instruments can be maintained through time and even be reinvented, since they contain crystallised human gestures (Simondon, 2005a, p. 340).

The technicity of technical objects allows for a continuous relationship of technical objects with their socio-technical networks and is rooted in Simondon's philosophy of individuation. Technical objects – as stated above – evolve and are conceived in analogy to organisms, that is, living beings. Living beings are, however, *open* systems and entertain a metastable relation with their milieus. This metastable relation allows for adaptation and transformation. It is – in terms of Simondon's philosophy – a condition of life to remain metastable. As long as an organism is metastable, it can transform itself and has not yet become stable; as such, it can create new 'stable forms' (Simondon, 2005b, p. 161) without being prone to being reduced to *one* stable form. Living beings are able to restructure themselves, they dispose of the potential of plasticity to adapt to new situations and even to restructure their bodily organisation (Simondon, 2005b, p. 161).

Even though a technical object cannot achieve the same level of self-organisation as a living being, its evolution strives towards this goal (Simondon, 2017, p. 50f.).[7] Simondon warns, however, that technical objects should not be identified with living beings. The aforementioned self-organisation of the living being cannot – according to Simondon – be achieved by technical objects due to their status as constructed beings: 'living beings [,] are concrete to begin with'.[8] However, as an expression *of* life (Simondon, 2014c, p. 321), technical objects are endowed with vitality through their relation with human beings. According to Simondon, it is *maintenance* that furnishes the technical object with a life in temporal and spatial networks:

> Similarly, the degradation of the technical occurs when the object is isolated in time (by the break that is the end of manufacture and the fall into the condition of venality) and in space (by the detachment that isolates the manufactured object from the conditions in which it could receive a perpetual regeneration maintaining it at the level of its full functional meaning/signification). Technicity degrades as it becomes objectified because the object, as a closed object, becomes archaic and degrades when it is no longer maintained in the network of technicity through which it was constituted. Technicity is a mode of being that can only exist fully and permanently in a network, both temporally and

> spatially. Temporal reticulation is made up of the object's revivals, in which it is updated, renewed, in the very conditions of its original manufacture. Spatial reticulation consists in the fact that technicity cannot be contained in a single object; an object is technical only if it operates in relation to other objects, in a network in which it takes on the significance of a key point; in itself and as an object, it possesses only virtual characteristics of technicity that are actualized in the active relationship to the system as a whole. Technicity is a characteristic of the functional set that covers the world and in which the object takes on a signification, plays a role with other objects.[9]

Technicity is, according to Simondon, the central aspect of the technical object that requires maintenance. Maintaining an object allows its technicity to enter into a network. Simondon points out that the production of objects merely to sell them creates closed objects. They become *things* (*une chose*) as soon as they leave the factory since they are only ever new and actual at this precise moment; after which they deteriorate and age, having lost contact with the actual reality (Simondon 2014b, p. 401). Maintenance thus calls for the 'opening' of objects in a literal and metaphorical sense: literally it can mean attempting to fix and maintain a technical object by opening it up and replacing components; metaphorically, it refers to a possible relationship with the technical object that goes beyond mere use and attempts to understand its operations and mode of being. Maintenance can thus restore the technicity of technical objects by reintegrating them into a network. This reintegration is aimed at the temporal condition of their genesis and the spatial condition that refers one object to others. Human acts of maintenance thus actualise the virtual technicity within a technical object. I, with Simondon, argue that the human technical interactions with technical objects while maintaining them, (re-)create signification by relating technical objects to each other in a technical ensemble:

> Man understands machines; for there to be a true technical ensemble man has to play a functional role between machines rather than above them. It is man who discovers significations: signification is the meaning [*sens*] that an event takes on with respect to already existing forms; signification is what makes an event have value as information.
> (Simondon, 2017, p. 150)

Maintenance can provide signification to the role of technical objects within an ensemble. Understanding technicity, that is, the spatio-temporal conditions of human beings, technical objects and technical ensembles in the world, creates information and meaning (*sens*).[10] Maintaining any tool, for instance, refers back to its past and relates it to future use. It is an attempt – as rudimentary as this might be – to understand the technicity and history of a particular object and to *continue* it.

The production of modern technical objects requires division of labour, standardisation and, to a certain degree, automation.[11] The changed conditions of production also bring a new way of relation to the produced objects – and thus also the changed form of maintenance. While the artisan holds the entire production process – so to speak – in his hands and both works on and understands the individual elements of the objects produced as well as ensuring that the object fits into the technical ensemble, in the factory these two aspects diverge. The manager and the worker of a factory each address only one level of the technical – either the 'ensemble', that is, the level of the technical object is integrated into the larger fabric of society and enters a network outside of the factory as a new whole, or on the level of the element, where workers assemble the technical object without having the perspective of the owner (Simondon, 2017, p. 80).

Both would tell very different stories about their relation to the technical object, since each of them has only a limited perspective which does not grant an understanding of the entire technicity of the technical object, but rather mere aspects of it. For Simondon, the ideal would be the engineer, who has simultaneously an understanding of the element and the ensemble.[12] Simondon was however well aware that already in his time, engineers had been reduced to a mere service function keeping the status quo of the functioning alive (Simondon, 2005a).

According to Simondon, concretisation also leads to different forms of maintenance: the maintenance of a tool differs from a concretised technical object. While honing a blade, for instance, acts directly on the material of the tool and requires body techniques that are in direct relation to the tool, maintaining a concretised technical object, a smartphone, for example, might only involve an update in order to keep the self-maintenance (*auto-entretien*) of the phone intact. This even holds when the phone is opened and the battery is replaced. However, all three activities, the honing of the blade, the software update and the replacement of a battery can all be points of departure for a deeper engagement with technical objects. They all share the temporal aspect of maintenance, that is, they provide a link between the past of the object, the present and the future. Information is inscribed in technical objects that can always be actualised and signifies a potentiality that remains present through time:

> The technical being realizes the summation of an availability that remains always present; the effort spread out in time, instead of dissipating, discursively constructs a coherent being that expresses the action or the series of actions that have constituted it, and keeps them always present: The technical being mediates human effort [...] The technical being

> is participatory; as its nature lies not only in its actuality, but also in the information it fixes and constitutes, it can be reproduced without losing this information; [...] it is open to any human gesture in order to use or recreate it, and is inserted in a universal elan of communication.[13]

It is precisely this 'universal elan of communication' that allows us to call the relation with technical objects 'hermeneutic'.[14] I understand hermeneutic here not in a methodological or historical sense, but merely as the possibility to discover and create and express signification with technical objects and ensembles. It can, but need not, use language in order to do so. This provides the basis on which to develop a conception of maintenance as '(hi)story-telling'. Human beings are in multiple relations with technical objects and must address the technicity of technical objects when maintaining them. That is, acts of maintenance require an understanding of the history of an object – what elements require maintenance, what are the conditions of *auto-entretien* and when are they met – as a prerequisite to keeping the objects up to date. The temporal aspect of maintenance allows for the revival and the constant actualisation of technical objects. In this process, technical objects do not remain the same but are transformed due to their temporality and also due to the effort of spatial reticulation, where the temporal aspect is put to the test, that is, the virtual characteristics of the technical object are actualised in a network.

This implies that the history of the technical object receives new signification by means of the active relation with the network.[15] Maintenance is thus the active continuation of the life of a technical object. Maintenance holds, as the German translator of Simondon's 'On Technical Mentality', Michael Cuntz puts it, technical objects in the present (Simondon, 2011). One could thus say that maintaining a technical object always begs the question: 'What kind of future does this object have?' Consequently, *how* an object is maintained cannot merely refer to the object itself, but must include a conceived virtual future. Actualising technicity in the present network may thus require adaptation or invention, making it necessary to:

> [...] know the language through which the human gesture that created it [the technical object] re-actualizes itself. [...] The technical being must be considered as an open, polarized being that seeks its complement, which is the working human, in the coincidence of the newly assembled whole. The user must take the place of the constructor. For this it is necessary that s/he falls at one with the essential schematism inscribed in the technical being, that s/he is able to think it, to understand it, to love it as if s/he had made it him/herself. The duality of human-nature is resolved in the functional unity of the working human. [16]
>
> (Simondon 2014h, 252–253)

Even if objects are built to sell in the modern capitalist mode of production, technical objects still contain technicity that can be accessed.[17] Yet, an ideal technical object would be in constant synchrony with the presence, which would facilitate the user's participation in the genesis of the technical object:

> On the contrary, if the object is open, i.e. if the user's gesture, on the one hand, can be an intelligent, well-adapted gesture, knowing the internal structures, and if, on the other hand, the repairer - who, moreover, can be the user - if the repairer can perpetually maintain the parts that wear out new, then there is no date, there is no aging. On a base which is a base of durability or at least of great solidity, one can install parts which will have to be replaced, but which, in any case, leave the fundamental scheme intact and which even allow it to be improved; because one can well think that at some point, if a better cutting tool is found for a machine intended for work involving cutting, this tool can be fitted, provided that it has the necessary standards, on the base, and thus the machine will progress with the development of techniques. This is what I call the open object.[18]

Even if the open object remains an ideal that might never or only seldom be produced due to economic interests, it reveals the central aspect of maintenance necessary for a harmonious relationship with technical objects and ensembles. On the level of use, the user's gestures already 'get to know' the internal structure of the object. Openness also refers to the external parts of the object, which can be improved and replaced if more efficient parts become available. The actualisation of an open object thus moves from the internal 'fundamental scheme' of functioning to the parts that can be replaced and connect the object with the contemporary ensembles of technical production (e.g., faster microchips, batteries, etc.). The fundamental scheme and the contemporary networks of production are thus in a recursive relationship with each other mediated by acts of repair and maintenance that keep the object up to date. This process creates signification since the object is positioned in the network and can enter relationships with other technical objects and ensembles.

In the foregoing, we observed that maintenance actualises the virtual technicity of a technical object. Technicity refers to the spatio-temporal integration in the technical ensemble and the practices of use. Maintaining technicity refers simultaneously to both the internal structure of an object as well as to its relation to other objects when functioning. Acts of maintenance create signification by relating the past of a technical object to its future operations, thus actualising its technicity. In the following, I will explore the notions of 'universal communicative elan', 'signification' and 'understanding of technical objects' by introducing the analogy of 'story-telling'. The communicative plan implied in the production, use and maintenance of technical objects allows users to participate in the process of becoming of the technical object. I will in Section 8.2 develop (hi)story-telling as a mode of participation with and maintenance of technical objects.

8.2 Simondon's Philosophy of Maintenance: Maintenance, Education and (Hi)story-Telling

I interpret Simondon's claim of needing to 'know the language' of the human gestures used in the creation and use of the object as well as the aim of the 'user becoming the constructor' in terms of story-telling. Even though Simondon himself only explicitly mentions the 'lack of technical poets' in an interview with the Canadian mechanologist Jean Le Moyne (Simondon, 2014b, p. 422), this does imply a reference to maintenance, since the poetic effort restores an original unity – the technicity – of outdated objects such as steam locomotives.[19] This 'poetry of techniques' recreates the history of an object and links it to the present and a possible future. It is a specific form of *poiesis* that gives meaning to an object and can be understood as a form of story-telling.

Like a great story, a technical object can remain perpetually actual and in contact with the presence; a potentially timeless connection between the story's 'audience', in this case, the technical object's users, the storyteller, the repairer and the story itself – the technical object, may be established and maintained. Even though the central message remains the same, the expectations of the audience, their social status, their listening skills and their knowledge of stories are all different and even the language they use is transformed over time. Some parts of the story may no longer fit and have been replaced and other parts may be completely omitted. Yet, in order to do so, the storyteller has not only to know the story itself but also its audience and the network in which the story is embedded. Both the story and its network are furthermore integrated into a broader network. This feature of the network does not only apply to external integration but must also be an internal aspect of the ideal open object. As stated above, it applies to both the story and the object.

In order to be actualised and maintained in a state of actuality, an object needs to be both: stable and transformable. The 'open object' consists thus of 'parts designed to be as close as possible to indestructibility' and of 'others ... in which the finesse of adaptation to each use, or wear and tear, or possible breakage in the event of a shock or malfunction, is concentrated'.[20] This constitution of the technical object signifies that it can 'allow the development of the technical mentality and [...] be chosen by it' (ibid).

To actualise and maintain a technical object it is necessary to grasp its technicity. To perform this task in its entirety, a form of abstract knowledge – that refers to the internal operations and understands the schemata of functioning – as well as the requisite skillsets to perform the tasks of maintenance are necessary. These two forms of knowledge go back to the distinction Simondon makes in *On the Mode of Existence of Technical Objects* between craftsmanship and engineering. Each refers to a specific mode of transferring knowledge. Craftsmanship is linked to the technical training of a 'master', while the training of an engineer is based on a transfer of knowledge via technical schemata that are laid out in an *Encyclopedia* (Simondon's example is Diderot and D'Alembert's *Encyclopedia*). On the one hand, craftsmen possess an almost mystical knowledge and intuition of the matter they are dealing with and act almost like magicians, whose knowledge remains a secret ability that cannot really be shared with others (Simondon, 2017, p. 106). On the other hand, a technical object in its proper sense requires a sort of abstract, generalised and rational knowledge, which can be represented in schemata and relies upon the formula and scientific knowledge, which belongs to the 'encyclopaedic order of knowledge' (Simondon, 2017, p. 123).

Even though Simondon describes these forms of knowledge in their extreme and claims that the teaching of the form of craftsmanship is aimed at the child, while the encyclopedism is aimed at the adult, the distinction between the skill taught by means of a body technique (Mauss, 2006) and a knowledge transmitted in an abstract form is still recognised today. However, even though craftsmanship and the encyclopaedic spirit may seem inverse to each other, both rely upon a quasi 'magical form of initiation' (Simondon, 2017, p. 112).

This ritualistic initiation allows us to develop the analogy of (hi)storytelling, as introduced earlier. According to each form of knowledge, a different story may be told: The craftsman's story is one of a magical initiation, an intimate relationship with matter and techniques that is passed down by a master to the apprentice; Encyclopaedic knowledge, however, may be seen as a story with a magical book at its core, which provides the key to a general knowledge bringing a variety of topics, matters and technical objects in relation to each other (Simondon, 2017, p. 112).[21]

The encyclopaedia generalises the initiation into technical knowledge – and as such it becomes a danger to the institutionalised craftsmanship in guilds, since if the encyclopaedia is clear and well understood, the initiation performed by the master craftsman becomes redundant (Simondon, 2017, p. 111). The difference between encyclopaedism and craftsmanship shows itself in the ways in which information and energy are distributed. In craftsmanship, energy and information are bound to the body of the craftsman. S/he is thus in direct contact with matter by means of the technical operation (Simondon, 2014f, p. 303). A machine, on the contrary, is not dependent on the human being in the same way. Energy can be provided by natural resources (oil, coal and wind) or by a technical ensemble, that is, an infrastructure. Once a machine is switched on, it performs its operations by itself, dependent on the human being only for information (Simondon, 2009, p. 20). The various scientific principles – the encyclopaedic schemata – are crystallised in the machine and thus inform the technical object. This process of informing cannot be reduced to the utilisation of the object. It is concerned with understanding the operation, providing it with direction and signification, which consists of maintaining its 'associated milieu', that is, its conditions of *auto-entretien* and its inner workings, and may also mean to transfer it into different milieus of use.

Comprehensive technological knowledge, however, requires both skills on the concrete performative level and an understanding of the schemata that allow for technical objects.

Technical objects crystallise a complex practice that emerges in its operation. The operation relates different structures to each other and leads to 'circular knowledge', that is, to a practical theory in its literal sense, which informs the theory of the practical synergy of heterogenous elements.[22] This line of argument shows not only the complexity of Simondon's philosophy but also a general point on the relation to technical objects in contemporary times. We are faced with objects that are extremely complex, which cannot be fully understood and maintained in their technicity by ourselves as users (at least not without some very specialised knowledge, which is simultaneously practical and theoretical, and not without access to very specific infrastructure, software and hardware components).

To elucidate a possible actualisation of maintenance, I would like to return to the idea of (hi)story-telling. Maintenance is conceived of as an ideal relation to technical objects, the technical ensembles and the infrastructure they rely on. Though it is difficult – near to impossible – to teach oneself the skills needed to maintain a modern digital object, it is certainly possible to enter into hermeneutic relations with technical objects. The encyclopaedic spirit and the spirit of craftsmanship serve as ideals for a technical education. For Simondon, the seeds of a technical culture, its corresponding education and the images required to create stories are planted in childhood. In childhood, an effort to understand technical operations takes the form of mimicry. This mimicry provides the images needed for storytelling: when children 'become' the airplane or the automobile and imitate the technical object at work, they acquire a glimpse of understanding of the technical object at an operational level. Obviously, this is not the encyclopaedic understanding of an engineer, but it provides a foundation for the creation of mediating images and bodily schema that can aid technological understanding in later life. Children grasp a technical scheme through play, which can later be conceptualised and objectified (Simondon, 2014g, p. 44). Initially, technical inventions are 'schemata of behaviour, of operation' (ibid.). Operations and images that served childhood play can be rediscovered and reconstructed to understand technical objects. These images and bodily schemata can be used to create stories that relate the human being and the technical object with each other at an operational level.

This relation – according to Simondon – exists not only in children but also on a higher level of understanding in other social groups that have an intimate relation with technical objects (Simondon, 2014g, p. 48ff.). Simondon uses the examples of the farmer, the housewife and the mariner to show that social images and prejudices are connected to social groups, the technical objects they utilise and the operations performed by the groups with these technical objects (ibid.). Such social images are used to implement stories about individuals, their social status in these groups and their use of technical objects. These stories may perpetuate stereotypes, which have to be countered by different accounts of how to relate and act in technical ensembles. Simondon shows that the woman in the kitchen, for instance, operates as a coordinator in the midst of technical objects (Simondon, 2014g, p. 47). She has skills and technical knowledge that go beyond the social image perpetuated by advertisements in the 60s. A contemporary example might be the hacker, who – even if s/he does not 'live' in the data – is by means of her/his programming skills a coordinator, embedded in the technical objects and ensembles. S/he can hack into systems and manipulate technical ensembles and infrastructures.

Going from these social images to the relation of the respective social group with the technical objects, a different picture can be drawn: the toys of children, the kitchen appliances of housewives and househusbands, the farmer's tractor and the mariner's vessel all reveal a participatory, non-dualistic relation with technical objects. The child *becomes* the airplane in play, the cook operates in the kitchen amidst the household appliances and orchestrates their interaction and the tractor is plurifunctionally used by the farmer to perform a wide variety of tasks (Simondon, 2014g, p. 49f.). The dualisms between body and mind, body and imagination, form and matter and human and machine are suspended in the act of playing, cooking, working with the tractor or navigating the vessel. The socio-cultural meaning of the technical object in these relations is still present because the child, farmer and cook correspond to this image precisely through their activity, but the relation itself consists of operations that produce norms that can in turn become part of a unified, technical culture.

This technical culture, where technical objects are treated as cultural objects, on par with works of art, etc., becomes possible, if one establishes a 'participatory observation' of technical objects beginning in childhood;

operations and images that one has played out in childhood can be rediscovered and reconstructed in order to create meanings which free the technical object from the norms imposed on it from the outside. The child observes the object and attempts to participate in its functioning (Simondon, 2014g, p. 43f.). The observed and played out schema gives access to the history of the technical object, to maintain it in the present and to actualise it in the later stages of life (Simondon, 2014g, p. 44). However, this does not mean that one can and should return to the supposedly innocent or natural state of the child. After all, understanding technical objects does not only consist of physically imitating operational schemata and empathising with them. Especially, when one has digital technologies in mind, this notion is downright absurd. While bodily techniques remain the basis of all techniques – including digital ones – since there is no strict, dualistic separation of the mental and the physical, and all bodily techniques are also imbued with mental operations, the understanding of technology points beyond itself and implies an understanding of matter that evokes theory and science.

The schemata of childhood are far from being sufficient to develop a 'technical culture'. They merely reveal a very basic notion of a possible relationship with technical objects that must then be fostered by proper technical education. Simondon developed such an idea of technical education in his work as a teacher. In two articles based on his own experiences, he explains how this education would be both adequate for the respective age of children and to enable new relationships with technical objects (Simondon, 2014d, 2014h). It further exemplifies how maintenance is understood not only as maintaining technical objects but also as maintaining (and regaining) a relationship to and with the natural world. Technical objects serve as media since the technical ensemble is the 'most stable and universal mix of the natural world and the human world'.[23] The goal of technical education is, according to Simondon, to create an intuitive understanding of the technical being:

> Our aim was the intuitive grasp of the technical being by the young intelligence. A child does not understand, in the deepest sense of the word, what a tree or an animal is. S/he can, however, understand, in the technical sense of the word, why a newly planted tree must be watered, why a tree needs light; this is because the child achieves an intuitive grasp of the organisation of the tree; s/he does not understand scientifically what assimilation and photosynthesis are, but s/he can *understand* what grafting or layering is. It is this kind of understanding, intuitive but not affective or animistic, which we call technical understanding. Between the primitive mentality and the scientific mentality, there is technical thinking.[24]

Here we can see precisely defined both the technical ensemble as a stable entity and technical thinking, which allows the development of maintenance skills through a specific kind of story-telling. Technical ensembles unite the natural and the human world, they are key to understanding human operations in the arena between the 'primitive' and the 'scientific mentality'. Maintenance in Simondonian terms refers to the temporal relationship with the concrete technical object, the natural world and the social world, co-produced with technical objects. Technical knowledge is therefore embedded within culture (Simondon, 2014d, p. 217). This is apparent from the position Simondon assigns to technical thinking in contrast to the scientific and primitive mentality.[25] Technical thinking may begin in childhood since it is rooted in socio-technical operations.

It is thus unsurprising that, as a teacher, Simondon considered it his task to contribute to the creation of a technical culture. In the classroom, he addressed the threefold dimension of nature, techniques as medium and culture and introduced his students to technical objects. The objects that were studied did not merely serve the function of being opened, repaired and tinkered with, but also to create a new, symmetrical social order in the classroom: the students worked in different groups and they had to work together in order to appreciate and understand the technical objects. Working in small groups provided the students with a sense of autonomy, initiative, invention and a sense of effort and solidarity (Simondon, 2014d, p. 214). Furthermore, Simondon did not understand himself as 'contremaître', but as a teacher, who monitors the progress of his students (ibid.).

The analogy of story-telling exemplifies how Simondon introduced the technical ensemble to his students. As stated above, Simondon played an analogous role to a storyteller: the goal was not *explaining* the scientific nature of the technical objects nor merely imitating the master's movements but rather to gain an *understanding* of the gist of the story; for the students should find a first intuition beyond an imitation of a technical object, but already pointing to its technicity and operations. Drawing the analogy of the storyteller further we could say that Simondon's story to his students addressed how to become a 'technician', which means to 'prefer the knowledge of a human operation to the knowledge of an objective structure', that is, not to see the technical object as something that requires explanation as a structure, but as something that operates and is the fruit of human hands and invention.[26]

Technical activity is always related to human activity that mediates between nature, society and the individual. It thus goes beyond 'mere work' and must be distinguished 'from alienating work' (Simondon, 2017, p. 255). It cannot be reduced to the use of the machine (ibid.), but rather consists of 'maintenance, adjustment, and improvement of the machine, which continues the activity of invention and construction' (ibid.). This in turn shows that maintenance is a human operation that does not merely operate on the technical object as an abstract and objective structure but relates to it as an entity at an intimate, spatio-temporal level.

This intimacy, this love for technical objects, is what a technician should aspire to. It is this message that Simondon conveyed to his students (Simondon, 2014h, p. 252). This relationship with the technical ensemble is exemplified in the telling image of complementary beings that fulfil a task together:

> The machine demands services from us and renders them to us, like a friend; the exchange of services, preferable to slavery, is not yet even the highest and most adequate relationship to the machine. It is necessary to 'pull the yoke' with her, to know her well, to work taking her neither as an end nor as a means, but as a fellow worker and as a complementary being [...]. This horizontal relationship must replace any vertical relationship.[27]

The story required for developing technical knowledge – which is in turn necessary to maintain technical objects and their relationship with the natural world – displays, at its core, a horizontal relation between natural, technical and human entities. The aim of a technical culture and consequently also of maintenance is to know and appreciate technical ensembles as such. They should not even enter into the logic of means and ends, but rather be ends in themselves that require care, since their well-being is necessary to fulfil a task together. Human beings and machines should work as an operative couple, each complementing the function of the other, instead of reducing their relationship to economic or energetic values (Simondon, 2017, p. 135).

The openness and opening of objects corresponds to the openness of the learning practice: the apprentice of a craftsman imitates operations and the object is presented merely as an occasion to fulfil standardised tasks (Simondon, 2014d, p. 214). Instead of a closed object, Simondon chose technical objects that were open to understanding (ibid.). The first experience with the object reveals a strangeness and complexity to the students (ibid.). This experience can be conceived of as an experience of a technical other that has an irreducible quality and is the point of departure for the horizontal, harmonious relationship with technical objects (Schick, 2021c). Maintaining this relationship means, as already mentioned above, becoming the constructor of technical objects instead of merely using them (Simondon, 2014h, p. 252). The image of this relationship goes even further. It does not suffice just to understand the technical being, but rather it is necessary to:

> coincide with the essential schematism inscribed in the technical being, s/he must be able to think it, understand it, love it as if s/he had made it.[28]

This coincidence with and love for technical objects does not mean that machines should be idolised, but it should prevent contempt and fear of technology and furnish the students with a 'healthy knowledge based on a careful attentiveness' for technical objects (Simondon, 2014h, p. 253). This becomes ever more relevant since the attitude towards an ever more rapidly increasing number of new technical developments oscillates between fear and idolatry. Especially, the robot or any automat – exemplified in today's digital realm by attitudes to artificial intelligence – is furnished by the imagination of its users with powers and abilities that have little to do with the actual technical object (Simondon, 2017, p. 16). To work against this common fear, Simondon introduced an automat to his students. It becomes quite clear that the basic notion of the automat still holds for contemporary computers and smart tools. Simondon defines any automat as follows:

> Every automaton, whether electronic or not, consists of four parts: a system of receptors, similar to sense organs; one or more amplifier or selector relays (the 'brain'); one or more effectors, similar to muscles; and finally, a source of energy (power supply).[29]

The main four 'organs' (sensory organs, brain, alimentary system and musculoskeletal system) correspond to the sensors recording movement and input of information (sensory organs), the selection of information on the motherboard (brain), the transformation of signals into movement and information as visual or audible output (muscles, movement) and the battery or power cable which powers the device.

To conceive the automaton in analogy to a living being with different 'organs' is a first step towards understanding the (hi)story of the technical object. Technical (hi)story-telling is, however, not merely about telling stories about technical objects, but rather about engaging with technical objects and seeing how signification can be created *with* technical objects. Maintenance can thus be conceived of as generating a (hi)story with the technical object, that is, to take the social relationship with the technical object seriously; while technical objects have a specific and genuine mode of existence, they are the result of human invention and a store of information and as such can always be actualised by them.

The (hi)story-telling relates human operations with the operations within technical objects. The operations that take place in the machine can thus be understood as operations that human beings also perform in a similar, but different way. Even if the machine itself is not opened, a relation to the machine becomes possible that understands it as something beyond its utility. However, this should not lead to the anthropomorphisation of the machine; its otherness must at all costs be preserved in order to realise the desired horizontal, harmonious relationship: since the love and friendship for technology takes the ontogenesis, that is, the genesis and evolution of the technical object as an individual into account. However, most technical objects are not constructed to be maintained by the user, which leads to alienation from the object's technicity (Simondon, 2017, p. 255f.).

We are faced with the question of how a concept of maintenance can be actualised in a world where almost any technical object is built merely to be sold and to be maintained, or more likely scrapped and replaced, by the constructor. The concept of maintenance, as I have introduced it in this chapter, can serve as a critical concept to tell different or new stories with technical objects. Maintenance as story-telling entails entering into a dialogue with technical objects and searching for their responsiveness. It implies an understanding of the history of the technical objects and the development of a perspective for their future use. A starting point is the re-evaluation of the relations with objects; understanding them as something containing human gestures and simultaneously irreducible to a specific human mode of operation, as they have their own mode of existence. Technical objects that correspond to the ideal of maintenance may yet become possible:

> they are conceived so that the different organs constituting them can be continually replaced and repaired over the course of use: maintenance is not separate from construction, it continues it, and in certain cases, completes it, for instance by means of breaking it in [rodage], which is the prolongation and completion of construction through rectification of the surface conditions during their functioning.
>
> (Simondon, 2017, p. 255f.)

The concept of maintenance itself is part of a story worth telling. A story that goes from concrete operations with technical objects in craftsmanship to the maintenance of the conditions of the *auto-entretien* of complex technical objects in the age of information. In this age, human gestures still organise the relationship with technical objects, but the 'crystallized human gestures' of the invention within the technical objects must be rediscovered. It points to an ethical relation with and in nature that is mediated by the technical ensemble: it also tells the story of a reciprocal relationship where maintenance of a technical object also entails the human being doing the maintenance keeps a relation to themselves as well as to the natural and social world.

8.3 Maintenance in the Networked Era?

How can maintenance allow for an intimate, participatory relationship with technical objects in the digital age? First, we must acknowledge that the task is too much to handle for an individual to tackle alone. Technical objects are constructed and sold as closed objects, users in the west are deskilled and the maintenance infrastructure has been outsourced to the global south (Graham and Thrift, 2007, p. 18f.). The genesis of a technical culture and an ideal of maintenance that keeps technical objects in the present and maintains a harmonious relationship with technology and nature presupposes a transformation of the attitude towards technical objects. It was the goal of this chapter to develop some aspects of how maintenance is integral to this transformation and how this different attitude can be developed. The different aspects of maintenance as (1) holding a technical object in the present, that is, to keep it up to date, (2) understanding its history, (3) being able to actualise its history in new and different situations as well as (4) maintaining a relationship of human beings with technical objects which is open and sees them as social beings, is crystallised in the notion of story-telling.

As Elizabeth Spelman points out, repair and maintenance is what keeps our social order together (2002, p. 1ff.) and is essential to human beings at all levels of their existence: the homo reparans repairs relationships, infrastructures and technical objects and restores natural habitats (ibid.). To understand these relations implies an understanding of not only the significance of the relations themselves but also of the mode of existence of the objects and entities we have relations with. Maintenance as (hi)story-telling provides a point of departure to relate to technical objects and technical ensembles as social beings. It probes into the object's past and relates it to the future. The temporal aspect of maintenance is intrinsically linked to the use of the object since it also implies the future perspective of what kind of relations with the object will be possible. A smartphone relates the user via the interface to its material structure, its internal, associated milieu that allows its functioning. The user maintains this internal functioning via updates which relate her to the technical ensemble of production. Using the smartphone, s/he interacts with other human and non-human entities (e.g., social platforms, satellites, etc.) spanning from other users close by to satellites in the orbit of Earth. The maintenance of a smartphone thus refers to the history of the infrastructure of production and use, the social construction of meaning with the smart tool as well as to the operations with the interface and the operations of functioning in the phone. The maintenance of a smartphone must thus take into account these aspects to develop a future relationship *with* the phone. Story-telling raises the questions of how to participate and cooperate with nature and technical networks The 'homo reparans' (Spelman, 2002) has thus in Simondon's case to be termed homo coordinans (Schick, 2021a), being in the midst of technical objects, able to maintain and provide the relations with nature, society and the technical ensemble with signification.

Notes

1 I would like to thank the anonymous reviewers, whose valuable comments pushed me to clarify my argument and who pointed me to the important notion of 'auto-entretien'. The publication of the chapter was supported by the German Research Foundation (DFG) – Project Number 262513311 – CRC 1187 Media of Cooperation, University Siegen.

2 Living beings can transform their internal structure – if necessary – while technical objects have to be maintained. Living beings are from the very beginning of their existence endowed with the aptitude to reorganise themselves. Artificiality, however, does not merely refer to the 'fabricated origin of objects' but to the engagement of human beings with artificial and natural objects that have been 'domesticated':

> As it evolves, this object loses its artificial character: the essential artificiality of an object resides in the fact that man must intervene to maintain the existence of this object by protecting it against the natural world, giving it a status of existence that stands apart. Artificiality is not a characteristic denoting the fabricated origin of the object in opposition to spontaneous production in nature: artificiality is that which is internal to man's artificializing action, whether this action intervenes on a natural object or on an entirely fabricated one; a flower, grown in a greenhouse, which yields only petals (a double flower) without being able to engender fruit, is the flower of an artificialized plant: man diverted the functions of this plant from their coherent fulfillment, to such an extent that it can no longer reproduce except through procedures such as grafting, requiring human intervention.
>
> (Simondon 2017, p. 49)

3 For an application of Simondon's philosophy to the digital, see, for example, Hui (2016); Rantala (2019).

4 This is also one of the few passages, where Simondon refers implicitly to his work as a teacher:

> Moreover, this is precisely how the functioning of thermal engines is explained to students in the classroom, each piece being isolated from the others like the lines that represent it on the blackboard in geometric space, *partes extra partes*. The old engine is a logical assemblage of elements defined by their complete and unique function. Each element can accomplish its own function best if it is, like a perfectly completed instrument, oriented entirely to accomplishing this function.
>
> (Simondon, 2017, p. 27)

5 The process of concretisation can be also applied to the digital phenomena. The development of AlphaGo to AlphaGo Zero evolves from a design where two different neural networks are each dedicated to a specific task (a 'policy network move that outputs move probabilities' and 'a value network that outputs a position evaluation' (Silver et al., 2017, p. 354)) to a design where both

tasks are integrated in a single neural network (Silver et al., 2017). An internal solution to the problem was found and improved the capabilities of the algorithm. Another example in biotechnology is the development of CAR-T cells (Schick, 2021b).

6 The other phase being 'religion'. Simondon's conception of 'technicity as a phase' is itself embedded in a complex 'phenomenological capture of the prehistorical process of hominization' (Bardin, 2015, p. 167). 'Phase' and 'phase-shift' are notions that stem from physics and signify the transformations within a system, where phases refer to each other (ibid., 173). It is not necessary for my argument to develop these notions further. For a more detailed account, see the article 'Technicity' in Jean-Yves Chateau's 'Vocabulaire de Simondon' (2008, pp. 103–7).

7 For a comprehensive discussion of the distinction between living and non-living beings, see DiFrisco (2014).

8 'What risks making the work of cybernetics partially inefficient as an inter-scientific study (which nevertheless is the objective Norbert Wiener attributes to his research) is the initial postulate concerning the identity between living beings and self-regulating technical objects. Yet the only thing we can say is that technical objects tend towards concretization, whereas natural objects, such as living beings, are concrete to begin with. One mustn't confuse the tendency towards concretization with the status of entirely concrete existence' (Simondon, 2017, p. 51).

9 All quotes are translated by the author if an original French text is referenced.

> De même, la dégradation du technique se produit quand l'objet est isolé dans le temps (par cette cassure qu'est la fin de fabrication et la chute dans la condition de vénalité) et dans l'espace (par ce détachement qui isole l'objet fabriqué des conditions en lesquelles il pourrait recevoir une perpétuelle régénération le maintenant au niveau de sa pleine signification fonctionnelle). La technicité se dégrade en s'objectivant parce que l'objet, en tant que fermé, s'archaïse et se dégrade, lorsqu'il n'est plus maintenu dans le réseau de technicité par lequel il a été constitué. La technicité est un mode d'être ne pouvant exister pleinement et de façon permanente qu'en réseau, aussi bien de façon temporelle que de façon spatiale. La réticulation temporelle est faite de reprises de l'objet en lesquelles il est réactualisé, rénové, renouvelé dans les conditions mêmes de sa fabrication première. La réticulation spatiale consiste en ce fait que la technicité ne peut être contenue en un seul objet; un objet n'est technique que s'il opère en relation avec d'autres objets, dans un réseau où il prend la signification d'un point-clef; en lui-même et comme objet, il ne possède que des caractères virtuels de technicité qui s'actualisent dans le rapport actif à l'ensemble du système. La technicité est une caractéristique de l'ensemble fonctionnel qui couvre le monde et dans lequel l'objet prend une signification, joue un rôle avec d'autres objets.
>
> (Simondon, 2014b, p. 82)

10 The temporal dimension of understanding technical objects even goes further in invention. According to Simondon, invention requires the imagination of a problem being solved. This implies that the future condition – where the problem is already solved – has effects on the present situation – where the problem is still to be solved:

> Only a thought that is capable of foresight and creative imagination can accomplish such a reverse conditioning in time: the elements that will materially constitute the technical object and which are separate from each other, without an associated milieu prior to the constitution of the technical object, must be organized in relation to each other according to the circular causality that will exist once the object will have been constituted; thus what is at stake here is a conditioning of the present by the future, by that which is not yet.
>
> (Simondon, 2017, p. 60)

11 Simondon claims that the process of concretisation made the 'production line' (Simondon, 2017, p. 29) necessary, since the elements of the technical objects have to fit precisely and to correspond with each other. There exists a recursive relation between the mode of production and the intrinsic values of technical objects, which results in standardisation:

> If technical objects do evolve toward a small number of specific types then this is by virtue of an internal necessity and not as a consequence of economic influences or practical requirements; it is not the production-line that produces standardization, but rather intrinsic standardization that allows for the production-line to exist.
>
> (Simondon, 2017, p. 29)

12 'It is rather difficult for a worker to know technicity through the aspects and modalities of his daily work on a machine. It is also difficult for a man who is the owner of machines and who considers them productive capital to know their essential technicity. It is the mediator of the relation between machines alone who can discover this particular form of wisdom. Such a function, however, does not yet have a social place; it would be that of the production planning engineer if he wasn't preoccupied by immediate output, and governed by a finality external to the operating system [régime] of machines which is that of productivity'. (Simondon, 2017, p. 160)

13 'L'être technique réalise la summation d'une disponibilité qui reste toujours présente; l'effort étale dans le temps, au lieu de se dissiper, construit discursivement un être coherent qui exprime l'action ou la suite d'actions qui l'ont constitué, et les conserve toujours présentes: l'être technique médiatise l'effort humaine [...] L'être technique est participable; comme sa nature ne réside pas seulement dans son actualité, mais aussi dans l'information qu'il fixe et qui le constitue, il peut être reproduit sans perdre cette information; [...] il est ouvert à tout geste humain pour l'utiliser ou le recréer, et s'insère dans un élan de communication universelle'. (Simondon, 2005a, p. 340)

14 Simondon does not use the term 'hermeneutical' himself. Language is not at the centre of his philosophical research and his ontology is quite different than the hermeneutics since nature is, for instance, 'not part of the discourse' (Bardin, 2015, p. 44f. & 105).

15 'Par contre, plusieurs élèves ont profondément compris et pratiqué cette prise de conscience de l'être technique, saisissant dans la machine l'histoire humaine déposée, et y ressentant la présence du monde' (Simondon, 2014d, p. 204f).

16 'Il faut connaître le langage par lequel se réactualise le geste humain qui l'a produit. [...] L'être technique doit être envisage comme un être ouvert, polarisé, qui appelle son complement qu'est l'homme au travail, dans la coincidence du tout recompose. L'utilisateur doit prendre la place du constructeur.

Il faut pour cela qu'il coincide avec le schématisme essential inscrit dans l'être technique, qu'il soit capable de le penser, de le comprendre, de l'aimer comme s'il avait fait. La dualité homme-nature se résorbe dans l'unité fonctionnelle de l'homme au travail.' (Simondon, 2014h, p. 252–253)

17 Stephen Graham and Nigel Thrift draw a similar conclusion: most contemporary technical objects (e.g., smartphones) are not built to be maintained but rather imply obsolescence, which goes hand in hand with a certain deskillment of the user (2007, p. 18ff.) They see possible solutions in 'fundamentally altering the nature of tool-being', which could change the problems of e-waste and the commodification of the relation towards technology by reintroducing the importance of maintenance and repair (2007, p. 19).

18 'Tout au contraire, si l'objet est *ouvert*, c'est-à-dire si le geste l'utilisateur, d'une part, peut être un geste intelligent, bien adapté, connaissant les structures internes, si d'autre part le réparateur – qui, d'ailleurs, peut être l'utilisateur –, si le réparateur peut perpétuellement maintenir neuves les pièces qui s'usent, alors il n'y a pas de date, il n'y a pas de vieillissement. Sur une base qui est une base de pérennité ou tout au moins de grande solidité, on peut installer des pièces qui devront être remplacées, mais qui, en tout cas, laissent le schème fondamental intact et qui même permettent de l'améliorer; car on peut bien penser qu'à un moment ou à un autre, si on trouve un outil de coupe meilleur pour une machine destinée à un travail impliquant la coupe, cet outil pourra être monté, à condition qu'il ait les normes nécessaires, sur la base, et qu'ainsi la machine progressera avec le développement des techniques. Voilà ce que j'appelle l'objet ouvert'. (Simondon, 2014c, p. 401f)

19 'Par ailleurs, on pourrait trouver aussi, par une plongée dans le temps, le pouvoir poétique de ce qui était extrêmement parfait et qui, un jour ou l'autre, sera détruit, est peut-être déjà détruit, par le cours d'une évolution, qui est extrêmement et très dramatiquement négatrice, de ce qui a été pourtant, un jour, une nouveauté: voyez les locomotives à vapeur, voyez les grands navires, que l'on met de côté parce qu'ils sont désuets. Ce que l'on appelle l'obsolescence, c'est une réalité économique mais, à côté de l'obsolescence économique, il y a une espèce de montée poétique qui n'a pas été, je crois, tout à fait suffisamment mise en valeur. Nous manquons de poètes techniques'. (Simondon, 2014b, p. 416) An example of such a poet is Jules Verne (ibid. p. 410), whose work allows for an imagination of the future and inspires the imagination of the adolescent (Simondon, 2014e, p. 191).

20 'Mais l'essentiel réside en ceci: pour qu'un objet permette le développement de la mentalité technique et puisse être choisi par elle, il faut qu'il soit lui-même de structure réticulaire: si l'on suppose un objet qui, au lieu d'être fermé, présente des parties conçues comme aussi près que possible de l'indestructibilité, et d'autres, au contraire, en lesquelles se concentre la finesse d'adaptation à chaque usage, ou l'usure, ou la rupture possible en cas de choc, de mauvais fonctionnement, on obtient un objet *ouvert*, pouvant être complété, amélioré, maintenu à l'état de perpétuelle actualité'. (Simondon, 2014d, p. 311)

21 An interesting example of such a story is *The Diamond Age* by Neal Stephenson (2011), where in a fictitious nanotechnological world an engineer invents an interactive, nanotechnological book to educate and foster the creativity of a child. In this world, standardisation and technical development have resulted in a new class system, where engineers correspond to the nobility of Victorian

times. However, creativity and the necessity to evolve and transform lack the saturated class of engineers. Stephenson's book shows multiple things: an imagined new socio-technical world, a new nanotechnological relation with the world and the effects technical objects have on social change.

22 Simondon provides an example from his time as a teacher:

> Cet exercice fait acquérir des notions ayant une valeur culturelle parce qu'il prépare la compréhension des connaissances scientifiques de thermodynamique (c'est l'allumage qui a permis la combustion interne dans les moteurs thermiques) sur le principe de Carnot, et surtout parce qu'il fait comprendre comment la synthèse de recherches au début complètement séparées, de thermodynamique d'une part, d'électricité d'autre part, ont donné naissance à un être technique viable. Le moteur à explosion est fils de la machine à vapeur et de l'eudiomètre. C'est une chaudière dans un cylindre.
>
> (Simondon, 2014e, p. 213)

23 'Il faut donc créer une nouvelle culture, un nouveau schématisme adéquat au monde humain et au monde naturel. Le mixte le plus stable et le plus universel du monde naturel et du monde humain, c'est l'ensemble des êtres techniques' (Simondon, 2014f, p. 241).

24 'Notre but était la saisie intuitive de l'être technique par la jeune intelligence. Un enfant ne comprend pas, au sens profond du mot, ce qu'est un arbre ou un animal. Il peut pourtant comprendre, au sens technique du mot, pourquoi on doit arroser un arbre qui vient d'être planté, pourquoi un arbre a besoin de lumière; c'est que l'enfant réalise une saisie intuitive de l'organisation de l'arbre; il ne comprend pas scientifiquement l'assimilation et la photosynthèse, mais il peut comprendre ce qu'est une greffe ou un marcottage. C'est ce genre de compréhension, intuitive mais non affective ou animiste, que nous nommons compréhension technique. Entre la mentalité primitive et la mentalité scientifique, il y a la pensée technique'. (Simondon, 2014e, p. 217)

25 To define technical thinking independently, especially of primitive mentality, is a critique of Levy-Bruhl's account of primitive mentality and its relation to techniques and scientific thinking. This goes back to the *homo faber* debate at the beginning of the twentieth century, where Durkheimians and followers of Bergson controversially debated the genesis of human intelligence (Sigaut, 2013). Durkheim, Mauss and other members of the Durkheim School were quite critical of the notion of primitive mentality since they argued that different sets of categories can be developed in different cultures without claiming that the mentality of the colonial or cultural other is essentially different (see the discussion between Mauss and Levy-Bruhl, 1923). In contrast to Levy-Bruhl, they argued that the genesis of the categories is a socio-technical process (Schick, 2022).

26 'Est technicien celui qui préfère la connaissance d'une *opération humaine* à la connaissance d'une *structure objective*' (Simondon, 2014e, p. 229).

27 'La machine exige de nous des services et nous en rend, comme un ami; l'échange des services, préférable à l'esclavage, n'est même pas encore la relation la plus haute et la plus adéquate à la machine. Il faut « tirer le joug» avec elle, la bien connaître, travailler en ne la prenant ni comme fin ni comme moyen, mais comme camarade de travail et comme être complémentaire [...]. Cette relation horizontale doit remplacer toute relation verticale'. (Simondon, 2014h, p. 253)

28 'L'utilisateur doit prendre la place du constructeur. Il faut pour cela qu'il coïncide avec le schématisme essentiel inscrit dans l'être technique, qu'il soit capable de le penser, de le comprendre, de l'aimer comme s'il l'avait fait' (Simondon, 2014f, p. 252).

29 'Tout automate, électronique ou non, se compose de quatre parties: un système de *récepteurs*, semblables à des organes des sens; un ou plusieurs relais *amplificateurs* ou *sélecteurs* (le «cerveau»); un ou plusieurs *effecteurs*, semblables à des muscles; enfin, une *source d'énergie* (alimentation)'. (Simondon, 2014e, p. 221)

Bibliography

Bardin, A., 2015. *Epistemology and political philosophy in Gilbert Simondon: individuation, technics, social systems, Philosophy of engineering and technology.* Springer, Dordrecht Heidelberg New York London.

Chateau, J.-Y., 2008. *Le vocabulaire de Simondon.* Ellipses, Paris.

DiFrisco, J., 2014. Hylomorphism and the Metabolic Closure Conception of Life. *Acta Biotheor* 62, 499–525. https://doi.org/10.1007/s10441-014-9233-9

Gillespie, T., 2014. The Relevance of Algorithms, in: Boczkowski, P., Foot, K., Gillespie, T. (Eds.), *Media Technologies.* MIT Press, Cambridge, MA, pp. 167–194.

Graham, S., Thrift, N., 2007. Out of Order: Understanding Repair and Maintenance. *Theory, Culture & Society* 24, 1–25. https://doi.org/10.1177/0263276407075954

Gray, J., Gerlitz, C., Bounegru, L., 2018. Data Infrastructure Literacy. *Big Data & Society* 5, 1–13.

Hui, Y., 2016. *On the Existence of Digital Objects*, University of Minnesota Press, Minneapolis, Minn. London.

Levy-Bruhl, L., 1923. La mentalité primitive. (Scéance du 15 février 1923 à la Société française de Philosophie, avec la participation de Mauss, Belot, Parodi, Ashbourne. Weber, Renoir, Piéron, Fauconnet, Gilson). Bulletin de la Société française de Philosophie 17–48.

Mauss, M., 2006. Techniques of the Body, in: Schlanger, N. (Ed.), *Techniques, Technology and Civilisation.* Durkheim Press/Berghahn Books, New York/Oxford, pp. 77–95.

Rantala, J., 2019. Blockchain as a Medium for Transindividual Collective. *Culture, Theory, and Critique* 60(3–4), pp. 1–14.

Schick, J.F.M., 2021a. The Potency of Open Objects: (Re-)Inventing New Modes of Being Human in the Digital Age with Bergson, Franco "Bifo" Berardi, and Simondon. *Techné: Research in Philosophy and Technology* 25 (3), pp. 279–406. https://doi.org/10.5840/techne2021723144

Schick, J.F.M., 2021b. "Mein fremder Killer in mir": Menschen, Organismen und Designer-T Zellen, in: Dzwidza, E., Speer, A. (Eds.), *Philosophische Anthropologie Als Interdisziplinäre Praxis. Max Scheler, Helmuth Plessner Und Nicolai Hartmann in Köln – Historische Und Systematische Perspektiven.* Brill/mentis, Paderborn, pp. 364–384.

Schick, J.F.M., 2021c. On Technical Alterity. *Foundations of Science* 27 (2022), pp. 515–520. https://doi.org/10.1007/s10699-020-09771-0

Schick, J.F.M., 2022. "Let us dare a little bit of metaphysics": Marcel Mauss, Henri Hubert and Louis Weber on the Categories of Causality and Technology, in: Schick, J.F.M., Schmidt, M., Zillinger, M. (Eds.), *The Social Origins of Thought: Durkheim, Mauss and the Category Project*. Berghahn, New York/Oxford.

Sigaut, F., 2013. Homo faber: documents d'un débat oublié. http://francois-sigaut.com/index.php/inedits/53-ouvrages-inedits/527-homo-faber-documents (Accessed 3 January 2024).

Silver, D., Schrittwieser, J., Simonyan, K., Antonoglou, I., Huang, A., Guez, A., Hubert, T., Baker, L., Lai, M., Bolton, A., Chen, Y., Lillicrap, T., Hui, F., Sifre, L., van den Driessche, G., Graepel, T., Hassabis, D., 2017. Mastering the game of Go without human knowledge. *Nature* 550, 354–359. https://doi.org/10.1038/nature24270

Simondon, G., 2005a. Notes complémentaires sur les conséquences de la notion de l'individuation, in: *L'individuation à la lumière des notions de forme et d'information*. Millon, Grenoble, pp. 331–355.

Simondon, G., 2005b. *L'individuation à la lumière des notions de forme et d'information, Krisis*. Millon, Grenoble.

Simondon, G., 2009. Technical Mentality. *Parrhesia: A Journal of Critical Philosophy* 7, pp. 17–27.

Simondon, G., 2011. Die technische Einstellung, in: Hörl, E. (Ed.), M. Cuntz (Tran.), Die Technologische Bedingung. Suhrkamp, Berlin, pp. 73–92.

Simondon, G., 2014a. Objet technique et conscience moderne (inédit), in: *Sur La Technique (1953–1983)*. Presses universitaires de France, Paris, pp. 363–365.

Simondon, G., 2014b. Entretien sur la mécanologie: Gilbertz Simondon et Jean Le Moyne, in: *Sur La Technique (1953–1983)*. Presses universitaires de France, Paris, pp. 405–445.

Simondon, G., 2014c. Culture et technique (1965), in: *Sur La Technique (1953–1983)*. Presses universitaires de France, Paris, pp. 315–329.

Simondon, G., 2014d. Place d'une initiation technique dans une formation humaine complète, in: *Sur La Technique (1953–1983)*. Presses universitaires de France, Paris, pp. 203–232.

Simondon, G., 2014e. Art et nature. La maîtrise technique de la nature (1980), in: *Sur La Technique (1953–1983)*. Presses universitaires de France, Paris, pp. 179–200.

Simondon, G., 2014f. La mentalité technique, in: *Sur La Technique (1953–1983)*. Presses universitaires de France, Paris, pp. 295–313.

Simondon, G., 2014g. Psychosociologie de la technicité, in: *Sur La Technique (1953–1983)*. Presses universitaires de France, Paris, pp. 27–129.

Simondon, G., 2014h. Prolègomenes à une refonte de l'enseignement, in: *Sur La Technique (1953–1983)*. Presses universitaires de France, Paris, pp. 233–254.

Simondon, G., 2017. *On the mode of existence of technical objects*. Univocal Pub, Minneapolis, MN.

Spelman, E.V., 2002. *Repair: the impulse to restore in a fragile world*. Beacon Press, Boston.

Stephenson, N., 2011. *The diamond age*, Reissued ed. Penguin, London.

Young, M.T., 2022. Maintenance, in: Michelfelder, D.P., Doorn, N. (Eds.), *The Routledge Handbook of the Philosophy of Engineering*. Routledge, London, pp. 356–368.

Part II

Ethics, Politics and Aesthetics of Maintenance

9 Maintenance of Value and the Value of Maintenance

Steffen Steinert

9.1 Introduction

While it's common to view maintenance practices as a means of ensuring proper functionality and usability of technology, this overlooks that technology is inherently value-laden. Technology can either facilitate or hinder the realisation of values. For example, roads and dams embody the value of safety, while artificial intelligence systems can promote transparency and fairness. Moreover, technology plays a significant role in shaping society and contributing to social development. Social media, for instance, can affect people's well-being and influence how individuals perceive and interact with each other. Maintaining our technology maintains not only technology's functional aspects but also its values.

In this chapter, we will expand upon existing discussions of value and technology by highlighting the importance of maintenance. While current discussions of value and technology typically centre on design and use, the role of maintenance is often overlooked. We can gain new insights and enrich our theoretical toolkit by bringing maintenance into the discussions of the relationship between technology and value. Looking at technology and value through the maintenance lens illuminates two crucial points often disregarded in these discussions. First, maintenance helps stabilise values by ensuring that technology continues functioning according to its intended purposes. Second, the maintenance lens reveals that technology has a life after the design stage and emphasises the importance of considering technology through time, including how embedded values may be transformed through maintenance. By focusing on maintenance, we can better understand the dynamic interplay between change and stability in technology and values.

This chapter provides an initial exploration of the intersection between maintenance and value and argues that the maintenance and value perspectives can greatly benefit from each other. While the value perspective sheds light on the role that values should play in maintenance considerations,

DOI: 10.4324/9781003316213-11

it is often underdeveloped in discussions of maintenance. Maintenance can help preserve social values by stabilising the values embedded in technology. Conversely, the maintenance perspective encourages us to consider technology over time and how the embedded values are stabilised and evolve through maintenance. This long-term perspective is essential to discussions of value and technology, as it challenges the common assumption that values are fixed in the design stage based on designers' intentions and that values remain stable. We can gain a comprehensive approach to technology maintenance and value by combining the maintenance and value perspectives.

The chapter is structured as follows: we will begin by characterising the maintenance perspective, and this characterisation will provide the foundation for our subsequent discussions of technology and value. Specifically, we will highlight three crucial aspects of the maintenance perspective relevant to these discussions. First, the maintenance perspective enables us to move beyond the design paradigm that characterises many accounts of value and technology. Second, the maintenance perspective shifts the focus from value disruption to considerations of value stabilisation, recognising that the maintenance of values is an ongoing process. Finally, adopting a maintenance perspective helps to reveal how values are transformed through maintenance, emphasising the dynamic and evolving nature of the relationship between technology and value.

With this groundwork established, we will turn to the examination of socio-technical systems through the maintenance lens. The maintenance lens reveals how the components of socio-technical systems and their embedded values are maintained and possibly transformed over time.

Finally, in the last part of this chapter, we will explore the relationship between the maintenance of technology and the maintenance of value. Here, we will consider how a maintenance account of technology and value could look that takes the entire life cycle of technology into account.

9.2 Maintenance: Beyond Design, Hardware and Conservation

Viewing technology from the maintenance perspective requires acknowledging that technology is not static but evolves over time. Therefore, a comprehensive approach that takes the dynamic nature of technology seriously involves examining the entire life cycle of technology, from its creation to eventual obsolescence (later, we will delve into pre-design and end-of-use considerations). In this section, we will briefly highlight three critical aspects frequently ignored in discussions of technology and value that become apparent when considering technology from a maintenance standpoint. In Section 9.3, these three aspects and the maintenance perspective will also guide the exploration of the relationship between value and technology in socio-technical systems.

9.2.1 Maintenance: Beyond Design

The maintenance perspective illuminates the crucial insight that there is more to technology than just design. Discussions about technology are often guided by what Mark Young calls the design paradigm (2020), which primarily focuses on a technology's design and the role of the designer's intention in creating it. This paradigm ignores the entire temporal dimension of technology, as it places design as the primary focus and views what happens to the technology afterward as secondary.

The maintenance perspective contrasts the design paradigm by inviting us to take the entire temporal spectrum of technology seriously. Technology is not finished after the design stage, and the designer's intention does not determine it. Instead, as many maintenance scholars have noted (Young, 2020; Weber and Krebs, 2021), technology has multiple temporal dimensions that require ongoing consideration. Therefore, we must recognise that technology is an ongoing process that extends beyond its initial creation and design phase. Many technologies can be considered *liminal* because they are continually reshaped, and their form must be negotiated (Suboticki and Sørensen, 2021).

Looking at technology through a maintenance lens can help us reevaluate and shift away from deeply sedimented ways of thinking about technology and value. One of these sediments is the design paradigm, which often dominates the discussions of value and technology. The focus on design and the designer's intention has resulted in approaches to value and technology that are often framed in terms of design and the intentionally designed properties of technical artefacts. For instance, some philosophers of technology, like Ibo van de Poel and Peter Kroes (van de Poel and Kroes, 2014; van de Poel, 2020), argue that technical artefacts embody values when their intentionally designed properties facilitate the realisation of these values. For instance, Michael Klenk (2020) has dubbed accounts that focus on design intentions 'intentional history accounts of value embedding'.

Furthermore, the design paradigm also influences accounts that do not focus on design intentions and their role in value. For example, although Boaz Miller (2020) recognises the long-term implications of values embedded in technology and technology's material longevity, Miller's treatment of value and technology focuses primarily on design. Although Miller briefly brings up redesign, practices of how longevity can be achieved, like maintenance, are neglected.

What can the maintenance perspective contribute to discussions of value and technology? By emphasising the longevity and stability of values, the maintenance perspective encourages a forward-looking approach that complements traditional design-focused accounts of value. Instead of thinking that design alone determines value, we must also consider the ongoing processes that maintain or transform values over the entire life cycle of technology. In this sense, the maintenance perspective challenges the narrow focus on the designer's intention that characterises many accounts of value and technology. As we will explore in Section 9.3, applying the maintenance lens to technology also has important implications for understanding socio-technical systems. But first, let us examine a second crucial aspect that adopting a maintenance perspective brings to the fore, and which will contribute to a more comprehensive understanding of the relationship between value and technology.

9.2.2 *Maintenance: Beyond Hardware*

Although hardware repair is a crucial part of maintenance, it is not the only aspect that deserves attention. Many maintenance scholars argue that maintenance encompasses the social dimension as well. For instance, Steven Jackson (2014) has emphasised that repair can help to maintain meaning. Maintenance and repair extend beyond material artefacts to our relationships, trust, health, honour, and status (Henke and Sims, 2020). Our desire to restore justice and repair our political system also falls under maintenance practices (Spelman, 2002). Therefore, the maintenance perspective highlights the interplay between the material and social aspects of maintenance and the broader implications of maintenance beyond the technical realm.

Maintenance studies have long recognised the symbiotic relationship between the maintenance of technology and the maintenance of social order. As Stephen Graham and Nigel Thrift (2007) suggest, it can be challenging to determine whether the focus of maintenance is the technical artefact itself or the social systems surrounding it. Andrew Russell and Lee Vinsel (2018) argue that the maintenance of technology often goes hand in hand with the maintenance and reproduction of social structures. Maintenance scholars also emphasise that the upkeep of social order requires continuous effort involving material maintenance and repair practices (Denis, Mongili, and Pontille, 2015, p. 7). Furthermore, Christopher Henke (1999) argues that while repair work fixes and maintains technology, it also repairs and restores social order in the workplace. Repair practices frequently involve reordering and fixing social relationships and beliefs, which Henke calls 'people repair' (p. 56). In their recent book, *Repairing Infrastructure*, Christopher Henke and Benjamin Sims (2020) argue that maintaining infrastructure can sustain social and political order. They coin the term 'socio-technical repair' to describe processes of repair that involve both material and discursive interventions to restore meaning.

Expanding our thinking about values and considering how values are maintained can inspire a new perspective on value and technology. The inclusive conception of maintenance that looks beyond hardware and acknowledges the maintenance of society can help us see that the values embodied in technology are not static or fixed but are continuously negotiated and maintained.

The maintenance perspective provides a valuable complement to many approaches in the philosophy of technology that prioritise the destabilising effects of technology on values. While these discussions frequently centre around the destabilising social impacts of technology use, such as social media's impact on mental health or sustainability concerns regarding energy systems, scholars have also developed analytical tools and frameworks for exploring how technology can transform morality. One such concept is techno-moral change (Swierstra, 2013; Kamphof, 2017), which highlights the transformative impact of technology on morally salient practices. Other scholars have explored the potential for technology to disrupt morality altogether (Nickel, Kudina, and van de Poel, 2022). In contrast to these approaches, the maintenance perspective offers a crucial counterpoint by emphasising how values are maintained over time, not just how technology destabilises or transforms values.

Although it is essential to recognise the potential of technology to destabilise morality and social order, it is equally important to acknowledge technology's role in stabilising and maintaining these values. Adopting a maintenance perspective reminds us that technology has a socially and morally disruptive potential and stabilises social and moral practices, institutions and orders. Stabilising technology by maintenance, then, also stabilises and maintains a particular social and moral order. As Bruno Latour (1990) famously argued, technology is society made durable. We can extend this idea and suggest that maintaining technology makes society and values more durable.

The stabilising effect of technology is especially evident with entrenched technologies, which are deeply embedded in the social fabric and can reinforce certain values and social practices. For example, the combustion engine and the infrastructure supporting it have stabilised a value regime centred on individual independence, freedom and car ownership as a status symbol. Car use and ownership are deeply socially ingrained in many cultures and discussions about them are often highly politicised. This is evident in some countries, such as Germany, where attempts to establish nationwide speed limits on highways face significant resistance from drivers and the automobile lobby.

Adopting a maintenance perspective helps us to gain a better understanding of the relationship between technology and value. While the design paradigm focuses on the creation and innovation of new technologies and their corresponding values, the maintenance lens highlights the ongoing work of preserving and sustaining technology and value over time. This lens sheds light on often overlooked aspects of technology and value, including the maintenance of social institutions and actors. At first glance, maintenance may appear to be a conservative force that merely sustains the status quo. However, as we will explore in Section 9.2.3, maintenance can be transformative.

The maintenance perspective provides a broader understanding of technology and value by highlighting the need to maintain hardware and social relations. This raises an important question: does the maintenance of technology also involve the maintenance of values? The answer, according to the maintenance perspective, is yes. Later, we will explore how this insight helps us better understand the maintenance of socio-technical systems.

9.2.3 Maintenance: Beyond Conservation

Many maintenance activities are forward-looking and focus on the stability or continuity of some technical artefact or system over time. However, maintenance is not always about restoring the status quo or conserving something. For instance, maintenance can also involve upgrading technology to meet changing demands or technological advancements. In addition, as discussed in Section 9.2.2, the maintenance perspective opens space for thinking about how maintaining technology relates to the maintenance of value.

In the literature on technology maintenance, there remains a tendency to emphasise conservation and preservation. For instance, Andrew Russell and Lee Vinsel argue that *maintenance* is all the work to preserve particular technical and physical orders (2018, p. 7). Similarly, Heike Weber and Stefan Krebs (2021) propose that *repair*, together with other forms of upkeep, is an intervention that prolongs the time that technology can stay in use. Furthermore, Gabriele Schabacher proposes that maintenance is a 'prospective routine procedure to prevent all forms of disorder' (2021). Also, Steven Jackson (2014) stresses order and proposes that repair involves acts of care that maintain order and meaning in socio-technical systems. This does not mean that these authors don't allow for transformation and change or consider maintenance exclusively in terms of conservation and preservation. Still, these aspects seem to be the primary focus.

Although there is a tendency to view maintenance as a force that conserves and sustains social order and values, scholars have pointed out that maintenance can be a creative, innovative and transformative process (Graham and Thrift, 2007; Russell and Vinsel, 2018; Vinck, 2019; Young, 2020). Rather than exact restoration, maintenance often allows for improvisation and innovation (Graham and Thrift, 2007). Furthermore, Henke and Sims (2020, p. 21f.) distinguish between repair as maintenance and repair as transformation, highlighting the potential for maintenance to enable change in existing social orders, including power structures and social practices. While repair as maintenance conserves and restores the status quo of practices and structures of power, repair as transformation can enable changes to these power structures and social practices. Therefore, maintenance should not be seen as solely a conservative force but also as a force that can bring change and transformation.

The maintenance perspective offers a valuable framework for discussing technology and values and goes beyond the traditional focus on value disruption. Not only does maintenance help maintain and stabilise existing values but it can also be a transformative process that leads to value transformation. This perspective encourages us to pay attention to value stability and transformation through maintenance practices. We should consider supplementing existing discussions about the relationship between technology and values with what could be called *maintenance as value transformation*. This concept acknowledges that values other than those intended by the designer can be embedded in technology through maintenance practices, potentially contributing to significant changes in societal values and norms.

In this section, we have highlighted how a maintenance perspective can provide insights into important but often neglected aspects of value and technology. We have identified three key features of this perspective. First, the maintenance perspective encourages us to go beyond the design paradigm to consider how values relate to technology after the technology has been created. Second, maintenance is not just about hardware upkeep but includes the maintenance and repair of social dimensions. The maintenance perspective emphasises the importance of value stability and conservation. Third, maintenance scholars have pointed out that maintenance is not merely a conservative force but can also be transformative. This opens possibilities for thinking about how values can be transformed through technology maintenance.

In summary, adopting a maintenance lens gives us a valuable perspective for analysing the relationship between technology and values. In the upcoming section, we will put the maintenance perspective to work. We will explore how the maintenance perspective can improve our analysis of socio-technical systems and value, building upon the three aspects introduced previously.

9.3 Maintaining Technology, Maintaining Value: The Case of Socio-technical Systems

This section will explore how the maintenance perspective can enhance our understanding of the relationship between values and technology by applying this perspective to socio-technical systems.

Taking a systems perspective on technology and focusing on socio-technical systems is warranted because even singular technical artefacts are embedded in broader systems. For example, a single bicycle is not an isolated technology but is embedded in the broader socio-technical transportation infrastructure, including roads, bridges, cars and public transportation. Furthermore, most highly impactful technologies, like transport infrastructure and artificial intelligence, are socio-technical systems that integrate social and technical elements. Therefore, by applying the maintenance perspective to socio-technical systems, we can better understand how the maintenance of technology and the maintenance of value and society are related.

Maintenance scholars have long recognised the importance of socio-technical systems, particularly infrastructure. For example, in their treatment of infrastructure repair, Henke and Sims (2020) describe infrastructures as socio-technical systems. Examining socio-technical systems brings attention to the interwoven relationship between technology and social factors, highlighting the links between technology and social structures, including power and privilege. This approach offers unique insights because it connects maintenance to value considerations and emphasises the crucial role of the maintenance perspective in the analysis of technology and value. By adopting a maintenance perspective that considers socio-technical systems, we can better understand how technology and values are maintained and transformed over time.

Before we go into the details of the relationship between maintenance, socio-technical systems and values, it is crucial to understand what socio-technical systems are. According to several scholars (Kroes *et al.*, 2006; Ottens *et al.*, 2006; van de Poel, 2020), socio-technical systems have three essential components. The first component is the material elements, which include technical artefacts and other hardware necessary for the system's functioning. The second component is the agents involved in the system's operation, which are primarily human but can also include artificial agents (van de Poel, 2020). Finally, socio-technical systems consist of social, legal and economic institutions, which are required for the system's functionality.

Institutions are structures of established rules, conventions or behavioural principles that structure social interactions (Fleetwood, 2008; Wilfred Dolfsma and Rudi Verburg, 2008). Social norms are institutions because they prescribe what to do in particular circumstances (Bicchieri, 2005; Brennan *et al.*, 2013). Institutions are enforced by social sanctions that discourage non-compliant behaviour. Laws and regulations are also institutions but in contrast to most social norms they are specified in writing and enforced by formal sanctions, such as fines. Understanding these components is essential to understand how maintenance can affect socio-technical systems and values.

To illustrate how the different components of a socio-technical system work together, we can turn to the Tokyo subway system, as described in Michael Fisch's ethnographic study (2018). This massive socio-technical infrastructure integrates hardware (such as trains and tracks), agents (including passengers and conductors) and institutions (such as social norms and municipal regulations) to ensure the functioning of the system. For instance, passengers have developed efficient boarding norms during rush hour that limit train delays and keep the system running smoothly. The example of the Tokyo subway shows how the different components of socio-technical systems are tightly interconnected and must work together seamlessly to enable the system to function properly.

All three components of the socio-technical system embody values, and the hardware of a socio-technical system comprises technical artefacts that embody values. How value embodiment occurs is subject to philosophical debate, and various accounts of value embodiments have been proposed, such as the historical-intentional account (van de Poel and Kroes, 2014) or the affordances account (Klenk, 2020). Regardless of one's preferred account of value embodiment, it is plausible that technical artefacts in a socio-technical system embody values.

Similarly, values are embedded in the institutional component of a socio-technical system. We can follow Ibo van de Poel (2020) here, who has suggested that values are embedded in the same way they are embedded in artefacts. He proposed that an institution embodies a value when it is conducive to this value because it has been designed for that value. Traffic rules are one example that van de Poel gives to illustrate his argument. For instance, the rule to drive on one specific side of the road embodies traffic safety because it was intentionally designed to facilitate this value. If all traffic participants follow it, traffic will indeed be safer.

Finally, human agents in a socio-technical system also embody values. While technical artefacts may be designed with specific values and functions in mind, human agents do not have such design features. Nonetheless, it is reasonable to say that human agents embody values because they endorse values, are motivated by them, and sustain them through their behaviour. For example, engineers who endorse the value of safety will think that safe technology is a desirable state of affairs and will consequently be motivated to design a safe product. They will feel a sense of satisfaction when they succeed in designing a technology they consider to be safe. Different approaches have been proposed to characterise individual values, including values as conceptions of the desirable (Kluckhohn, 1951), values as abstract trans-situational goals or ideals (Schwartz, 2015; Maio, 2016) or values as patterns of relatively robust attitudes, such as emotions and desires, that provide reasons for actions (Tiberius, 2018). Despite the various accounts of values, it is widely accepted that human agents hold or endorse values.

Now let us discuss how maintenance relates to socio-technical systems and values. To ensure the proper functioning of the entire system, it is necessary to maintain all three components of a socio-technical system – hardware, agents and institutions. Furthermore, a maintenance perspective can enhance our understanding of values and technology by recognising that maintaining the components of a socio-technical system also entails preserving values embedded in them. The maintenance of these components and their embedded values are illustrated in Figure 9.1.

Let us now consider the maintenance of the components one by one, beginning with the maintenance of hardware.

When it comes to maintaining a socio-technical system, the hardware component is crucial. Technical components inevitably break down or wear out over time and thus must be repaired or replaced to ensure the system's proper functioning. Take, for instance, the complex socio-technical system of an airport. It comprises various hardware components, including airplanes, airport buildings, air traffic control equipment and runways, all requiring ongoing maintenance to keep the airport operating smoothly.

The maintenance perspective reveals that maintaining the hardware is not solely about replacing parts or repairing broken components. It also

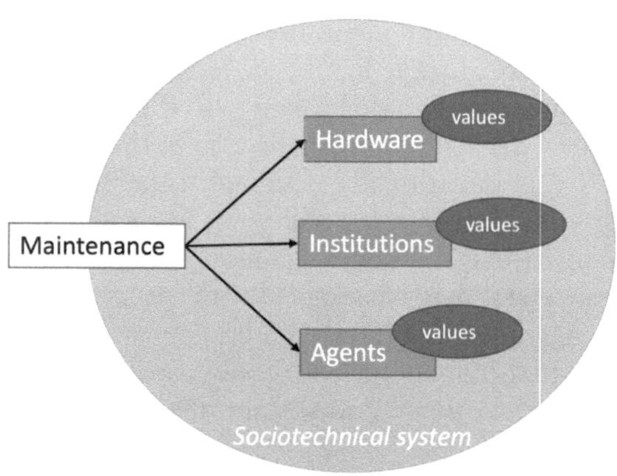

Figure 9.1 Maintenance and socio-technical system.

involves maintaining the values that are embodied in the hardware. Depending on which account of value embodiment one finds most convincing, the maintenance of values will take different forms. For example, the affordances account of value embodiment (Klenk, 2020) asserts that artefacts embody values because they have affordances, which are relational properties that create action possibilities for the agent and make some actions likelier than others.[1] Technology affords or enables some valuable actions more than others. Following the affordance view of value embedding, maintaining technology would entail maintaining its affordances and, hence, its values.

In contrast to the affordance view, on the function account of value embodiment (van de Poel and Kroes, 2014), an artefact embodies values due to the functional properties it was designed for. If the maintenance of the hardware keeps the technology function stable, then the embodied value is also preserved. Section 9.4 will explore the maintenance account for other value domains besides design values.

The debates about value in the philosophy of technology often overlook the fact that the values of technology are not static but rather require ongoing maintenance to remain stable. The predominant focus on design has led to a neglect of the role of maintenance practices in upholding values. By contrast, a maintenance perspective highlights the importance of maintaining values through active practices. This perspective provides a valuable counterbalance to the prevalent emphasis on value disruption and destabilisation in the philosophy and ethics of technology. By highlighting that values are stabilised through maintenance, it offers a fresh angle for exploring and understanding the complex relationship between technology and values.

However, the maintenance perspective recommends attending to conservation and transformation, especially focusing on transformation through maintenance practices. By applying this perspective to value and technology, we can open up intellectual space to consider how values can be transformed through maintenance. This approach challenges the implicit assumption in many debates that values embodied in technology are fixed in design. In reality, not only are embodied values maintained but they can also be transformed. Through maintenance practices, we can transform the values embodied in technology, thereby enabling it to meet new or changing requirements. For instance, to better align artificial intelligence with our value of justice, the maintenance of an algorithm may change the system components responsible for producing biased results while maintaining the rest of the algorithm.

To illustrate how maintenance as value transformation is viewed from the standpoint of different accounts of value embodiment, let's take the example of the abovementioned affordance view. According to this view, technology embodies values by virtue of the affordances it provides, which are the potential actions or interactions it allows or invites. Hence, changing the affordances of technology can result in a transformation of its embodied values. For example, consider a social media platform designed to maximise user engagement, but it has been found to lead to negative social consequences such as addiction, polarisation and disinformation. By changing the platform's affordances, such as reducing notifications, it becomes more aligned with the value of promoting healthy social interactions. In this way, maintenance practices that modify the affordances of technology can lead to value transformation.

Let us now turn to the other critical elements of a socio-technical system. As we have seen, taking the maintenance perspective seriously involves recognising that maintaining the entire system requires more than hardware maintenance. All components must be maintained to ensure that the system functions correctly. Therefore, companies, governments or society must consider maintaining all the components to maintain a socio-technical system.

For example, to maintain a socio-technical transportation system like an airport, it is necessary to maintain social institutions like air traffic control regulations, which govern the system and enable it to operate. Similarly, to maintain a road traffic system, it is crucial to ensure that hardware components like roads and signs are in good condition, but it is equally important to maintain institutions like traffic rules.

Above, we have highlighted that institutions, like artefacts, embody values. However, the maintenance of institutions is often overlooked in discussions of technology and value. Taking a maintenance perspective requires us to consider the maintenance of institutions and how it relates to maintaining embodied values. This perspective also highlights the importance of paying attention to how institutions are maintained and stabilised over time. Maintenance scholars have stressed the relationship between maintenance and social order, emphasising the significance of institutional maintenance for the stability of values (Graham and Thrift, 2007; Russell and Vinsel, 2018; Henke and Sims, 2020). Adopting a maintenance perspective can help us to better understand how institutions maintain embodied values.

Let us briefly consider how institutions can be maintained. Institutions, such as social norms and traffic rules, can be maintained through various means, including positive and negative reinforcements. Negative reinforcement includes sanctions, which can take different forms, from public disapproval to verbal or physical discipline. In some cases, the state usually enforces physical sanctions through the police and courts. Positive reinforcement of institutions includes incentives to encourage people to adhere to norms and other means of ensuring compliance. For instance, drivers' education ensures that drivers internalise and follow traffic rules. At this point, the maintenance of institutions overlaps with the maintenance of actors, which we will discuss below.

The maintenance of enforcement mechanisms is indeed crucial for the successful enforcement of institutions. However, it is important to note that maintaining a socio-technical system does not just mean maintaining these mechanisms but also the broader social and technical components that enable these mechanisms to function effectively. For example, maintaining a traffic enforcement system not only involves hiring and training police officers but also ensuring that the road infrastructure is in good condition, traffic signals are functioning properly and drivers are educated about the rules. All these components work together to create an effective traffic enforcement system.

Furthermore, it is important to recognise that we should not ignore the connectedness of different socio-technical systems. The maintenance of one system can impact the maintenance of another system, and the failure to maintain one system can have ripple effects on other systems. Therefore, a comprehensive understanding of socio-technical systems and their interconnectedness is necessary for a comprehensive understanding of maintenance, but we will leave this complexity aside here.

To fully embrace the maintenance perspective, we must move beyond the idea that maintenance preserves the status quo. Maintenance can be a tool for intentional change, including the transformation of institutions and values embedded in them. In the case of socio-technical systems, institutions can be changed purposefully to embed new values. For example, if there is a demand to make shipping more sustainable, the maintenance of the system may require revising shipping routes to achieve this goal. The new shipping routes will then embody the value of sustainability, reflecting a deliberate effort to transform the institution to align with new values (Künneke *et al.*, 2015). This kind of intentional change of institutions is an important aspect of maintenance that is often overlooked in discussions of technology and value.

Another example of maintenance as institutional change has occurred due to the rise of sophisticated language models like ChatGPT. In response to concerns about how these models may reshape education and scientific integrity (Cotton, Cotton and Shipway, 2023), many educational organisations have sought to adapt their institutions, like policies and guidelines, to maintain specific educational values, such as scientific integrity. By revising their rules and guidelines for scientific integrity, these organisations have adapted to the new reality and ensured that their values remain embedded in the educational system. This educational change highlights how the maintenance of socio-technical systems often requires the purposeful transformation of institutions to embed new values.

Before we move on, it is important to note that the above examples of institutional change, such as the revision of shipping routes and changes to scientific integrity guidelines, involved changes in formal rules and regulations. However, not all institutions in a socio-technical system are formal. Some institutions, like social norms, are informal, and their change can often be unintentional.[2]

The maintenance perspective considers the temporal dimension and acknowledges that institutions are not fixed. In addition to focusing on the maintenance required to stabilise institutions and their embedded values, the perspective also highlights the potential for institutional transformation. By recognising this dynamic aspect of socio-technical systems, the maintenance perspective provides a more nuanced understanding of these systems.

Finally, let us turn to the maintenance of agents as components of socio-technical systems. Besides external factors like sanctions, agents follow institutions because they have internalised them through processes like socialisation and habituation (Fleetwood, 2008). Because agents likely internalise certain institutions during their education and training, maintaining internalisation processes contributes to maintaining institutions. Thus, it is crucial to recognise that the maintenance of agents involves ensuring a steady supply of agents and the continuous development and training of existing agents. This can involve providing professional development opportunities, updating knowledge and skills and promoting a culture of ongoing learning. The maintenance of agents also encompasses recruiting and retaining a diverse range of agents to reflect the diverse needs of the socio-technical system and its users.

Moreover, the maintenance of organisations that shape agents involves not just universities and vocational schools but also other institutions that support agents, such as professional associations and networks. These organisations provide resources, mentorship and networking opportunities vital for agents' ongoing development and support. The maintenance of these organisations requires a sustained effort to ensure they remain relevant, responsive and accessible to the agents they support.

The link between agents and institutions highlights the importance of maintaining internalisation processes. Agents' internalisation of social institutions, like norms, rules and conventions, underpin the socio-technical system. To maintain these internalisation processes, organisations must promote a culture that emphasises the importance of following social institutions and provides agents with opportunities to practice and reinforce these institutions regularly.

However, the agents in socio-technical systems include more than just operators. Users are another significant group of agents. For instance, when focusing on travel infrastructure, train passengers and pedestrians should be considered. A maintenance perspective should pay attention to this group because, as we have seen, it looks beyond design and considers what happens after technology leaves the factory. For example, in the case of the Tokyo subway discussed previously, how commuters use a transportation system, including norms of smooth entry and exit, can significantly impact the system's operation.

Earlier, we discussed how agents embody values. However, the question of how agents' values are maintained is often overlooked in the literature on technology and values. This is a critical oversight because the maintenance of values is relevant to technology's smooth functioning and stability. Adopting a maintenance perspective can help correct this oversight by emphasising the importance of maintaining values in addition to other components of the system.

To illustrate the role of values in the maintenance of a system, let us consider the example of traffic infrastructure. Drivers who endorse the value of safety will behave in a way that does not harm the material parts of the infrastructure. For instance, in certain countries with colder climates, valuing safety means changing tyres between winter and summer, which avoids damaging the road[3]. Similarly, a plane passenger who values efficiency will likely behave in a way that contributes to quick boarding, which helps with the smooth operation of the flight.

Now that we have established that the values of agents, operators, engineers and users are relevant for maintaining technology, we must ask how the values embodied in agents are maintained. A part of the answer lies in considering organisations that shape agents and their values. First, let us briefly consider the organisations that shape operators and engineers and then turn to the topic of maintenance of users' values.

To fully understand the maintenance of values in socio-technical systems, it is important to consider the role of various organisations in shaping the values of the agents involved. While it may not be possible to provide an exhaustive list of all the factors that influence the values of these agents, certain organisations, such as universities and vocational schools, play a particularly significant role in shaping the values of operators and engineers. During their education and training, these agents internalise values crucial for maintaining the socio-technical system, such as efficiency, safety and sustainability. These values, because they translate into behaviour, like designing an efficient product, are conducive to maintaining the socio-technical system. Organisations play a key role in shaping and maintaining the values of users because they influence the internalisation of norms. For instance, a driver's education program is mandatory in many countries, and drivers must pass a test before receiving their license. This education likely instils values such as safety and responsible driving in the prospective driver. In addition, formal rules and regulations and official sanctions, like fines, help maintain drivers' values and keep their behaviour in line with established institutions crucial for maintaining the socio-technical system.

The interdependence of the components of a socio-technical system is an essential feature of such systems. As we have seen, agents' values are closely related to institutions and maintaining one component can contribute to maintaining the other. However, it is also important to note that the components can be maintained independently, and the intention to maintain one component does not necessarily imply the intention to maintain the whole socio-technical system. For instance, a government may wish to maintain certain rules and regulations regarding the use of technology without the intention to maintain the entire socio-technical system. For example, a government may enforce strict data privacy laws without the intention to maintain the entire digital infrastructure. Similarly, the European Union may wish to maintain its rules about the use and design of cars without the intention to maintain the socio-technical traffic system in its member states.

So far, we have examined socio-technical systems from a maintenance perspective, which proved to be a powerful tool for identifying key components of the connection between technology and value previously overlooked. However, our treatment was more like a snapshot view. For this reason, our considerations were not entirely faithful to the dynamism embraced by the maintenance perspective. To fully align with the maintenance perspective, we must consider technology through time, particularly how values are maintained throughout a technology's lifespan. Section 9.4 will focus on technology and maintenance over time.

9.4 Technology through Time: Maintenance and Value Domains

The maintenance perspective shifts our attention from the initial design of technology to its ongoing trajectory through time. It encourages us to focus on how technology is stabilised and transformed. In contrast to accounts that emphasise the importance of design, the maintenance perspective highlights the ongoing importance of maintenance, repair and adaptation. While design is crucial, it is only the beginning of technology's life cycle. In this section, we will see how adopting a maintenance perspective can give us insights into how values are maintained in different domains throughout the life cycle of technology.

At this point, it is crucial to introduce how a maintenance account of value and technology would consider the whole life cycle of technology. To provide such an account, let us revisit the concept of embodied values we discussed earlier. We previously noted that some philosophers argue that values are embodied in technology through design. However, it is essential to recognise that design is only one of many domains that contribute to shaping the values embodied in technology. Many philosophers of technology also emphasise the importance of the use domain and its associated values[4] (van de Poel and Kroes, 2014; van de Poel, 2020).

The values related to design are *intended values*, which are the values that designers or developers aim to embody in the technology they create. For instance, the designers may intend for the technology to embody values such as sustainability or safety. In addition to intended values, we can distinguish *realised values*. These are the values that are actually realised in the use of technology. Notably, intended values and realised values may not always align. For instance, it may turn out that the technology is not sustainable or safe once people use it. Finally, there are *embodied values*, which we have previously discussed. Embodied values are intentionally embedded in the technology and realised in use because the technology has been designed for these values (van de Poel, 2020). Let us now turn to a maintenance account of value and technology, which can provide a broader analytical perspective than existing accounts of value and technology.

Distinguishing between design and use is a good start, but to fully capture the relationship between technology and value, a maintenance account would emphasise the full temporality of technology. Adopting a maintenance perspective means looking beyond just design and use and considering the value domains that span the entire life cycle of a technology.

The life cycle of technology involves more than design and use; it also includes a domain that could be called *discard*. While discard may appear simply as a form of use, it is prudent to distinguish between use and discard because what happens to technology after it becomes obsolete may not necessarily affect how people use it or how creators intend it to be used. For instance, how to discard technology may not even be considered in the use plan. A use plan describes an intended use of technology, including a set of manipulations of the technology so that it can adequately fulfil its function (Houkes *et al.*, 2002). This is not to say that discarding a technology is never part of a use plan or that designers cannot translate their intentions regarding how technology is discarded into design requirements. For instance, if designers and engineers intend technology to be recyclable, they can make devices that are easy to disassemble. Nevertheless, an outlook that focuses on design and use, like the predominant paradigm in the philosophy of technology, invites us to do, risks ignoring the entire life cycle of technology. Adopting a maintenance perspective, which means focusing on technology over time, helps to prevent this limited view of technology.

It is important to acknowledge that discarding technology is not a value-free process because it is connected to value-sensitive issues such as sustainability, justice and safety. For example, many discarded clothes and computer technologies end up in markets or landfills in the Global South, where they are often disassembled or repurposed. Therefore, the discard domain can realise specific values like the use domain. The values realised in the discard domain are related to the intended values designers aim to achieve when the technology is discarded. Designers can significantly promote sustainability, safety and other values during the discard phase by making the technology easy to disassemble and break down into recyclable parts.

Besides the domains of design, use and discard, there is yet another domain that we should consider. In taking the social embeddedness of use and design seriously, our thinking needs to reflect that the intended values of design do not arise from a social vacuum. Therefore, adding pre-design as a fourth domain to our analysis is crucial.

The pre-design domain encompasses the social and cultural factors that contribute to the values intended by designers. For example, the values that engineering students internalise during their education play a crucial role in shaping the intended values of the technology they develop. However, the values they learn in their education are influenced by larger social contexts, such as cultural norms and expectations.

Additionally, the domain of pre-design includes the demands of employers and purchasers of technology. These demands are not isolated from larger social values and expectations and are often shaped by them. For instance, a company may demand that its technology be designed to minimise environmental impact or prioritise user safety.

A maintenance account of value and technology requires us to adopt a comprehensive perspective that considers the entire social life of technology. To achieve this comprehensive view, we must focus on the four domains introduced earlier, namely design, use, discard and pre-design and examine how maintenance relates to values in each domain. The central idea is that maintaining each domain helps maintain the associated values. The relation of maintenance, domains and values can be illustrated in Figure 9.2, which I adapted from van de Poel and Kroes (van de Poel and Kroes, 2014; van de Poel, 2020).

In the remainder, we will elaborate on the maintenance of pre-design, design, use and discard. It should be noted that there may be an overlap between the types of maintenance, and my analysis is not exhaustive. Furthermore, maintaining one domain can contribute to the maintenance of multiple values.

To elaborate on the *maintenance of pre-design*, it is important to consider the various social domains that shape the intended values that go into the design of technology. As mentioned, engineering education plays a significant role in shaping students' values and beliefs. In their courses, students are often confronted with tasks, materials and projects highlighting the importance of instrumental values related to efficiency and cost reduction.

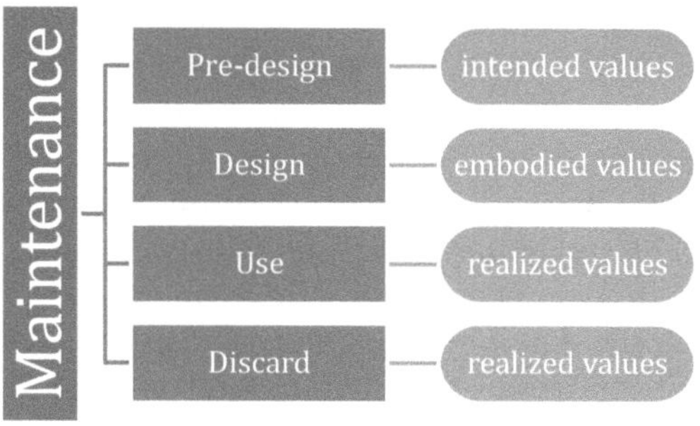

Figure 9.2 Maintenance of value domains.

Students internalise these values, later shaping their design decisions about intended values. Maintaining the status quo of engineering education that features courses and curricula that mainly focus on some values and not others, the university contributes (unintentionally) to maintaining some intended values and not others. This need not be the case, however, and universities may opt for what Henke and Sims call 'repair as transformation' (2020, p. 21), which means finding ways to repair while changing some practices, instead of simply repairing to conserve the status quo. For instance, by including ethics education, students may also internalise moral and social values that can inform the design of their future projects.

To further illustrate the maintenance of pre-design, consider the impact of social and cultural norms on the intended values of technology design. For example, gender roles and stereotypes may be deeply ingrained and could influence the intended values of technology design. Maintaining a status quo that perpetuates these norms would likely result in technologies that reinforce and perpetuate gender biases and inequalities.

Maintenance of design relates to the maintenance of social factors that influence the design of technology and its embodied values. Embodied values are the values for which technology is intentionally designed to facilitate. To illustrate, let us consider a previous example of rules and regulations that prescribe which values should be intended and promoted in technology design, such as rules concerning safety and sustainability standards. By motivating designers to create technology in a certain way, these institutions make it likely that the technology furthers specific values. Maintaining institutions that ensure the technology is adequately designed then maintains embodied values.

Recall that intended and embodied values can misalign, for instance, when the technology is not designed correctly, is misused or has unintended consequences. Some practices, such as the tendency or even requirement to choose the cheapest design option to save costs, can contribute to this misalignment since designing affordably may come at the expense of other values, such as the health and well-being of users. Thus, maintaining these practices through economic incentives perpetuates the mismatch between intended and embodied values.

Maintenance of use concerns the conservation of factors that affect technology use, thereby maintaining realised values. Like pre-design and design, social institutions such as customs, laws and regulations are crucial in maintaining specific technology use. Maintaining certain norms of technology use through incentives or sanctions can ensure that technology is used according to its intended purposes and values. For example, laws that govern the use of vehicles and organisations that enforce sanctions, like the police and courts, aim to ensure that cars are used safely. The government may also reward specific technology use by providing tax breaks.

When technology is not used as intended, and intended values and realised values misalign, society and designers must find ways to maintain specific technology use. Municipalities, for instance, may create new regulations to increase the safe use of E-scooters by restricting where they can be ridden. Besides social maintenance of use, there is also technical maintenance of use. Designers have tools to maintain specific technology use, such as apps needed to operate the scooter, which can restrict where the scooter can be parked or limit the maximum velocity in certain areas of the town, realising the value of safety in traffic. Designers can also achieve technical maintenance of use by designing technology with specific affordances (Klenk, 2020; Steinert and Dennis, 2022), which makes particular forms of technology use more likely, thereby maintaining realised values.

There is some overlap between the different forms of maintenance, which means that maintaining one domain, and its accompanying values, can contribute to maintaining other values. This is particularly evident in the maintenance of use. When social institutions or affordances maintain a specific use of technology, it maintains the realised value and embodied value because the latter is intended by the designers and realised in use. In other words, by maintaining a specific use of technology, designers and society can ensure that the technology's intended values are realised.

Finally, the *maintenance of discard* concerns the factors that influence how technology is treated after it loses its usefulness and hence, which values are realised in this treatment. Discard shares many similarities with use and could be viewed as a form of use. Therefore, the aforementioned points about the maintenance of use applies equally to the maintenance of discard. Social institutions, such as norms and laws concerning recycling, contribute to how technology is discarded, and maintaining these institutions maintains the accompanying realised values, such as sustainability. Providing the necessary infrastructure to discard technology can also be a way of maintaining the values associated with discard. For example, providing convenient opportunities for people to discard their old electronic devices so that they can be collected and appropriately recycled facilitates the realization of values like sustainability and safety.

9.5 Conclusion

The chapter's main purpose has been to demonstrate how examining technology and values from a maintenance perspective can enhance discussions of the relationship between value and technology in the philosophy of technology.

We have explored three fundamental aspects of a maintenance perspective, which have helped us to understand the relationship between maintenance and value more fully. First, taking a maintenance perspective enables us to move beyond the design-centric approach and ask how value stability is maintained after the initial design. Second, the maintenance perspective on technology and value allows us to recognise that technology can sustain values rather than merely destabilise them. Third, maintenance is not limited to preservation but can also be transformative. Adopting a maintenance perspective allows us to consider how the maintenance of technology can contribute to the transformation of values.

The maintenance perspective offers a novel lens to examine technology and value. To demonstrate the usefulness of this perspective, we applied it to socio-technical systems, highlighting how maintaining such systems preserves the values embedded in their components and the relationships between social and technological maintenance. The maintenance perspective also allows us to explore value considerations in the whole life cycle of technology, thereby opening up new avenues for philosophical inquiry into technology and value.

An upshot of a maintenance account of value and technology is that extant perspectives on technology and value should be expanded. Adopting a maintenance perspective can also inspire new investigations within philosophy and ethics of technology. For example, if maintenance practices contribute to the stability of values, then responsible maintenance should supplement responsible design approaches, such as value-sensitive design or design for values (van den Hoven, Vermaas, and van de Poel, 2015; Friedman and Hendry, 2019).

Responsible design aims to facilitate responsible practices of technology design by considering technology's impact on people, the environment and society. As a supplement to responsible design, responsible maintenance would extend these value considerations to technology maintenance.

Responsible maintenance differs from frameworks like design for maintenance and maintainability approaches (Dhillon, 1999). Design for maintenance aims to anticipate and plan for future maintenance needs to ensure that the technology remains in good condition or continues to promote safety and health (Martinetti and Singh, 2019). Design for maintenance refers to processes during the design phase of technology, while responsible maintenance focuses on the impact of maintenance. Although design for maintenance is relevant and can include important value considerations, responsible maintenance supplements design for maintenance in that it infuses maintenance with value considerations by considering the impact of maintenance on people and society. For instance, by questioning what is maintained and by whom and which values are sustained and transformed through maintenance?

Furthermore, to further develop a maintenance account of value and technology, a possible next step is to investigate the mechanisms of value stability and transformation. For instance, it would be valuable to consider when and under which conditions maintenance practices contribute to stabilising or transforming values. This could involve analysing specific case studies of technology maintenance and their impact on values or developing theoretical frameworks to guide such analyses. Additionally, exploring the ethical implications of maintenance practices and their effects on values could be a fruitful area for further investigation.

One motivation for the chapter was to convince readers that the maintenance perspective can make valuable contributions to the philosophy of technology. A maintenance account of value and technology, which highlights the crucial role of maintenance in sustaining and transforming values, helps to shake loose common assumptions in the field. More work is necessary to establish a complete maintenance account of value and technology. It is key that we maintain our efforts to develop and deepen our understanding of the relationship between maintenance and value in technology.

Notes

1 The idea of affordances originated in ecological psychology and was developed by psychologist James Gibson. According to Gibson, affordances are what the environment 'offers the animal, what it provides or furnishes, either for good or ill' (1979, p. 127). Affordances can be characterised as the relation between the abilities of the organism and the features of the environment (Chemero, 2003). As relational phenomena, affordances depend on the properties of the environment and the features of the agent interacting with it. Design scholar Don Norman applied the idea of affordances to human–technology interactions. Norman proposed that affordances are '[...] perceived and actual properties of the thing, primarily those fundamental properties that determine just how the thing could possibly be used' (1988, p. 9). For instance, a doorknob affords grasping whereas a computer mouse and a trackpad afford particular ways of handling. The idea of affordance helps to analyse what technology does and how it does it. Recently, Jenny Davis (Davis, 2020) has presented

mechanisms of affordances that help to discern how technology shapes social behaviour and its political implications.

2 Further discussion of these types of norm changes can be found in Brennan *et al.* (2013, p. 107ff.).

3 I would like to thank the editor for suggesting this example.

4 This is an oversimplification and the phases need not be strictly compartmentalised. There are feedback loops, and the use can inform the design (see, for instance, van de Poel (2020)).

References

Bicchieri, C. (2005) *The Grammar of Society: The Nature and Dynamics of Social Norms*. Cambridge: Cambridge University Press.

Brennan, G. et al. (2013) *Explaining Norms*. 1st ed. Oxford: Oxford University Press.

Chemero, A. (2003) 'An Outline of a Theory of Affordances', *Ecological Psychology*, 15(2), pp. 181–195. https://doi.org/10.1207/S15326969ECO1502_5

Cotton, D.R.E., Cotton, P.A. and Shipway, J.R. (2023) 'Chatting and Cheating: Ensuring Academic Integrity in the Era of ChatGPT', *Innovations in Education and Teaching International*, pp. 1–12. https://doi.org/10.1080/14703297.2023.219014820

Davis, J.L. (2020) *How Artifacts Afford: The Power and Politics of Everyday Things*. Cambridge, Massachusetts: The MIT Press (Design thinking, design theory).

Denis, J., Mongili, A. and Pontille, D. (2015) 'Maintenance & Repair in Science and Technology Studies', *Tecnoscienza Italian Journal of Science & Technology Studies*, 6(2), pp. 5–15.

Dhillon, B.S. (1999) *Engineering maintainability: how to design for reliability and easy maintenance*. Houston, TX: Gulf Pub. Co.

Fisch, M. (2018) *An anthropology of the machine: Tokyo's commuter train network*. Chicago; London: The University of Chicago Press.

Fleetwood, S. (2008) 'Institutions and Social Structures', *Journal for the Theory of Social Behaviour*, 38(3), pp. 241–265. https://doi.org/10.1111/j.1468-5914.2008.00370.x

Friedman, B. and Hendry, D. (2019) *Value sensitive design: shaping technology with moral imagination*. Cambridge, Massachusetts: The MIT Press.

Gibson, J.J. (1979) *The ecological approach to visual perception*. Boston: Houghton Mifflin.

Graham, S. and Thrift, N. (2007) 'Out of Order: Understanding Repair and Maintenance', *Theory, Culture & Society*, 24(3), pp. 1–25. https://doi.org/10.1177/0263276407075954

Henke, C. (1999) 'The Mechanics of Workplace Order: Toward a Sociology of Repair', *Berkeley Journal of Sociology*, 44, pp. 55–81.

Henke, C. and Sims, B. (2020) *Repairing Infrastructures: The Maintenance of Materiality and Power*. Cambridge, Massachusetts: The MIT Press (Infrastructures series). Available at: https://direct.mit.edu/books/book/4962/Repairing-InfrastructuresThe-Maintenance-of

Houkes, W. et al. (2002) 'Design and Use as Plans: An Action-Theoretical Account', *Design Studies*, 23(3), pp. 303–320. https://doi.org/10.1016/S0142-694X(01)00040-0

van den Hoven, J., Vermaas, P.E. and van de Poel, I. (2015) 'Design for Values: An Introduction', in J. van den Hoven, P.E. Vermaas, and I. van de Poel (eds) *Handbook of Ethics, Values, and Technological Design*. Dordrecht: Springer Netherlands, pp. 1–7. Available at: http://link.springer.com/10.1007/978-94-007-6970-0_40 (Accessed: 15 October 2018).

Jackson, S.J. (2014) 'Rethinking Repair', in T. Gillespie, P.J. Boczkowski, and K.A. Foot (eds) *Media Technologies*. The MIT Press, pp. 221–240. Available at: https://academic.oup.com/mit-press-scholarship-online/book/14976/chapter/169335302 (Accessed: 10 August 2022).

Kamphof, I. (2017) 'A Modest Art: Securing Privacy in Technologically Mediated Homecare', *Foundations of Science*, 22(2), pp. 411–419. https://doi.org/10.1007/s10699-015-9448-5

Klenk, M. (2020) 'How Do Technological Artefacts Embody Moral Values?', *Philosophy & Technology* [Preprint]. https://doi.org/10.1007/s13347-020-00401-y

Kluckhohn, C. (1951) '2. Values and value-orientations in the theory of action: an exploration in definition and classification', in T. Parsons and E.A. Shils (eds) *Toward a General Theory of Action*. Cambridge, MA and London, England: Harvard University Press.

Kroes, P. et al. (2006) 'Treating Socio-technical Systems as Engineering Systems: Some Conceptual Problems', *Systems Research and Behavioral Science*, 23(6), pp. 803–814. https://doi.org/10.1002/sres.703

Künneke, R. et al. (2015) 'Understanding Values Embedded in Offshore Wind Energy Systems: Toward a Purposeful Institutional and Technological Design', *Environmental Science & Policy*, 53, pp. 118–129. https://doi.org/10.1016/j.envsci.2015.06.013

Latour, B. (1990) 'Technology is Society Made Durable', *The Sociological Review*, 38(1_suppl), pp. 103–131. https://doi.org/10.1111/j.1467-954X.1990.tb03350.x

Maio, G.R. (2016) *The Psychology of Human Values*. Routledge. https://doi.org/10.4324/9781315622545

Martinetti, A. and Singh, S. (2019) 'Design for Maintenance of Infrastructures: The Lesson of the Morandi Bridge', in S. Singh et al. (eds) *Transportation Systems*. Singapore: Springer Singapore, pp. 43–54. Available at: http://link.springer.com/10.1007/978-981-32-9323-6_4 (Accessed: 16 June 2023).

Miller, B. (2020) 'Is Technology Value-Neutral?', *Science, Technology, & Human Values*. https://doi.org/10.1177/0162243919900965

Nickel, P.J., Kudina, O. and van de Poel, I. (2022) 'Moral Uncertainty in Technomoral Change: Bridging the Explanatory Gap', *Perspectives on Science*, 30(2), pp. 260–283. https://doi.org/10.1162/posc_a_00414

Norman, D.A. (1988) *The Psychology of Everyday Things*. New York: Basic Books.

Ottens, M. et al. (2006) 'Modelling Infrastructures as Socio-technical Systems', *International Journal of Critical Infrastructures*, 2(2/3), p. 133. https://doi.org/10.1504/IJCIS.2006.009433

Russell, A.L. and Vinsel, L. (2018) 'After Innovation, Turn to Maintenance', *Technology and Culture*, 59(1), pp. 1–25. https://doi.org/10.1353/tech.2018.0004

Schabacher, G. (2021) 'Time and Technology: The Temporalities of Care', in A. Volmar and K. Stine (eds) *Media Infrastructures and the Politics of Digital Time*. Amsterdam University Press, pp. 55–76. Available at: https://www.degruyter.com/document/doi/10.1515/9789048550753-004/html?lang=de (Accessed: 10 August 2022).

Schwartz, S.H. (2015) 'Basic individual values: sources and consequences', in T. Brosch and D. Sander (eds) *Handbook of Value*. Oxford University Press, pp. 63–84. Available at: http://www.oxfordscholarship.com/view/10.1093/acprof:oso/9780198716600.001.0001/acprof-9780198716600-chapter-4 (Accessed: 19 June 2019).

Spelman, E.V. (2002) *Repair: The Impulse to Restore in a Fragile World*. Boston: Beacon Press.

Steinert, S. and Dennis, M.J. (2022) 'Emotions and Digital Well-Being: on Social Media's Emotional Affordances', *Philosophy & Technology*, 35(2), p. 36. https://doi.org/10.1007/s13347-022-00530-6

Suboticki, I. and Sørensen, K.H. (2021) 'Liminal Technologies: Exploring the Temporalities and Struggles in Efforts to Develop a Belgrade Metro', *The Sociological Review*, 69(1), pp. 156–173. https://doi.org/10.1177/0038026120918166

Swierstra, T. (2013) 'Nanotechnology and Technomoral Change', *Etica & Politica*, 15(1), pp. 200–2019.

Tiberius, V. (2018) *Well-being as Value Fulfillment: How We Can Help Each Other to Live Well*. Oxford, New York: Oxford University Press.

van de Poel, I. (2020) 'Embedding Values in Artificial Intelligence (AI) Systems', *Minds and Machines*, 30(3), pp. 385–409. https://doi.org/10.1007/s11023-020-09537-4

van de Poel, I. and Kroes, P. (2014) 'Can technology embody values?', in *The Moral Status of Technical Artefacts*. Dordrecht; New York: Springer, pp. 103–124.

Vinck, D. (2019) 'Maintenance and Repair Work', *Engineering Studies*, 11(2), pp. 153–167. https://doi.org/10.1080/19378629.2019.1655566

Weber, H. and Krebs, S. (2021) 'The Persistence of Technology: From Maintenance and Repair to Reuse and Disposal', in S. Krebs and H. Weber (eds) *Science Studies*. 1st edn. Bielefeld, Germany: transcript Verlag, pp. 9–26. Available at: https://www.transcript-open.de/doi/10.14361/9783839447413-002 (Accessed: 19 July 2022).

Wilfred Dolfsma and Rudi Verburg (2008) 'Structure, Agency and the Role of Values in Processes of Institutional Change', *Journal of Economic Issues*, 42(4), pp. 1031–1054.

Young, M.T. (2020) 'Maintenance', in *The Routledge Handbook of the Philosophy of Engineering*. Routledge, pp. 356–368.

10 An Eco-Ethics for the End of the Anthropocene

Finding Ethical and Sustainable Paths through Consumerism, Disposability and Planned Obsolescence

Simon Penny

10.1 A Storm Is Blowing in From Paradise

From where we stand, two decades into the twenty-first century – in the closing act of the Anthropocene we might say – we look back over the detritus of 250 years of industrialism, colonialism and commodity capitalism. We might be reminded of Walter Benjamin's meditation on history, prompted by a small painting by Paul Klee:

> A Klee painting named 'Angelus Novus' shows an angel looking as though he is about to move away from something he is fixedly contemplating. His eyes are staring, his mouth is open, his wings are spread. This is how one pictures the angel of history. His face is turned toward the past. Where we perceive a chain of events, he sees one single catastrophe which keeps piling wreckage and hurls it in front of his feet. The angel would like to stay, awaken the dead, and make whole what has been smashed. But a storm is blowing in from Paradise; it has got caught in his wings with such a violence that the angel can no longer close them. The storm irresistibly propels him into the future to which his back is turned, while the pile of debris before him grows skyward. This storm is what we call progress.
>
> (Walter Benjamin 1968)

That 'progress' has created greater storms. If we look ahead, we survey the prospect of ever-increasing climate catastrophes and calamities: a sweltering planet, a great extinction, dying oceans, wars over water in some places, battles against water in others and apocalyptic fires and storms. In 2022, Pakistan was disastrously inundated. New York experienced monsoons, Hurricane Ian laid waste to large parts of Florida. California experienced droughts and floods simultaneously. The Colorado River is drying up and the great Ogallala aquifer that feeds the US grain belt is nearly

DOI: 10.4324/9781003316213-12

empty. The forests of western USA and Canada, Siberia, Australia and Amazonia erupt into fiery conflagrations. Enormous chunks of Antarctica and Greenland ice are breaking off. Less obviously, arable land diminishes, aquifers salinate, oceans acidify, coral reefs die and fish populations plummet.

The beginnings of global government, the League of Nations and then the United Nations, were catalysed by the horror of the 'world wars'. In comparison, the climate crisis seems diffuse and abstract and the immediacy of local social, economic and political issues, not to mention simple profit incentives, pushes the larger crisis into the background. It may take more dramatic disasters to galvanise nations into concerted action, but the awareness of impending existential crisis is, at long last (and perhaps not too late), growing across diverse communities and especially the youth.

One aspect of this end of the Anthropocene is species extinction. We understand that it is ecologies as much as individual species that must be saved. It is easy to evoke sympathy for turtles snagged in plastic but not so easy for leeches or hagfish or mites. Likewise for grand forests or coral reefs but not so much for saltmarshes or thawing tundra. It is a testament to rising consciousness that awareness of previously obscure(d) aspects of the economy, like the *waste stream* is growing. Food waste, surprisingly, has become a political issue. The re-use of municipal greywater is a topic of public interest. Fast food, *fast fashion* and *single-use plastics* (*SUPs*) have come under scrutiny. Recycling itself has been exposed as a sham. Terms like *Anthropocene* itself, along with *Circular economy*, *lifecycle planning* and *carbon footprint*, are more commonly used and understood.[1]

The idea of sustainability has attained a new urgency, but 'sustainability', like so many overused catchphrases, is ripe for critique. Greenwashed home and body products and appliances exploit sustainability concerns.[2] Purchasing a new, 'green' appliance or smugly sporting clothing containing recycled fibres is evidence of being duped into further consumption. In most cases, not being responsible for further extraction and manufacture is the best choice. Keep the old thing and fix it. The 'right to repair' movement and 'repair cafés' attest to new interest in maintenance and repair – the sad irony being that many appliances and consumer goods are inherently irreparable.

Any thinking person recognises the urgent need for change – political, social, technological and environmental – but few are ready to willingly embrace deprivations. It is sobering to reflect that any middle-class westerner has at their disposal more energy than a king or emperor of the pre-industrial world. The period of fossil fuel use is but a blip on the timeline of humanity, but we in western culture are as addicted to fossil fuels as a junkie is to heroin. As nuclear electricity was once touted as 'too cheap to meter', we are naturalised to the assumption of cheap, convenient energy-consuming infrastructure: we flick on lights, we expect water to flow from the faucet, thermostats run central heating or air conditioning to keep us cosy or cool. Without thought we jump in the car to get groceries, taking a ton of steel with us at speeds once thought deadly for humans (5000 lb electric cars are still cars).

A central question of this chapter concerns the ethical reconciliation of the global with the personal. We are all already compromised to the extent that we take part in contemporary western culture. We travel in planes, with some shame, but we still do it. The industrialised nations we live in have profited from the production of greenhouse gases, that less-industrial nations now suffer from the effects of. As Derrick Jensen argues in his essay 'Forget shorter showers: why personal change does not equal political change' (2015),[3] ethical behaviour on the personal level and effective (political) action are different things. Change at the personal level will not reconfigure the practices of major culprits of climate crime – industries, corporations, cities and states.

So how should one conduct oneself in one's daily, domestic life in order to at least minimise hypocrisy? What kinds of activities ought one engage and which eschew? In what follows, I offer suggestions for ways of reconciling the global and the personal – the result of a lifetime of striving for eco-ethical awareness in lived practice. This essay is written in two different registers: in Part II, I map out a big-picture view of the production, distribution and disposal of contemporary commodities in historical, economic, regulatory and technological dimensions (I live in the United States; many of my examples reflect that context). Parts III and IV are autobiographical, even confessional, outlining my idiosyncratic approach to living that attempts to satisfy, on a personal level, the ethical implications of Part II.[4]

10.2 Swimming in the Global Waste Stream

We exist in a consumer commodity economy that is predicated on continual re-purchasing of intentionally short-lifespan products. Industry and marketing are oriented towards continuous manufacture, continuous (re) consumption of roughly equivalent things and the untroubled assumption of an endless waste stream. This is most obvious in clothing, automobiles, domestic commodities and appliances. Big-Ag has embraced this logic with successive new generations of pesticides and GMO crops, while the food distribution industry deploys 'sell-by' dates that encourage unnecessary disposal. Designed-in use-by dates have different contributing causes: technological change, materials lifetimes and the bald-faced market manipulation we call *fashion*. 'New! Improved!' was a cliché of advertising in the mid-twentieth century – this year's model is just somehow sexier. The exploitation of psychological insecurities to boost sales was pithily captured long ago in the lyrics of the Rolling Stones song 'I can't get no satisfaction'. Manipulation of tastes in cosmetics, clothing, cars and appliances has been a fixture in media for a century or more – in evidence since the early days of the automobile industry – managed by the advertising industry, that became an economic force in the early nineteenth century.

Planned obsolescence, like the growth economy itself, is a twentieth-century notion. The term originated in the USA in the 1920s. It is achieved in three main ways – goods become obsolete, unfashionable or irreparable. As Rivera and Lallmohamed explain:

> Creating goods with a limited lifetime led to increased consumption. It was a business strategy to create mass consumption, which the country needed in a time of economic crisis. Producers who once were producing quality products started to find ways to make goods either more fragile, or difficult to repair so that people would be forced to replace the older version sooner. Processes and devices have been created with a predetermined lifetime to maximize economic outputs.
>
> (Rivera and Lallmahomed 2016)

We are naturalised to a mindset that ownership of a commodity is inherently short term and transitory – from 10 minutes in the case of SUP packaging to 10 years for cars (with clothing and appliances in between). The awareness of the fundamental problem of sustainability is growing. But the global economy is a giant ship – turning it around will necessarily take time. The reconfiguration of cities and lifestyles to lower energy consumption and carbon neutrality will necessarily be a slow and incremental process. All manner of infrastructure and livelihoods will have to be reconfigured and this reorientation will have deep ramifications for businesses, for the economies of states and for the health and welfare of citizens.

10.2.1 Technological 'Progress'

Those of us in a position to read this text are naturalised to relentless technological change, and the rhetoric thereof. Solid-state electronics, plastics, chemical and pharmaceutical industries and chemical agriculture all emerged or grew substantially in the post-WW2 period – all tightly integrated into the growth of the larger petrochemical infrastructure. In the same period, digital computing developed from an obscure research idea to a ubiquitous technology – and a major consumer of energy globally.[5] In my lifetime, the number of artificial satellites orbiting the Earth has gone from zero to around 8000 (the number continues to expand rapidly).[6] Pushing tons of metal up beyond the atmosphere must be the most grandiose display of energy use ever conceived – but who mentions the carbon footprint of rockets?

Among the industrially mass-produced artefacts and rapidly changing technologies of commodity capitalism, *improvement* implies *obsolescence*.[7] Commercialisation of technological developments implies commodities of short lifespan: in periods of technological 'revolution' entire systems rapidly become obsolete: even if you could find a dial phone, there's nowhere to plug it in anymore. No one would use a 20-year-old desktop computer. We have become accustomed to the assumption that Moore's law will extend indefinitely.[8] Unavoidable in the first decades of the computer era, many developments in digital technologies are now seen as self-perpetuating and sometimes frivolous.[9] As often as not, 'this year's model' is dressed-up with fashionable shapes and colours while little has changed 'under the hood'. Bloated code demands faster processors and more memory. Constructed desires drive technological escalations that lead to the naturalisation of expectations. Technological peccadillos become infrastructure or rights: people stream movies onto tiny screens of their phones while on the train to work – oblivious of the vast, energy-consuming technological infrastructure brought into play. In a commodity capitalist culture, the only limit to the existence of such travesties is profitability. The perpetrators of this nonsense should be called out as the environmental criminals they are.

10.2.2 (Fast) Fashion

Today, many consumer commodities are designed to have a short cultural relevance or desirability. Internet niche marketing and sophisticated digitally coordinated manufacturing have moved the process into a hyper-rapid temporal space – China-based international fast-fashion giant Shein is a prime example. This fashion churn drives unsustainable production in a number of ways – aside from stimulating the creation of petrochemical-based fibres and textiles (nylon and polyester), the expected short cultural lifespan of such goods predicts the use of cheaper materials that wear out or break quickly. In addition, the vagaries of markets create environmental outrages like the mountains of unsold fast fashion in the Atacama Desert, where the high UV turns toxin-exuding synthetics into microplastics (Duong 2021). If you make a good pair of shoes, I will buy one pair in maybe 10 years. If you make shoes that fall apart, and cannot be fixed, I will be forced to buy a new pair every year. You will get my money 10 times. Who can argue with that? Certainly not the bosses of industry nor Nike stockholders nor the planners of national economies. The fashion-conscious will not wear last year's sneakers anyway.

10.2.3 Things Fall Apart

In the contemporary consumer commodity system, things are designed to break, thus requiring (regular) repurchase and resisting both repair and recycling (Slade 2007). Chemical and materials engineering is sophisticated; materials are designed at a molecular level. An entire branch of the discipline of materials science is dedicated to making things break on time – manufactured materials with a designed-in lifespan that are engineered to decompose at a certain age, seldom into Earth-friendly detritus – timed to fall apart in accordance with the calculated buying power of the customers for that product. The plastic that comprises semi-structural parts in your car will crumble, fracture and decompose into toxic dust in 10 years (give or take for ambient UV). Tyres, hoses, wiring insulation, interior panels, upholstery, window seals and 'disposable' air and oil filters – when those materials begin to lose their structural integrity, there is no fixing them.

Composite and synthetic materials are cynically designed to fail, with reliability and precision, at a particular age, not unlike the replicants in Blade Runner (or indeed, like all living things). We have succeeded in creating materials that fall apart that no known organism can metabolise.

The global plastics crisis – nonbiodegradability and the impossibility or nonviability of recycling – has resulted in the global ubiquity of microplastics and a recognition of their dire health impacts on individual organisms, species and entire ecosystems. Petrochemically derived synthetic materials dubbed 'forever chemicals' are particularly egregious: they enter the waste stream but are not reintegrated into cycles of biological metabolism. Acronyms for deadly chemicals such as BPA, PFAS and EDCs (endocrine-disrupting chemicals) are now used in daily language.

Humans have battled the destructive qualities of weather, rodents and micro-organisms on organic materials for millennia – wood rots, leather gets hard and cracks and wool is chewed by moths. But that destruction is also salvation, micro-organisms turning detritus into bioavailable molecules to be truly recycled, in the great ecological chain. Planned obsolescence need not imply irreparability nor unrecyclability nor non-biodecomposability, but it usually does.

10.2.4 'Disposable' Packaging and Single-Use Plastic

The infamous 1960s 'Crying Indian' ad campaign was a classic case of corporate cynicism in response to the 'litter crisis' created by disposable packaging – much of it the new technology of plastics.[10] This faux PSA (public service announcement) ad campaign produced by an organisation called Keep America Beautiful in association with the AdCouncil shamed the public and made cleaning up 'litter' a civic responsibility (Summers 2019). If the inherent cynicism was not enough, the 'Indian', Iron Eyes Cody, was actually a Hollywood extra of Italian descent who claimed some indigenous heritage.[11] The 'Crying Indian' campaign put what should have been a governmental or corporate responsibility onto the shoulders of consumers. The Indian is still crying.

The litter crisis was a result of the use of 'disposable' packaging components that supported the new drive-through junk-food industry that emerged in parallel with the rise of car ownership. Automobile-based dining – only in America! Seen as a convenience half a century ago, disposable packaging and SUP food containers (polystyrene foam, compressed styrene, polythene, plastic-impregnated paper, etc.) have become a global environmental disaster. Some things are necessarily consumables – my food, the energy I use to cook it and the water I drink. But why should I buy a new cup with every cup of coffee, just to discard it when the coffee is gone? This is a kind of consensual socio-economic insanity. People all over the world use and have used 'disposable' food containers – in Oceania, banana leaves, and in Mexico, lightly fired terracotta – but crucially, these materials bio-decompose.

Ludicrously, most of the plastic bottles we consume exist to move water. Domestic and personal cleaning products – laundry soaps, floor cleaners, shampoos and so on – that line the shelves of supermarkets are mostly water, diluted for 'convenience' and no doubt for profit. So vast quantities of water are trucked around the country in specially made plastic bottles. The solids in a large bottle of shampoo would not fill a shot glass and a bar of soap will make a gallon of 'liquid soap'. Taking the water out would substantially reduce packaging, retail floorspace and diesel consumed in transport.

10.2.5 There Is No Unscrambling Some Eggs

Composite materials, combined with planned obsolescence, make a mockery of pretenses of recycling, maintenance and repair. Appliances and furnishings – refrigerators, office chairs, printers, phones and synthetic carpets – are all diabolical cocktails that defy demanufacture. The amalgams of materials in our commodities make them virtually impossible to recycle and reclaim. There is no way, or no economical way, to tease out and sort those molecules back into their categories. Such diabolical composites are everywhere: combinations of metals and plastics in microelectronics, combinations of paper and plastic in the humble milk carton, laminations of different plastics in 'disposable' plastic water bottles or the mixtures of synthetic and organic fibres in a polycotton shirt. In the humble sneaker, the sophistication of manufacturing processes and the bonding of materials makes them irreparable and inherently unrecyclable *because they are composites*. Miniaturisation and encapsulation of subcomponents prohibit repair. No one repairs a microprocessor – there is no fixing such things. In such contexts, repair amounts to pointing a diagnostic tool at the system, throwing away a faulty component and ordering the replacement component by the serial number in an Internet-linked database.[12] It is ironic that the most sophisticated forms of technology that humans have made cannot be repaired.[13]

10.2.6 Recycling – The Big Lie

Corporate support for recycling policies can be interpreted as a way to reduce consumer resistance to consumption of 'disposable' materials by assuaging the potential consumption-guilt – business, apparently, knows no shame. But 'recyclable' almost never means 'will be recycled'. The percentage of plastic recycled in 2022 is estimated at 5%, recently downrated from the 9% estimate of recent years.[14] While we dutifully clean and sort our recycling, most of the waste we put in the recycling bins ends up in landfill.[15] The remanufacture implied in recycling is ecologically fallacious – the energy and material consumption of recycling is unsustainable, ecologically and economically. What little recycling does occur consumes energy and new toxic materials (plasticisers and the like), creating toxic waste and liberating potent greenhouse gases. Recycling is often predicated on the availability of cheap energy. The processes involved in recycling other materials we purport to recycle – aluminium, paper and steel – are barely economically viable, due in part to high labour and huge inputs of materials and energy.

Much so-called recycling is down-cycling. The 'recycling' of glass by my local trash service amounts to crushing it for roadmaking – turning food-grade glass into gravel hardly constitutes recycling. I know how much fuel it takes to turn broken glass into new bottles. I had a friend who was a craft glassblower, he had to save for three months to afford gas to fire his kiln and lehr for one work session.[16] After 10–12 hours firing the kiln, burners roaring, a few kilos of glass was molten, so hot it glowed like the sun. Having brought the glass to working temperature, he had to work non-stop for 36 hours. Working on such a dangerous material while sleep deprived was always a fraught proposition. Then, he would sleep while the red-hot glass objects underwent controlled cooling (annealing) for days (or they would explode).

When I was a child, our milk was home-delivered. Not so many years before, the milkman had a horse-drawn cart – the horse knew to walk slowly along the street, keeping pace with the milkman as he went back and forth from cart to doorstep (who needs self-driving cars?). When the milkman left the milk, he would take our empty milk bottles. At the dairy, they would be washed and refilled. Some of our milk bottles were burnished and scratched, evidence of hundreds, perhaps thousands of cycles of cleaning and refilling. How many bottles can you sterilise with the energy it takes to remake a bottle out of melted crushed glass? Thousands?

10.2.7 *Growth and Degrowth*

Consumer spending currently comprises about 70% of US GDP.[17] The economy we know is predicated on repeated consumption, of the same things, or the new model, over and over again – often driven by 'fashion', deployed as a driver of consumption. *Buy it today, buy it again tomorrow.* Incessant production, incessant consumption of resources and power and incessant creation of waste: this engine of the economy is constructed around the assumption of continuous, repeated consumption and continuous disposal. Goods that wear out with no potential for repair. The waste stream is the skeleton in the closet of commodity capitalism.

If we bought only what we needed, bought things that did not wear out or become obsolete, local, national and global economies would crash. To abandon disposable products and inbuilt obsolescence, to abandon consumerism, would destroy the economy as we know it. Yet to continue on this path is to ensure environmental destruction. This puts us in an awkward position: the global commodity economy itself resists 'maintenance'. The growth economy is unsustainable.

What should be sustained? If the continued functioning of a system depends on the consumption and throughput of resources that continuously enter the waste stream, unreusable and unrecyclable, is that system sustainable? Viewed this way, the commodity economic system is like a voracious animal, consuming and excreting. But unlike the cowpat, the diesel particulates or the used toner cartridges are not input resources for another (biological) system. Externalities be damned! In a closed system, there are no externalities – the biosphere is a closed system. Choosing to manufacture the disposable, the unmaintainable is, in terms of eco-ethics, unacceptable.

What is ethically sustainable? The fallacy lurking at the heart of much sustainability discourse is the assumption that the profligate consumption we are used to is in any way sustainable. Just because we are accustomed to taking a ton of steel with us to get the groceries or leapfrogging vast oceans in a few hours, or for that matter, leveraging global digital networks so we can watch an episode of an old sitcom when we might otherwise be looking out the window of the bus, does not mean that these can or should be sustained.

10.2.8 *The End of the Anthropocene*

Increasingly, I feel like I am living in the final act of a global Shakespearian tragedy. We stand on stage, appalled and aghast as the bodies pile up. The longer-term results of impetuous avaricious behaviour are revealed.

Except perhaps for the very rich, luxurious lifestyles predicated on an abundance of low-cost energy, access to information, freedom of movement by land, sea and air, sophisticated medicine, disposable wealth, even access to water, seem likely to abate. The global poor will, inevitably feel it first and worst; indeed, they already do. Populations of climate refugees will inevitably grow. Storms, floods, fires and droughts will contribute to the failure of infrastructure.

We live immersed in highly interconnected systems that have built upon other systems over generations. My home Internet comes on repurposed cable TV infrastructure that literally hangs off power infrastructure installed 100 years ago. At some point, the interdependence of the systems we depend on to facilitate new projects will begin to degrade and the economic load of maintenance will itself become unsustainable. Components of subsystems will fail, bringing down entire systems. In some US cities, water infrastructure is failing – due in part to inadequate or deferred maintenance. Restoring century-old piping now lying under roads and buildings is a gargantuan engineering task. Some things are so broken that there is no fixing them.

Surviving with trucked and bottled water and porta-potties in anything but the short term is unsustainable. Industries and services and stores will shut down. The quality of life in those places will plummet and they will depopulate. Collapse of some kind seems inevitable. The best we might hope for is to mitigate, and hopefully survive, climate crisis and engineer more parsimonious modes of living that maintain health, safety and dignity. There is good reason to think that even this humble aspiration may not be possible – that climate pressures will outpace us.

Russell Hoban published Riddley Walker (Hoban 1980) at the height of the global fear of nuclear apocalypse – miraculously (that time), we escaped the nuclear annihilation that had so many of my generation living in nihilistic dread. The novel is set in southern England a century or two after a nuclear apocalypse: humanity has regressed to a kind of iron-age culture. Radioactive bogs are dredged for C20th iron detritus – bits of buildings, railways and factories. The scenario that Hoban laid out may as well depict human life in a post-Anthropocene world, adapted to a depopulated and detechnologised normal.

10.3 Bricolage Aesthetics: Maintenance, Repair and Salvage

Maintenance, at its most basic, implies continuation, as in maintaining a vigil. Mark Thomas Young offers '*a preliminary conception of maintenance as an activity through which we provide stability to something over time*' (2020). This stability clearly includes less-tangible things: institutions, traditions, relationships and 'the peace'. In the context of this anthology and similar concerns, 'maintenance' is often associated with large-scale technological systems, civic infrastructure, industries and the built environment (Henke &Sims 2020). My concern here is with tangible things, artefacts and skills, in a more humble, domestic sense. Sweeping and dusting and vacuuming and doing laundry: such (often gendered) tasks do constitute maintenance. Maintenance implies managing wear, the relentless assaults of entropy and decay. It seems to imply stasis: to bring the artefact back to the as-new condition, as closely as possible – as often as not, these days, with OEM replacement parts purchased online via a part number. In some contexts, that might mean yet another coating of paint or the replacement of a broken part with a new one.

Sewing on a button, sharpening the kitchen knives, re-grouting the tile-work and pruning the fruit trees and grapevines – these maintenance tasks blur into 'repair'. The transition from maintenance to repair usually involves replacement or addition. Repair implies know-how, skill and experience: 'I know how to fix that'. At the end of repair that is furthest from maintenance, the work has more creativity to it – improvisation and bricolage. Re-creation might not be an overly pretentious descriptor (restoration is something else again: a fetishised, obsessive maintenance tinged with nostalgia) (Figure 10.1).

Figure 10.1 Kettle repaired (by the author) with a piece of rosemary wood that had grown in his garden, bound with marline.

Photo: S. Penny.

What happens when something is irreparable? A less-aware person will consign whatever it is to the waste stream and the more aware will *repurpose* – if we must buy yoghurt, we can at least use the plastic containers for something. Some well-meaning but misguided souls ardently craft plastic waste into nasty handcrafted sustainable folk art – plastic bottles as plant pots and so on – as if gilding such trash with folksy aesthetic somehow redeems it from being the poison it remains: no silk purses from sow's ears.

A more extreme form of repurposing is *salvage* – the territory of scavengers, gleaners, lowly rag and bone men, scrap metal merchants, trash-sorters and dumpster divers. My neighbour replaced her fence, the bottoms of the palings were rotted. They were left in a roadside pile, destined for landfill. I took the planks and trimmed them. I clad a shed with the perfectly usable, if weathered, lumber. The offcuts became firewood or mulch.

10.3.1 Salvage Eco-ethics

Beyond, or beside, an ethics of maintenance, is an ethics of repair, and beside those, an eco-ethical responsibility to salvage. Permaculture, the radical environmental philosophy developed by Australians Bill Mollison and David Holmgren in the 1970s is mostly known as an organic gardening and regenerative agriculture philosophy. A central concept in permaculture is that there is no 'waste'. Any output of any subsystem is the input to another subsystem – even sewage. *Salvage eco-ethics* aspires to a condition in which there is no waste. It is intentionally parsimonious in the sense that it is always conscious of economising energy and material resources.

Self-evidently, in a sustainable world, making things once, robustly, with a long life, is the way to go. If they break, they can be repaired. If they come to the end of their useful life, their parts are salvaged and repurposed. An ecological ethics of poverty, of scarcity, is an appropriate antidote to our naturalisation of the (illusory) overabundance of the fossil-fuel-driven orgy of commodity economics. Why would anyone straighten a used nail when nails are so cheap? But nails are *not* cheap by any real accounting. Iron ore is mined on one continent with giant trucks and high explosives, shipped across vast oceans in diesel-powered mega-ships, smelted and refined and manufactured in factories with giant furnaces and then shipped again. No, an eco-ethics demands that I repair and reuse that nail if I can.

10.3.2 An Autobiographical Intermezzo

When I was about 12 years old, I had an epiphany regarding the finiteness of the planet: *you can't throw anything away*! You just put things somewhere else. This must have been about 1967, before the iconic 'blue marble' images of our planet from space that jump-started the environmental movement and adorned the cover of the *Whole Earth Catalog*. Who knows how this idea popped into the psyche of a boy in suburban Sydney in the mid-1960s? Did some person, some book, put it in my head? I don't think so – more likely a Daffy Duck cartoon. A few years later, as a teenager, I would wander the rocks of the estuary near where I lived and find tangles of fishing line, which I would take home and painstakingly untangle. I never used the line, I wasn't that interested in fishing. On a backpacking trip as a young man, I spent several days in the yard of the hostel I was staying in, tidying accumulated detritus in a large garden into organised categories – rocks, dead leaves, branches, bits of metal and junk. I was driven by a deep, anti-entropic desire to reorder. I'm still not sure why I did it. I thought myself an artist at the time – I thought it was a kind of artwork.

At art school, I gravitated towards sculpture – there was a workshop and tools. I had the great good fortune to become a student of one Owen Broughton: sculptor, foundrymaster, anthropologist and tool collector. The profundity of what he taught me has only grown with time. He had a seemingly infinite knowledge of tools, artisanal practices and manufacture. He had been Henry Moore's foundry foreman, responsible for successfully pouring those elephantine bronze castings. He could tell you how in India they quarry granite with driftwood, how the Romans cast the bronze soles of centurion's sandals, how Australian colonialists surveyed land on horseback. He had a collection of barbed wire, meticulously set in hardboard pages of an album 3 feet high. If you bothered to ask, he could tell you how, when and where each sample was made and on what kinds of machines. I learned toolmaking at the forge from him. We annealed old files and reworked them into chisels and drawknives – a rusty file is still a useful billet of carbon tool steel, after all.

Throughout my career as a sculptor, I consistently *upcycled* materials, decades before the term was coined. I scavenged scrap metal yards and came away on my bicycle with useful materials and treasures – memorably, on one occasion, towing a 20'-long bundle of aluminium tubing along heavily trafficked roads: a bicycle semitrailer. I raided county dumps in the hills for tree trimmings that would be fashioned into finely articulated structures. I made paper out of dryer lint and dust vacuumed from the floor – the very definition of waste. As a child, in somewhat impoverished conditions, I cobbled together rudimentary vehicles from bits of wood and tin can and wheels scavenged from old prams found in dumps, then careened down dangerous slopes, narrowly avoiding culverts, cars and serious injury (little boys can get themselves into no end of trouble). Have I embroidered an ecological and theoretical justification around the psychic effects of that impoverishment? – quite possibly. Be that as it may, as long as I can recall, the work of repair has been present in my psyche as a *responsibility*. The discarding of almost anything is difficult because everything, presumably, has a use.

I live in an increasingly desertified country, but I have less guilt about water use because I know the water I bathe in positively contributes to the plants I grow and what doesn't percolates down into the depleted aquifer. My bath drains into the garden, where the trees enjoy the phosphates in the organic soaps I use. The trees, while photosynthesising, provide shade, keeping the soil cooler and reducing surface evaporation. They provide fruits containing some of that water, transformed into nutritious juices, which I then consume and contribute my urine back to the soil (urine, unsurprisingly, is an excellent 1,1,1 fertiliser). In this little universe, everything has value, is used and is transformed. The yard of my current house was arid when I found it. Now it has inches of mulch and a thriving soil microbiome – a bustling ecosystem of bugs and worms and spiders. Worms are turning up in places they never were before, and birds are coming after them. Organic matter has never left the property – even the coffee-grounds and used paper towels are not 'trash' – but the makings of mulch. Lizards and hummingbirds snap at the fruitflies rising off the compost heap. It is a humble triumph. Every week my neighbours line their curbs with enormous trash and recycling rolling bins, filled, often with lids open, overfilled. Offerings in the cult of the waste stream. Our bins are always almost empty (we are compelled to discard some things – mostly toxic or plastic things that have no use to any organism).

10.3.3 Shoes and Socks

Now I worry about old socks, worn at the heel or toe. Socks made from sophisticated polymers derived from the residues of fossil forests, dispersing festoons of toxic microfibres. What am I going to do with them? I do not want to throw them away to add to the 95% of all plastic ever manufactured, now discarded, dispersed in the waste stream that now encompasses the entire planet. With the best will in the world, it may take months before I darn them. So they sit, not usable but not waste, taunting me with their poisonous existence. They can't be recycled, at least not without the consumption of more energy (probably more fossil fuels) and not without the addition of more toxic chemicals. I can't burn them – God forbid! They won't rot in the compost heap – no known creature will feed on them. They are microscopically immortal. Their parts will divide to invisibility, but they will continue to invade bodies and wreak metabolic havoc. So must I keep these socks, and plan to one day, darn them? But who darns socks when you can buy a new pair for a dollar? These damned cheap colourful plastic socks with the fashionable logo – dragged across giant oceans in giant iron ships burning bunker fuel at hundreds of gallons per hour, belching toxic fumes. What a bargain! I will wear them around my neck, like a dead albatross, forever, to remind us of our sin as a culture, of making things that are foreign to the Earth: a betrayal of the planet and our responsibility to it.

And these shoes, all these shoes, bonded in industrial autoclaves that cannot be repaired. I've tried, but glue and stitching and rivets just don't jive with industrial vulcanisation or plastics that are designed to fall apart. The sad fact is that, even with my library of tools and encyclopaedic skills, I cannot fix them, as much as I feel the responsibility to try. I have a pair of work boots, with good solid soles with plenty of tread left. They have well-made leather uppers, with good eyelets and durable laces. But between the insole and the sole is a layer of what was some kind of foam, that has decomposed into toxic goo. There's no fixing that. And there is no obvious pathway to making the boot usable again. Irreparable. An eco-ethical disaster. What am I to do with them, these lingering spirits of commodities, ghosts of footwear, casualties of a war of price-cutting and market share?

10.4 Doing the Eco-ethics of Salvage

In Section 10.4.1, I describe idiosyncratic personal solutions to the question of ethically sustainable lifestyle. I set out to be this person. For decades, I have strived to understand the composition and manufacture of things. The knowledge or 'literacies' I draw upon are at least as much know-how as know-that. As an artist and artisan, a latecomer to academia, I am increasingly aware that my references and resources are different from my primarily book-learned colleagues. I draw upon my extensive hands-on experience in metal fabrication, electronics and electromechanics, boatbuilding, home renovation, precision machining, sailing and gardening. I know what it takes to design and make a machine that works, a structure that stays up and a hull that withstands often violent natural forces. I know the qualities of iron and steel, of timbers hard and soft, of rock and concrete and of plastics and composite materials. I know how to put them together, with fasteners and adhesives and molten metals, and how to take them apart again, with blowtorch, wrench, saw or sledgehammer. I know the kinds of concentration involved in such work, and the fatigue of the body, the calluses on the hands, the regular superficial wounds: cuts and scrapes and bruises, splinters of metal and glass and wood. There is know-how, and there is also a pragmatic, artisanal know-that – significantly unlike scholarly know-that.

10.4.1 Life in the Workshop – Repair, Bricolage and Improvisation

Pursuing salvage ethics implies skill and experience with (and access to) tools and materials – a kind of intimacy. It also calls for a kind of patient humility. Such a path is not for everyone – some are just disinclined, some are intimidated and some resist the accumulation of the necessary stuff. It has its psychic rewards, not simply assuaging guilt, but, on good days, a successful experience confers a sense of creative competence.

For better or worse, I see value in material detritus that most people would toss without a thought. I straighten used nails I pull from wood and I'll recut a thread on a burred bolt. The broken drill bit – a precision hardened shaft of a specific diameter. The usefulness of a used paint can or milk carton does not escape me nor do scraps of wood, down to tiny pieces – wood scraps are always useful. Ditto string and cordage and rope: when you need a small piece, why cut a big piece? Even if an appliance is trash, there are salvageable cables and fasteners and hose clamps and a switch to scavenge. In my line of work, there's no telling what will come in handy – that scrap of wood may be just the thing to raise up a workpiece to level or support a piece under the drill or mix the paint or spread the glue. After which use, it might be even more useful or find another use. And if they're no good for that, they're firewood, sawdust or mulch, of course.

I fix things. I do have a fondness for tools and restoring tools, not in some fetishistic collector way, but to bring them back to a good usable state. A friend once picked up a hammer in my shop. Its handle, cracked near the head, I'd bound with copper sheet and bronze wire. That's love, he said. I guess he was right. I do feel a responsibility to make things better. Especially things that have originally been built with skill and care and have perhaps, seen hard use or abuse. I do not feel a compulsion to lavish care on a damaged Ikea flatpack bookshelf. That material cannot be repaired. But a chair that has pieces that were identifiably once part of a tree, perhaps some joinery made with skill and care – that deserves respect.

10.4.2 Taxonomy, Organisation and Sorting in the Shop

Maintaining the material resources for general repair is a not-insubstantial undertaking. Having accepted the responsibility to maintain and repair, and to accumulate relevant parts and tools according to salvage ethics, maintenance of order is implied. There's little more frustrating than poking about trying to find that thing, or tool, you know you've got, somewhere. Worse – infuriating – is when you can't find it and go out and buy one, only to find it a week or so later. It's important to know what you've got and what you might need to get more of, so you don't run out at a crucial moment in a project. You need to know where it is when you need it. It's not useful if you can't find it or have forgotten you have it. And it's amazing how quickly you forget – not *that* you have it but *where* it is, so part of the work is regular reinforcement of memory. This occurs organically in the regular working and ferreting around: 'so *that's* where I put that collection of scraps of piano-wire – in the small chain and lightweight cable box!'.

There is always order, and there is always what looks like clutter and chaos – the things in the process of sorting, the salvaged whatevers: a pair of hinges, an electric motor, a sheet of glass – a standing reserve of other people's trash – have to be stored, in the right place in this demented Wunderkammer. How well sorted is sorted? – must every screw be identified, labelled and stored in a knowable place? No, that would be going too far. But everything in a big pile: that would be useless. Each individual item requires some consideration – which category is for it? Do the categories have to be altered? And categories are always mutable. Some systems are absurd – only a fool would organise their bookshelf according to the colour of the spines.

The contents of the workshop are too contingent and fluid to be exhaustively notated. Like any library or warehouse, the system of organisation has to have a certain spatial logic to it and a certain granularity. To sort is to define categories at a pragmatic level of detail. But total organisation is a Borgesian impossibility because things are like and unlike other things in infinite ways. The organisation of animals in the *Celestial Emporium of Benevolent Knowledge* related by Jorge Luis Borges is a wry epistemological joke. It divides all animals into these categories:

Those that belong to the emperor
Embalmed ones
Those that are trained
Sucking pigs
Mermaids (or Sirens)
Fabulous ones
Stray dogs
Those that are included in this classification
Those that tremble as if they were mad
Innumerable ones
Those drawn with a very fine camel hair brush
Et cetera
Those that have just broken the flower vase
Those that, at a distance, resemble flies

This list (so admired by Michel Foucault), appears in Borge's short essay on John Wilkins, the Baroque taxonomist. Borges observes 'there is no description of the universe that isn't arbitrary and conjectural for a simple reason: we don't know what the universe is' (1942).

10.4.3 Pathological Maintenance

About a year ago, I went to an estate sale to buy an old machine lathe. I found the residue of the working life of a man who, for 30 years or more, had built model boats. Not my thing really, grown men making small-scale models of boats they might have known or simply fantasised about as kids. But I could read his backyard shop. In its disarray was the history of a life dedicated to a craft. Scattered all about, customised tools, scraps of careful drawings, hand-made cardboard patterns, jars and little drawers of parts – a chronicle of a life and a working process. Each little thing, down to the scars on the edge of the workbench, spoke of purposive practice. I saw the little scraps, the offcuts saved, and (mostly) I knew why they'd been kept and what they might come in handy for. I felt like a knowledgeable detective or archaeologist.

I made an offer and bought it all. All that junk. Ostensibly, because I knew that amongst it all might be parts for the lathe that might be otherwise unobtainable. I had a complex feeling, a combination of respectful empathy and a little mania, or at least compulsion. I drove away feeling simultaneously that I had a treasure trove in the back of the truck; doubting my own sanity; and dreading the task I had now committed myself to. Then, for over a year, sporadically, I sorted and categorised broken tools, outdated technological components and tiny fasteners, that may or may not have value for me or anyone. The decision-making often outweighs the value of the object – the work becomes demented and obsessive. I am wasting my time, but I cannot throw it away. I am yoked to this labour, not Sisyphisean, but obsessive and of dubious value. What is it that against my better judgement, persuades me that this is good work that needs to be done? What dues am I paying and to whom? I became, by a combination of intention, accident and predilection, the custodian, the archivist, of an eccentric, encyclopaedic collection of diverse fasteners. I'm sure I could write a long and boring book just describing their variety just as an entomologist would describe beetles – a taxonomy of threads and head shapes, exotic, special purpose and common. Some people organise dead beetles and butterflies and some organise postage stamps or coins. I organise nuts and bolts – ostensibly for utility, not so much in the interests of science or history.

My fetishism is not that of the 'collector', looking for that elusive magazine issue, a 'mint' 'NIB' action figure or the missing piece to complete the set of whatever.[18] Neither is it the lure of the nostalgic novelty – the old tin toy or the bakelite comb. To say that my collecting is utilitarian and motivated by ecological ethics is only half of the story. I have an abiding interest in the way materials fit together and the way forms can be coordinated in movement: the taxonomy and topologies of mechanism, especially ingeniously simple ones: did you know you can use a coil spring to constrain a shaft to rotate in only one direction? This collection is a library of mechanical ideas.

I once roomed with a friend in his deceased grandfather's suburban house. At the bottom of the garden was a shed, untouched more or less, since the old man demised. The great depression left its mark on that generation – the shed was cluttered, full of stuff, much of it in labelled reused wooden cheese boxes: bits of old leather for shoe repair and parts for garden hoses. A dusting of fine dust covered it all. My friend maintained it as a kind of museum for the old man. At the time, I found it quirky and anachronistic. Now I have become that man.

John Haugeland once boasted that he had more nuts and bolts than any other philosopher. The remark endeared him to me, but philosopher though he was, in the world of nuts and bolts, he was a child paddling in the shallows. Wordsworth found infinity in a grain of sand. There is an infinite variety of nuts and bolts: the thread size and type, the length, the material and the kind of head. Eleanor Rosch's 'principles of categorisation' apply, but there are always 'boundary objects' (1988).[19] This organisation is a kind of Goedelian game where contingencies and exceptions undermine reassuring logical structures. There are threaded shafts that are not quite bolts but bear a family resemblance. This taxonomical task is fascinating precisely because it seems to exceed nomenclature. *The order of these things remains resident in the relationships between the things themselves, and not in some symbolic descriptive order by which they are described.*

10.4.4 The Obsessive Bricoleur

My mania is practical. It begins, usually, with a thought: 'that could be useful for…'or 'with a little work I could fix that'. I look at the lace hooks on the pair of old hiking books whose soles were somehow defective and cannot be repaired: I could reuse those lace hooks, they are good lace hooks. So begins a design and planning process – how to detach them without breaking them, how to reattach them and what special tools would be needed – salvage ethics at work. Once this thought process is underway, the tools and materials are marshalled, first mentally, then physically – the rasp or the file or the wire brush, the sandpaper or the steel wool or the abrasive pad, the pry bar, the chisel, several mallets and hammers, and so it goes. The process depends on a pre-existing, more or less encyclopaedic array of tools for working different materials, at different scales, from the size of a coin to the size of a house.

I found myself regretting that I'd thrown away a plastic toilet brush that must have cost almost nothing new – another nasty, short-lifespan, mass-produced plastic object made to wear out and be tossed and replaced with an equally diabolical piece of crap a year or two later. Not SUP in the usual sense, but in geological time it's all the same. But my regret was geometrical: the handle was conical, finely tapered from about 3 cm to about 1.5 cm. A very handy object when you need to connect, say, slightly different diameters of hose. So who would want that? Only someone who does that sort of thing from time to time, like me (these machine-made commodities are paradoxical – they are worthless, but exhibit a level of precision that would have been regarded as miraculous barely a century ago). It is an open question whether this work is a blessing or a curse. It is handy to have the thing you need when you need it.

There is something deeply satisfying about making some broken thing – whatever it is, a sapling with a broken stem, a rusty tool, a worn piece of clothing – whole again or at least serviceable. There is some self-care in this, combined with a compulsion about it that at times seems unbalanced. Oh, if only, sometimes I could just throw something away. How I dream of being so oblivious that I could simply dump the old printer in the trash. How I wish I believed that recycling actually worked. But I know too much. I wear my inability to throw things away like some penitent. It's not that I want to save it, but it's not useless – that t-shirt with a hole or two. That old sneaker – if it goes into landfill it will be there...forever? How can I be responsible for that? I only keep things that are useful. But, unfortunately, everything has a use. It's just a question of whether you have the knowledge to 'see', and if you have the skills, tools, time and inclination. It is my misfortune to have all five. I was bemoaning my condition recently and some well-meaning type said – you're an artist – you could make junk sculpture. I responded, with wry self-reflection, that this was not possible because nothing is 'junk'.

10.4.5 Hoarding, Collecting, Exhibiting and Purging

I knew a guy who could not throw away newspapers. His house was stacked with piles of broadsheet – I was amazed the floors supported its weight. Reading Katie Kilroy-Marac on collectors and hoarders, I was heartened that (contrary to some suspicions) I do not appear to possess Hoarding Disorder, as clinically defined (2018, pp. 20–38). Hoarders, who tend to be lower income (at least when depicted on reality TV), are considered misguided or even ill. Yet, 'collectors' who tend to be more wealthy are tolerated as eccentrics or admired (especially if they are potential philanthropists). Some kinds of collections – archives museums – are positioned at the very pinnacle of culture. Bennett (1988) describes an 'exhibitionary complex':

> ...history and natural science museums, dioramas and panoramas, national and, later, international exhibitions, arcades and department stores – which served as linked sites for the development and circulation of new disciplines (history, biology, art history, anthropology) and their discursive formations (the past, evolution, aesthetics, man) as well as for the development of new technologies of vision.
>
> (Bennett, 1988, p. 73)

In this juxtaposition of hoarder and collector, a third, possibly equally pathological category – the *purger* – goes un(re)marked. While, superficially, embodying the virtue of tidiness (Marie Kondo style), the purger presumes the waste stream, takes for granted the huge hidden industry of disposal, which literally sweeps our trash 'under the rug', bulldozing a skin of soil over vast toxic landfills. Immersed in commodity capitalism as a fish is in water, we inevitably position ourselves, intentionally or unintentionally, in relation to the obscene accumulation of things and stuff that characterises our culture.[20] The conditions of the hoarder, the collector and the purger bear an uncanny but perhaps unsurprising similarity to the array of eating disorders. The lesson in both cases is the same: too much and too little are both undesirable. Tread lightly on the Earth and tread lightly on oneself as well.

10.5 Conclusion

In the foregoing, I have sketched the outlines of an argument for eco-ethics, not in terms of a philosophical argument, but in relation to the climate crisis, commodity capitalism and environmental sustainability. I've traced an argument from the global: laying out the realities of the waste stream and inbuilt obsolescence to the idiosyncratically autobiographical: describing the way I have come to conduct and position myself with respect to the social, biological and technological aspects of my world. On a personal level, my solution is simple – a permaculture eco-ethics: consume as little as possible (especially of non-biodegradable materials); repair and repurpose; be hyper-conscious of energy and resource use and minimise waste: a mindful salvage-ethics.

It is important to recognise that this is a two-way flow – the personal also informs the political. My day-to-day grappling with tools and materials informs and grounds my activism. But we are all (already) compromised, enmeshed in systems that often have us doing things we would prefer not to. We ought not to let the perfect be the enemy of the good.[21] As Sami Grover has noted, we are all climate hypocrites, but we must not let that stymie us (2021). Given that modern economies run on consumer spending, Anna Lappés remark 'Every time you spend money you're casting a vote for the kind of world you want' (2003) is a recipe for quietly reconfiguring the big picture. But voting with your wallet is not enough. We should not imagine that ethical personal practices or lifestyles are a substitute for socio-political action.

Notes

1 The term carbon footprint, ironically, was coined by British Petroleum (BP) in the early 2000s. It was based on the concept of 'ecological footprint' by William E. Rees and Mathis Wackernagel (University of British Columbia, 1992).
2 Markets, apparently, will cynically work any societal trend for profit. Perhaps, the most absurd example I've seen is a benchtop blender for food waste – shiny new white plastic with electric motors that seems to confuse simple maceration with bioactive composting.
3 This essay has been made into a powerful short film by Jordan Brown (2015).
4 I am not qualified to position this text in the context of the academic ethics discourses, so with apologies, I leave that for my ethicist colleagues.
5 Different estimates suggest that power consumption of the Internet, globally, exceeds that of all commercial aviation or all automobiles.
6 Of which nearly 5000 are 'active' and, as of mid-2023, over half belong to Elon Musk.
7 We assume there will be no (new) internal combustion cars in 20 years. But some things are less susceptible to such obsolescence: paper is paper, bricks are bricks and knives and forks remain knives and forks. In old-world cities, people commonly live in houses and walk on roads (and send their waste down sewers) built hundreds of years ago.
8 Gordon Moore was Intel's longest-serving CEO and chairman. In 1965, he proposed that roughly every two years, the number of transistors on microchips would double and hence computer performance would constantly increase. Now known as Moore's Law, this held roughly true for 50 years. More recently, there have been arguments that we are reaching the physical limits of such miniaturisation.
9 We do not need yet another style of miniature power plug. In 2022, the EU approved a 'common charger directive'. By the end of 2024, all mobile phones, tablets and cameras sold in the EU will have a USB Type-C charging port.
10 On corporate cynicism and advertising, see Oreskes and Conway (2010).
11 A national nonprofit public education organization named Keep America Beautiful, Inc. (KAB) was formed in 1953 with the mission of 'engaging individuals to take greater responsibility for improving their local community environments'. KAB's first PSA focused on litter prevention. It partnered with the Ad Council in 1960 to produce a campaign focused on the harmful environmental effects of litter and other forms of pollution. Ten years into the KAB-Ad Council partnership, in 1971, an Italian actor playing the part of a Native American (who became known as 'The Crying Indian') or 'Iron Eyes Cody' appeared in an anti-litter commercial. As he looks over a polluted landscape and sheds a tear, a voice-over says: 'People start pollution. People can stop it'.

This powerful commercial won many awards, including being named one of the top 100 advertising campaigns of the twentieth century by Advertising Age. Its success inspired other environmental messages from other groups as well. https://www.psaresearch.com/a-brief-history-of-public-service-advertising/ Accessed 20 June 2023. In a curious recent wrinkle to this story, on 23 February 2023, the NCAI announced 'Today, Keep America Beautiful, Inc. announced the transfer of ownership of the well-known "Crying Indian" anti-pollution public service announcement that first aired on Earth Day in 1971, to the National Congress of American Indians (NCAI) Fund' https://www.ncai.org/news/articles/2023/02/23/national-congress-of-american-indians-acquires-rights-retires-crying-indian. Accessed 29 June 2023. The NCAI refused this author rights to reproduce any image of the crying Indian for this chapter.

12 In November 2021, Apple reversed its policy that opening an iPhone voided warranty and is allowing owner repair. But this 'repair' is limited to replacing a defective component.

13 The evaluation of the public good of computing and the Internet (as opposed to its role as a vehicle of advertising and the manipulation of public opinion) and the idea of Internet access as a 'right' far exceeds the scope of this chapter and would demand at least a book-length treatment (see, for instance, Broeders (2016)). Likewise, strategising a greener information sector – assessing energy use, materials extraction, cost and cleanliness of manufacture, designing for recycling and waste management – is a vast topic, but see *Anatomy of an AI system* (Crawford and Joler 2018).

14 'Plastic recycling was estimated to have declined to about 5–6% in 2021, down from a high of 9.5% in 2014 and 8.7% in 2018. At that time, the United States exported millions of tons of plastic waste to China and counted it as recycled even though much of it was burned or dumped'. https://www.greenpeace.org/usa/news/new-greenpeace-report-plastic-recycling-is-a-dead-end-street-year-after-year-plastic-recycling-declines-even-as-plastic-waste-increases/ accessed 23 January 2023. See also https://www.greenpeace.org/usa/reports/circular-claims-fall-flat-again/. Accessed 23 January 2023.

15 See, for instance, https://www.kqed.org/news/11901288/you-cant-recycle-your-way-out-californias-plastic-problem-and-what-to-do-about-it. Accessed 24 January 2023.

16 A Lehr is an annealing kiln where blown glass objects are very slowly cooled, over days.

17 This figure had been repeatedly cited in economic commentaries over recent years. See, for instance, https://www.whitehouse.gov/cea/written-materials/2023/07/27/the-advance-estimate-of-second-quarter-real-gdp/#:~:text=Consumer%20spending%2C%20which%20amounts%20to,the%20strong%20U.S.%20labor%20market. Accessed 23 August 2023, and https://www.ceicdata.com/en/indicator/united-states/private-consumption--of-nominal-gdp. Accessed 23 August 2023.

18 NIB: 'New, In Box' (ebay jargon).

19 See also Bowker, GC & Star, SL 1999. *Sorting things out: classification and its consequences*, MIT Press, Cambridge.

20 I read once that an Amazonian indigenous culture identified only 300 different kinds of artefacts – baskets, cutting tools, bows and arrows and so on. We have 300 in every room, and thousands more in the garage, the car, the supermarket and the workplace.

21 An aphorism which is commonly attributed to Voltaire, also Montesquieu.

References

Benjamin, Walter. 1968. On the Concept of History IX. Illuminations. Harcourt Brace Jovanovich.

Bennett, T. 1988. 'The exhibitionary complex', *New Formations*, no. 4, pp. 73–102.

Borges, J.L. 1942. *The Analytical Language of John Wilkins*, *La Nación*, Argentina.

Bowker, G.C. & Star, S.L. 1999. *Sorting things out: classification and its consequences*, MIT Press, Cambridge.

Broeders, D. 2016. *The Public Core of the Internet: An international Agenda for Internet Governance*. Amsterdam University Press.

Brown, Jordan. 2015. *Forget Shorter Showers*. https://www.filmsforaction.org/watch/forget-shorter-showers/ accessed 23 January 2023.

Crawford, Kate & Vladan Joler, 2018. Anatomy of an AI system. https://anatomyof.ai

Duong, T. 2021. 'Chile's Atacama desert: where fast fashion goes to die', *Eco Watch*, 15 November, accessed 26 June 2022, https://www.ecowatch.com/chile-desert-fast-fashion-2655551898.html

Grover, Sami. 2021. *We're all climate hypocrites now*. New Society pub.

Henke, Christopher & Benjamin Sims. 2020. *Repairing Infrastructures, The Maintenance of Materiality and Power*. MIT Press.

Hoban, R. 1980. *Riddley Walker*, Jonathan Cape, United Kingdom.

Jensen, Derrick. 2015. Forget Shorter Showers https://orionmagazine.org/article/forget-shorter-showers/

Kilroy-Marac, K. 2018. 'An order of distinction (or, how to tell a collection from a hoard)', *Journal of Material Culture*, 23(1), pp. 20–38. https://doi.org/10.1177/1359183517729428

Lappé, Anna. 2003. O Magazine (The Oprah Magazine) June.

Oreskes, Naomi, Erik Conway. 2010. *Merchants of doubt: how a handful of scientists obscured the truth on issues from tobacco smoke to global warming*. New York: Bloomsbury Press.

Rivera, Julio L. & Amrine Lallmahomed 2016. Environmental implications of planned obsolescence and product lifetime: a literature review, *International Journal of Sustainable Engineering*, 9(2), pp. 119–129.

Rosch, E. 1988. 'Principles of categorization', in A. Collins & E.E. Smith (eds), *Readings in Cognitive Science, a Perspective from Psychology and Artificial Intelligence*, Morgan Kaufmann Publishers, pp. 312–322.

Slade, Giles. 2007. *Made to Break: Technology and Obsolescence in America*. Harvard University Press.

Summers, M. 2019. *The shocking true story behind the 'crying Indian' commercial*, accessed 26 June 2022, https://www.ranker.com/list/crying-indian-commercial-true-story/megan-summers

Young, Mark Thomas. 2020. "Maintenance" in Dorn & Michelfelder (eds) *The Routledge Handbook of the Philosophy of Engineering*. New York: Routledge pp. 356–368.

11 Aesthetic Values in the Maintenance of Urban Technologies

Sanna Lehtinen

11.1 Introduction

Urban technologies and their maintenance provide an interesting and even somewhat unexpected angle to study the city and its aesthetic dimensions. This perspective is strongly influenced not only by the way in which we define technology but also by how the city in question is imagined: whether it is an actual place or the idea of the city in general. As a new topic for the philosophy of technology, maintenance opens up new perspectives within the existing discourse on urban aesthetics by refocusing our attention on technologies over time. In order to begin bridging the approaches of urban technological aesthetics and their maintenance, it is important to emphasise that urban technologies are understood quite broadly in this context. This means that the focus is not only on digital technologies or the latest smart city developments which have been largely driving the latest technology-inclined studies of the city but also on more rudimentary technologies. The definition of technology used here follows that utilised in recent work delineating a specific field of study for the philosophy of urban technologies (Nagenborg et al. 2021) and thus probes more broadly into the existing and fast-developing plurality of human–technology relations in cities. The benefit of this is a broad understanding of urban environments and especially their aesthetic implications. Further analyses of different types of technologies can build upon these insights by providing a more detailed understanding of the contemporary urban lifeworld.[1]

It is possible to approach cities themselves as technological artefacts. This is because they are, to a large extent, produced by human intentional activity and considered prime examples of complex systems. The focus of this chapter, however, is on technologies of varying scales which one can encounter perceptually and experientially in cities either directly or through their use and impacts on urban surroundings. Most of the time, these types of technologies have some sort of perceivable presence and their implementation, use and upkeep have an impact on how the city looks and feels

DOI: 10.4324/9781003316213-13

to its inhabitants and visitors alike. This type of experiential emphasis typical of philosophical urban aesthetics has strong phenomenological underpinnings. It is further brought together with a postphenomenological[2] understanding of different types of human–technology relations that characterise urban everyday life. This broader take is necessary in order to show that the aesthetic dimension of cities affected by urban technologies is not only a matter of design or a step in the early phases of their realisation but an iterative, ongoing process and a topic for an open deliberative debate for which studying the basic tenets of the aesthetic experience of the city is of crucial importance. In the philosophy of technology, there has been a disproportionate focus on design which has led to the neglect of maintenance and its further implications as a lucrative topic for the field.

Aesthetics has for the most part of the twentieth century been strongly associated with the arts and the philosophy of art, especially in the Anglo-American context. Topics such as intentionality and interpretation have been developed in the specific context of western art forms. Although there is an obvious link to the theme of maintenance through designers' intentions which already take a stance towards how the designed artefacts or places are to be maintained, the central insights of contemporary aesthetic theory have not been applied as broadly outside the sphere of the arts as they could be. Since the turn of the twenty-first century, the overlooked questions about how things should and could look like and how their aesthetic value is linked to other values and practices have, however, been proven increasingly relevant as a much broader take on the aesthetic dimension of human activities has been developed in the form of urban and everyday aesthetics. Ecological, environmental and everyday-focused approaches in philosophical aesthetics have become prevalent and subjected to systematic analysis since the 1960s. The increasingly complex problems and grand challenges that humanity is facing have shown the need to understand human aesthetic values far beyond the limited, albeit fascinating context of the arts. Motivated by growing environmental concerns, this broadened academic interest has increasingly been directed to the influence and aesthetic impact of human activity on various types of environments. Urban environments have gained more relevance in academic discussions in the few last decades, when cities started to be perceived not only as the unavoidable by-products of societal development but also as places of central aesthetic relevance for humans. This shift in the mindset is important for the interest in the aesthetics of urban technologies, as it leads to a more all-encompassing idea of what constitutes quality of life in urban environments of various types (Lehtinen 2021a).

In its current orientation, as a relatively new branch of philosophy, urban aesthetics sheds light on how humans not only perceive and pay attention to but also further assess and evaluate the sensory and experiential features of contemporary cities globally. In traditional aesthetic theory, aesthetic judgements are linked to aesthetic values that have usually strong collective and societal importance and that come up in public discussion on how cities normatively should look and feel like. The focus of inquiry can be on either a more distanced, conceptualised perspective of contemporary cities or the so-called inhabitant's perspective, which emphasises lived everyday experience. Whichever perspective is taken as the main starting point, it is obvious that the realm of technology and the extent to which it becomes a quintessential part of the specifically urban experience, has not been traditionally at the centre of attention in philosophical aesthetics. (Lehtinen 2021a) In the context of contemporary cities globally, it is increasingly important to address this omission as we are faced with questions concerning 1) how urban technologies are impacting the overall look and feel of the city, 2) how different types of technologies make different values in the city perceivable and 3) how to understand changes in the technological realm, that is, what actually changes in the urban lifeworld when new technologies enter the picture?

This chapter has been developed around two central arguments. The first one is that the maintenance of urban technologies is an aesthetic issue in the first place, not just a purely practical or functional issue. The second argument goes further to claim that the maintenance of urban technologies is of key importance in showing care for future generations. This goes to show how flexible aesthetic values can and could be when societal priorities change. The first argument is made in order to broaden the disciplinary landscape of the emerging intersection of urban aesthetics, the philosophy of the city and the philosophy of maintenance as part of the philosophy of technology. The second argument is developed further to show how the care and aesthetic attention with which humans tend to urban technologies, can provide insight into how we may care for and maintain the planet in a less destructive way for various human-affected ecosystems. The end goal is to show that care in itself does not compromise aesthetic values: on the contrary, it is crucial to ask what type of values and ideals are the main drivers behind current maintenance practices and in what ways they may need to be rethought and adjusted in order to achieve a future we desire.

11.2 Traces of the Past, Traces of the Future

Urban technologies can be defined in various overlapping or even conflicting ways. For the purpose of looking into their aesthetic character and capacities, they are sometimes part of a larger architectural ensemble or at least implemented in a way that is indiscernible from buildings themselves as objects. They can be built in from the beginning, as in the case of smart architecture, or implemented later on. The latter strategy is referred to as retrofit practices, in which compromises in design solutions need to be made in order to fit new technologies into the already existing structures of the built environment. Oftentimes technologies are also distinctly separate and physically clearly detached from buildings, sometimes in order to ease the implementation of future upgrades. This implies already a temporal disconnect between the basic building stock of the city (which is approached in aesthetic terms commonly as 'architecture') and the more anonymous technologies that are either hidden or just bluntly attached for the time being to the facades or other surfaces of buildings. Often urban technologies are implemented in ways which suggest they are regarded as uninteresting or as having low aesthetic value. The assumed anonymity and replaceability of technology, for example, culminates in the way more recent digital technologies are designed into the processes of the city. The implementation of 5G antennas and small cell towers into urban neighbourhoods is an example of this, showing a whole breadth of different strategies of aesthetically 'dealing' with them, ranging from camouflage to full-on takeover of public space both visually as well as spatially (Lehtinen & Fantini van Ditmar 2022).

What makes digging into the implicit aesthetic assumptions regarding 'urban technologies' difficult is the common, shared understandings of the definition and range of technology needs to be always specified in relation to the particular task in question. Not all specifically urban forms of technology are accessible to everyone and often their functioning is not understood or consciously paid attention to when engaged in activities they enable. There is sometimes significant overlap with different categories of urban technologies which makes analysing them difficult. Recent philosophical accounts of urban technologies take into consideration their broad and varied nature and their impact on urban life. Epting (2021), for example, provides a taxonomy of urban technologies that is based on their role in facilitating sustainability transitions and securing a livable environment for citizens. According to Epting, urban technologies can be thus either 'wicked' or 'saving', depending on what their main impacts on cities are and whether they are ethically justifiable. Nagenborg et al. (2021), on the other hand, offer a view of new and emerging urban technologies that focuses less on judging their moral character and more on how they function and are used to mediate the relationship between citizens and their environment. This mediation can be seamless and unperceivable or more articulated and impactful in shaping our experience of the urban environment. It is advisable to keep both ways of defining technologies in mind, although the human-centric mediation role tends to be more central for the aesthetic repercussions discussed further in this chapter.

To investigate the aesthetic dimension of urban technologies and their maintenance in particular, one can start with the existing elements of the built environment. On the level of individual buildings, the scale in a city varies from the smallest micro houses to high-rise tower blocks and skyscrapers, which could not be built without the very existence of multiple technologies and innovations, for example, those enabling mechanical vertical transportation in the form of the elevators or escalators (Graham 2018). Contemporary buildings, even in their most simple forms are highly technologised artefacts produced by human intentional design to complement the current form and ideas of what a city is or should be. Buildings as advanced technological objects are thus not detached from urban imaginaries and the aesthetic forms they often take. It is tempting to speculate that these imaginaries have a far longer lifespan than what concrete urban plans and designs specify which is to a great extent reliant on technological possibilities, whether presently at hand or imagined towards the future.

The construction phase of an entirely new building attaches certain meanings to the urban environment which becomes always to some extent rearranged by the new building and its subsequent uses. How built places are then further maintained and cared for, gives some insight into the values of the community that uses and maintains them. Concrete maintenance practices can be telling of how a community relates itself to their technological surroundings in cities. Showing care to buildings, either in the form of everyday maintenance practices such as cleaning and painting or through symbolic acts through more speculative or artistic means, is a very concrete way of making attachment and ownership of a place tangible (Lehtinen 2020). To what extent manifesting this care extends to the details and varied examples of the technological realm, has been less easy to trace and depends again on what types of technologies are in the focus. Tending and caring for the technological realm and specifically their aesthetic range, however, represents a potentially significant shift in thinking. This can have repercussions in moving towards more sustainable technological practices, in which continuous upgrading is not a goal in itself.

Even in cities, all technologies do not work in direct contact with architectural constructs: technologies are to be found also in a more independent form in urban environments: we find examples of this on the streets and in the so-called in-between spaces in cities. Infrastructural technologies are often hidden, either completely or partially and this affects to which extent they draw attention or stimulate imagination through their aesthetic qualities. Technological constructs of the particularly detached type, such as electrical grids or telecommunications-related constructs, often take up space from non-human species when they are installed in places in which animals could potentially find shelter or in which plant ecologies are affected by human intrusion. Technologies within urban ecology can also offer refuge to animals, although this is most often an unintentional consequence. Bees or birds can build their nests under the coverings used to shield technologies against the elements or spiders weave their webs in places where electricity produces extra warmth emanating to the surface of structures. Technology thus sometimes provides habitats to animals in urban environments, but most often technology and natural ecology in cities are kept intentionally apart. The presence of animals and vegetation is often seen as an uncontrollable element in the maintenance of technologies, and the long-term impacts of urban technologies on non-human ecologies are not fully known either (Hernan & Ramirez-Figueroa 2022). There is a deep ontological divide in thinking about cities in these terms: the natural and the human originating are easier to manage when kept apart whenever possible. This goes to show, that human decisions on maintenance affect other-than-human ecologies even in cases where there is an especially strong intention to keep them separate on purpose.

Despite common tendencies to hide urban technologies from view, many moments of everyday life in cities nonetheless involve a recognition of urban technologies in use. Not all technologies are perceptually discernible as the aforementioned example 4G and 5G networks show but the installed structures and parts still have a noticeable presence in the urban environment. This perceptual dimension requires attention, recognition and interpretation in order to be correctly linked to particular technology 'in action'. In the fourth wave of industrialisation and through the expansive digitalisation of cities, the act of linking the perceivable structure of urban technologies and their 'hidden' and often converged uses is becoming exceedingly difficult and often requires a high degree of technological expertise. This is especially pertinent in cases of breakdown which reveal, not only the vulnerability but also the opaqueness of increasingly complex technological systems (Graham & Thrift 2007).

The selected examples from the technological realm have had a clear point of implementation to use and most likely will have also an expiration date in the urban sphere. These types of perceptually discernible technologies can be found on many scales of the city, ranging from traces of industrial founding phases of urban expansion to communication technologies that have had significant shifts throughout the twentieth and twenty-first centuries alike. It is also important to consider the future impact of the implementation of such technologies. This goes to underline that the aesthetic impacts of maintenance decisions do not concern only how the appearance of technologies is kept as they have been designed in the first place but in thinking and rethinking how they could and should look in the future.

11.3 Aesthetics of Care in Maintenance Practices

Maintenance, repair and upkeep are areas of human activity in which human values become explicit and gain a perceivable, even tangible, dimension of some type. This dimension is temporally important as well, as maintenance is not only an act but a continuous cycle with consequences that reach far in time: it is tempting to state that, to a certain degree, maintenance gives perceivable presence and temporal longevity to human values. Maintenance practices mean taking an active stance towards keeping something – a more or less networked artefact of some scale – in good material and functional condition, looking and working like it is generally expected to. Sometimes maintenance might even take a form of 'curated decay', in which the ultimate end goal is to align maintenance into collaborating with ecological processes rather than working against them as the logic of maintenance usually dictates (DeSilvey, 2017). In societies under chaotic conditions, this might be a prolonged reality, when resources are directed only to the bare essentials. I would argue, that even then the aesthetic value of urban technologies does not vanish but takes on another meaning for the citizens or from the perspective of maintenance decisions. The ingenuity with which citizens maintain living quarters in conditions of scarce resources can develop into original forms of creativity. This is visible, for example, in the aesthetically rich favela culture of Rio de Janeiro in which the creativity of the inhabitants often stems from a severe lack of resources. The solutions employed for extending buildings, for example, or accommodating for a lack of infrastructure, produce aesthetically distinct forms and a layered overall look for entire neighbourhoods.

To what extent different maintenance practices take the aesthetic dimension into account varies according to the maintained object, artefact or process. Depending on the extent the technology in question is considered already an integral part of the make-up of the building, block or neighbourhood, its aesthetic value might not be contested. Very visible (and audible) historical mobility technologies are a good example, such as the street elevators in some Lisbon *bairros* or trams in many other historic downtown locations. Despite their very noticeable presence (or precisely even because of it), old, creaky and rattling tram cars have become emblematic of the city and soundscapes of particular corners in cities such as Lisbon. This type of established and culturally recognised aesthetic value is also tied to the heritage value of mobility technologies, which captures the complex interplay between current and past values implicit in human–technology interactions in cities.

Recognised cultural heritage value leads to certain expectations towards the object's aesthetic value that characterise some large-scale infrastructural technologies in cities. Bridges, for example, are often considered a part of the rich, vibrant heritage of cities and thus cultural as well as aesthetic values are more obviously a part of the discussion on their maintenance. It is relevant to point out that even cultural heritage is not enough to pinpoint the aesthetic values attached to human artefacts, as the aesthetic dimension implies the possibility of a more deeply personal take on how a particular artefact is perceived and how it contributes to its surroundings and the ever-changing lifeworld in flux (Giombini 2020). Aesthetic values can be a cause of extreme disagreement between community members, whereas heritage value tends to be something that is generally agreed based on shared historical and cultural perspectives and thus can often function as a foundation for long-term maintenance plans. The relative flexibility of aesthetic value depending on variations in personal experiences, interpretations and nuances makes it more difficult to grasp. It is fair to assume, that this is one of the reasons why the idea of aesthetic value has been difficult to grasp in negotiations over the maintenance of such urban elements as technologies.

Recent years have witnessed an increasing interest in 'aesthetics of care' in philosophical and applied aesthetics, especially in relation to environmental and everyday aesthetics. One of the proponents of this new approach, Yuriko Saito, has proposed bridging Western philosophical aesthetics and Japanese aesthetics for a more comprehensive take on aesthetic sensibility. In her recent work, Saito links aesthetic interest and values with contemporary perspectives in care ethics (2020, 2022). The strong feminist underpinnings of the care ethical tradition direct attention to the underappreciated work and practices that enable the functioning of a society which are often performed by underprivileged social groups. This perspective is useful for discerning dimensions of aesthetics that have previously been left in the dark due to the common tendency to valorise innovation or the design phase in many cultures (Young 2020).

Care can be understood in this context at least from two different yet often complementary perspectives. It can refer to 1) care towards material condition in the form of 'attentive care' that is shown through maintenance and repair practices or 2) care towards future generations, whose living conditions are altered and maintained through decisions made today on the basis of current knowledge. These two modes of care are closely connected in most cases. Some important nuances can, however, be distinguished in concrete maintenance practices with the help of this distinction. One way of determining which mode of care is more central is by considering the relationship with environmental harm and overall sustainability in particular maintenance decisions. If certain maintenance decisions and practices are likely to harm the living conditions of future generations of humans and non-humans alike, they cannot be considered to show care towards future generations. An example of this is the use and replacement of carbon-intensive materials, such as concrete or glass, in building maintenance, which are now understood to contribute to a warming climate which threatens the livelihood of future generations. The realisation of this type of temporal influence has recently led to a significant cultural shift which is still ongoing in many of the fields in which long-lasting decisions over maintenance are made. The increased recognition of our obligations and influence on the living conditions of future generations has encouraged a re-evaluation of current maintenance practices. Followed as a strict principle in various maintenance practices, many of the previously made decisions would not be feasible anymore.

The perceivability of care itself as a practice, and how it becomes visible and sensed is another important perspective to understanding maintenance practices and decisions from the care ethical and aesthetic perspective. It has thus been easier to focus on the first mode of care as understanding the material conditions and ramifications of maintenance practices is perhaps the clearest point of entry to upkeeping elements of the urban sphere. It would be, however, ultimately important to link these two instances of care: focusing first on what is possible with what exists already and then understanding what the further consequences of these actions will be in times and locations far beyond one's immediate reach. Articulating more clearly, this relation between the different dimensions of care will help in re-assessing further some of the maintenance choices that have previously been considered obvious and uncompromised. Through the lens of intergenerational aesthetics, in particular, a significantly new perspective can be gained on debated issues such as the renewal of parts of nationally significant built heritage (Capdevila-Werning & Lehtinen 2021).

With the intergenerational focus in mind, the goal of maintenance, repair and care can sometimes be to make material conditions to some extent better than before in clearly acknowledgeable ways. Perhaps faults have been identified in the original design or new ways have been found to improve on existing structures or systems. Instead of merely aiming at maintaining an existing state, this kind of incremental amelioration also often changes the appearance of the maintained artefact. An example of this could be changing the façade material of a building into something entirely new, after discovering that the original material has not proved durable or sustainable enough for the long-term use of the building (Capdevila-Werning & Lehtinen 2021). In such cases, even a significant change can be justified if it enables a building to avoid demolition and replacement. However, if the change in appearance is significant, it can often be important to communicate to stakeholders the reasons motivating the change in appearance or the overall aesthetic status quo. Sometimes even the slightest changes in well-known or otherwise familiar elements of the city can produce discontent among the urban dwellers, especially in cases where the decision-making processes have not been transparent enough. As pointed out by Mark Thomas Young in his chapter in this volume, the Golden Gate Bridge is a good example of this – the debate about whether to install suicide barriers often centred on the aesthetics of the bridge. Many worried that the addition of nets or raised guard railings would ruin the iconic look of the bridge, considered to be emblematic and beautiful in its unchanged appearance throughout the decades.

As concerns over the ecological and social sustainabilities of many human practices have grown, understanding the deeper consequences and implications of aesthetic choices in design have become increasingly important. This in turn has led many to scrutinise the aesthetic dimension of decision-making implicit in maintenance and repair decisions (see e.g. Young 2021). Aesthetics opens a way to assess not only the chosen methods and goals but also the deeper layers of values that have directed the plans and designs. The aesthetic layer of human world-making does not refer merely to surface qualities that are to be taken into consideration as the last part of the design process. Conceptualising the care aesthetical perspective and implementing it in the decision-making processes would be taking a step further in acknowledging the aesthetic importance of the various elements that create the urban lifeworld.

11.4 Deliberation of Aesthetic Values

It is safe to say that aesthetic choices of some calibre are made always when maintenance decisions are made. Often, they are by-products of some other choices but nonetheless, repercussions to the aesthetic realm will take place when decisions are put into effect. Further on, there are many detailed moments in the process of maintenance, which will result in some type of aesthetically relevant outcome. While these outcomes are often nuanced they can be more significant than we may often think. With repainting a house, for example, nuances in colour choice might seem insignificant but the change in perception can nonetheless be very significant when the newly repainted house is contrasted with the colours of the adjacent buildings and the details of the streetscape. When it comes to architectural preservation and restoration, there are rules and guidelines as the preservation and maintenance of architectural heritage is already an established field in itself. The consideration of urban technologies, however, has a much broader scope as it concerns a wider spectrum of different types of technologies than just those directly linked to architectural solutions. However, much like architecture, many of these technologies also have a significant effect on how the urban environment looks and feels to its various users.

Yet, while any changes to the appearance of a city risk disrupting its aesthetic character, preserving cities as museums of outdated technologies is not a feasible goal either. So how should we keep track of and value past uses of the city in a way that does not hinder and even supports the current needs of the citizens? One way of approaching this question appeals to conceptions of 'authenticity' which have recently been utilised to analyse the way cities are perceived and characterised (Wittingslow 2021). In the context of philosophical aesthetics, however, this concept can be risky to use, as it implies the existence of some original, true-to-character form and identity of the city. Yet, many cities seem to genuinely have centuries-long histories and traditions that constitute a uniqueness to their character. Implementing new technologies into these kinds of environments can be especially difficult, as the layers of history often quite literally do not have space left for new layers to emerge. Consider, for example, the mobility problems currently faced by many historical European cities, in which new metro lines are nearly impossible to build because of the constant discovery of archaeological findings in the excavation sites of planned routes.

It is important to remember in this context that the aesthetic layer of cities should not be considered only in the broad sense but from the so-called macro perspective with a strong visual orientation (Lehtinen 2021a). When decisions are made, the city or a part of it is often approached in imagination, based on impressions and ideas about the place and its aesthetic qualities. Even with detailed visualisations and plans, the so-called micro aesthetic layer is easily overlooked in these deliberations. Understanding how the aesthetic character of a place opens up through multisensory, street-level experience, in the midst of the hustle and bustle of everyday urban life, is of crucial importance when making aesthetically informed maintenance decisions. This is, however, also very difficult in practice because such processes often unfold on different timescales. In planning or architectural design processes, participatory and citizen-oriented approaches have become more prevalent in recent years, but the described tension between the macro and the micro layers of aesthetics cannot be resolved only by making the inhabitants participate more in the processes. One solution to alleviate the friction in favour of more quotidian experiences is to expand the temporal scale of the discussions and to take them into the locations themselves, during many different kinds of conditions (seasons, hours of the day and so on).

Approaching urban technologies as processes that require ongoing care and value deliberation to maintain, changes the perspective slightly and enables us to see these technologies beyond their most immediate use value. Another aesthetic dimension important to the continuous cycles of upkeep comes from the creativity that is required to tackle the challenges that time itself presents in the process. This applies both to the material world and the social sphere of the use of urban technologies. Materials decay and age in unexpected ways, especially newer plastic-based and composite materials that do not endure changing conditions, which complicates the task of maintenance. The aesthetic appreciations of individual people often change during their lifetimes and even societal aesthetic norms are prone to evolve. Because it is considered fickle and prone to change, aesthetics has often been considered a difficult area to tackle as a value domain. Such perceptions arise at least in part through the influence of the long-standing idea of the autonomy of aesthetic value, based on the Kantian notion of aesthetic appreciation as disinterested. However, recent work on environmental and everyday aesthetics has often challenged the notion of disinterestedness and tends to put forward an idea of aesthetic interest and appreciation as deeply intertwined with ethical and social values (Lehtinen 2021b). At the very least, it should be acknowledged that the level of disinterestedness differs when considering the aesthetic value of something that we use and interact with on a daily basis. In comparison to artworks which form the traditional focus in the study of aesthetic judgement, aesthetic values in the realm of the everyday are part of a value network in which different values are interlinked in more complex ways than in the context of art.

Something that once was fashionable might once again come into vogue. Such fluctuations in aesthetic trends might be a tempting cue for determining long-term aesthetic value. It is at least useful to postpone aesthetic judgement until a fuller picture of the significance of the artefact in question has been gained. Large-scale infrastructural technologies provide interesting examples of this. In lower Manhattan, New York City, dilapidated railroad tracks were turned into a hugely popular urban park, the High Line. In Paris, sections of the *La Petite Ceinture*, an inner city circular railway, were also rehabilitated and transformed into recreational areas. The development of both of these sites stems primarily from a commitment to saving the main traces of a city's past. Yet in addition to forming parts of notable post-industrial landscapes, the unused railroads also exhibited unusual biodiversity. Demolishing these structures would have therefore meant not only erasing parts of the historical layers of each city but also erasing places that now support urban nature and function quite literally as breathing spaces within the city.

These examples show in significant ways how infrastructural technologies, as examples of urban technologies more broadly, are not discernible from the spaces and environments in which they have been built. They merge into new kinds of places that evolve towards new meanings as their past uses become largely forgotten and obscured. In the case of the High Line, such processes are employed in the design of the new park as the old rail tracks are left visible and untouched, sometimes rendered into newer, stylised design elements of the trail. Deliberation of aesthetic values within the community takes place on multiple levels simultaneously and traces of past technologies act as cues for these deliberations. However, whereas the aesthetic dimension can be directly addressed, often it is merely implied by discussions on other topics. Similarly, decisions about aesthetics are often masked as something else. This is recognised at least in the sphere of architecture where the buildings that are considered ugly are more likely to be demolished than repaired. The decisions have aesthetic repercussions, so the supported and maintained state of urban technologies manages to make other central values perceivable.

11.5 Conclusions

When discussing the range and influence of urban technologies in relation to the use and overall look and feel of cities, aesthetic values are still an understudied element. This chapter has aimed to demonstrate how aesthetic value intersects importantly with other value considerations, having, for example, further ecological and social consequences through the ethical values that the appearance of urban places is interpreted to represent. The aesthetics of cities or elements of the built environment are not to be treated separately from other values but are deeply tied to how these elements are valued and what type of decisions are made over their fate. Technological artefacts, networks or their perceivable parts should not be considered an exception to this. On the contrary, the maintenance of various types of urban technologies can be seen as indicative of aesthetic values more broadly extending to how well the society in question adapts to changing conditions for the human lifeform. It can serve even as a testing ground for increasing the flexibility of societally acceptable values.

Taking into consideration the ongoing shifts and tensions in aesthetic values would help to develop the burgeoning field of the philosophy of maintenance further in order to better address the complex reality of the contemporary world. Especially, in relation to urban environments and their development, aesthetics is rarely only about aesthetic values but also intersect with social and ethical values. What there is to be sensed, perceived and experienced in the first place is indicative of societal values more broadly. This extends also to what is missing from the cityscape, as in the example of the many technologies which are hidden or simply erased when they become obsolete for most city dwellers. This implies that looking at cities over time might reveal new dimensions to established problems and issues in aesthetics.

While the ruins of the future might not consist solely of technological constructs even in the most highly technologised imagination, the appearance and perceptual qualities of human-designed technological artefacts form a great deal of the legacy that we leave for future generations. These should not be approached merely as traces of the past from a heritage perspective or as updatable elements from a technical perspective but also through their impact on the overall aesthetics of cities. The aesthetic perspective undermines the continuity of human–technology relations over time and also facilitates experimentation with contested aesthetic values and new possibilities. Understanding what is aesthetically pleasing, negative, interesting or even merely tolerable could have further consequences on how to appreciate cities as complex systems, in which there is 'something for every taste' and which conform and adapt to the increasing requirements for sustainability transformations.

Notes

1 A concept derived from Husserlian phenomenology, for the use in the specifically urban sense, see Madsen and Plunz (2002).
2 Post-phenomenological referring here to the views in philosophy of technology based on Don Ihde's work, see, for example, Rosenberger and Verbeek (2015).

References

Capdevila-Werning, Remei & Sanna Lehtinen (2021). "Intergenerational Aesthetics: A Future-Oriented Approach to Aesthetic Theory and Practice", *Philosophical Inquiries*, 9(2), pp. 175–198.

DeSilvey, Caitlin (2017). *Curated Decay: Heritage beyond Saving.* Minnesota University Press.

Epting, Shane (2021). *Saving Cities: A Taxonomy of Urban Technologies.* Cham: Springer.

Giombini, Lisa (2020). "Reconstructing Heritage: Places, Values, Attachment", *Contemporary Aesthetics*, vol. 18 article 9. Available at: https://digitalcommons.risd.edu//liberalarts_contempaesthetics/vol18/iss1/9

Graham, Stephen (2018). *Vertical. The City from Satellites to Bunkers.* London: Verso.

Graham, Stephen & Nigel Thrift (2007). "Out of Order: Understanding Maintenance and Repair", *Theory, Culture and Society*, 24(3): 1–25.

Hernan, Luis & Carolina Ramirez-Figueroa (2022). "A Luminous Abstraction: 5G Narratives and Futuring Otherwise", *Mediapolis: Journal of Cities and Culture*, Dossiers no. 3, vol. 7. https://www.mediapolisjournal.com/2022/09/luminous-abstraction

Lehtinen, Sanna & Fantini van Ditmar, Delfina (2022). "Introduction: The Rise of 5G", *Mediapolis: Journal of Cities and Culture*, Dossiers no. 3, vol. 7. https://www.mediapolisjournal.com/2022/09/rise-of-5g

Lehtinen, Sanna (2021a). "Urban Aesthetics and Technology", in *Oxford Handbook of Philosophy of Technology*, ed. S. Vallor. Oxford: Oxford University Press, pp. 449–469.

Lehtinen, Sanna (2021b). "Aesthetic Sustainability", *Situating Sustainability: A Handbook of Contexts and Concepts*, eds. R. Toivanen & P. Krieg, Helsinki: Helsinki University Press, pp. 255–268. https://doi.org/10.33134/HUP-14-18

Lehtinen, Sanna (2020). "Buildings as Objects of Care in the Urban Environment", in *Aesthetics in Dialogue: Applying Philosophy of Art in a Global World*, eds. Z. Somhegyi & M. Ryynänen, Berlin: Peter Lang, pp. 223–236.
Madsen, Peter and Richard Plunz (2002). *The Urban Lifeworld: Formation Perception Representation*. New York & London: Routledge.
Nagenborg, Michael, Taylor Stone, Margoth González Woge and Pieter Vermaas, eds. (2021). *Technology and the City: Towards a Philosophy of Urban Technologies*. Dordrecht: Springer.
Rosenberger, Robert and Peter-Paul Verbeek, eds. (2015). *Postphenomenological Investigations: Essays on Human–Technology Relations*. Washington, DC: Rowman & Littlefield.
Saito, Yuriko (2022). *Aesthetics of Care: Practice in Everyday Life*. London: Bloomsbury.
Saito, Yuriko (2020). "Aesthetics of Care", in *Aesthetics in Dialogue: Applying Philosophy of Art in a Global World*, eds. Z. Somhegyi & M. Ryynänen, Berlin: Peter Lang, pp. 187–201.
Wittingslow, Ryan (2021). "Authenticity and the 'Authentic City'", in *Technology and the City: Towards a Philosophy of Urban Technologies*, eds. M. Nagenborg, M. González Woge, T. Stone, & P.E. Vermaas. Dordrecht: Springer, pp. 253–270.
Young, Mark Thomas (2021). "Now You See It (Now You Don't): Users, Maintainers and the Invisibility of Infrastructure", in *Technology and the City: Towards a Philosophy of Urban Technologies*, eds. M. Nagenborg, T. Stone, M. González Woge, & P.E. Vermaas. Dordrecht: Springer, pp. 101–120.
Young, Mark Thomas (2020). "Maintenance", in *Routledge Handbook of the Philosophy of Engineering*, eds. Diane P. Michelfelder & Neelke Doorn. London: Routledge, pp. 356–368.

12 Negotiating Visions of Waste

On the Ethics of Maintaining Waste Infrastructures

Joost Alleblas and Benjamin Hofbauer

12.1 Introduction: Waste, Ethics and Maintenance

The ethics of waste has become an important academic and societal topic (e.g., Thompson 1979; Strasser 2000; Hawkins and Muecke 2002, 2010; Hird 2022). The growing literature on this subject shows the many ways in which waste has evolved as a normative category to control and form peoples, bodies and identities. Literature also shows how waste has become a global problem in which responsibility and accountability are difficult to assign. However, the relationship between the ethics of waste and the maintenance of waste infrastructures in urban contexts is still underdeveloped, despite the ongoing urbanisation of the global population and the ever-increasing amounts of improperly managed municipal waste. Accordingly, this chapter seeks to highlight how cultural understandings of waste have become embedded in waste infrastructures, becoming instructive for their design, and explore what role maintenance plays in the continued embedding of these understandings. Furthermore, a collective reimagination of waste is crucial, especially in High-Income Countries [HIC] (UN-Habitat 2010). Support for the behavioural and lifestyle changes deemed necessary to deal with the current environmental crises is only possible through a collective confrontation with waste in these urban contexts. An analysis of the socio-technical systems of waste and the ways in which these systems are maintained allows us to better grasp how emergent, new interpretations of what constitutes waste and how to best deal with it are hampered by the design of these systems and by a static conception of their maintenance and repair.

Previous chapters in this volume have discussed how the imagination, design and construction of systems and artefacts are intricately related to their maintenance and repair and how maintenance can actually be conceptualised as an extension of the design process (see also Young 2021a, 2021b; Edwards 2003). However, regarding the combination of maintenance and waste, this idea seems counterintuitive. Maintenance and waste

DOI: 10.4324/9781003316213-14

are temporally related, but do not necessarily overlap: on the one hand, maintenance seeks to uphold the functioning of a system or artefact, potentially reimagining it in the process.[1] Waste, on the other hand, represents the end of a system or an artefact's life cycle. What is wasted is no longer of use or its usefulness has been neglected (Hawkins and Muecke 2002). Waste, therefore, often marks an end of what has been maintained, the end of value. Furthermore, though dirt and waste are distinct concepts, improperly managed waste can lead to the creation of dirt. Dirt has been amply discussed in the maintenance and repair literature (e.g., Dant & Bowles 2003; Harmer, Cooper, Fisher, Salvia & Barr 2019; Lejeune 2019). Here, dirt stands in an antagonistic relation to the maintenance and repair of artefacts, such as cars, vacuum cleaners and cameras, bridges and server parks, but also urban infrastructures.

Thus, three perspectives on waste and maintenance emerge: 1) waste as the discarded, disvalued object or material (solid, liquid or gaseous, hazardous or not); 2) waste as that which, if improperly managed, endangers the functioning of machines and instruments, processes and systems, leading to more maintenance and repair and more hazards; 3) maintenance and repair of artefacts as a counterforce in consumer societies, keeping things from turning into waste. The effective management of waste then is a critical service in a society. Waste management infrastructures appear as critical for the maintenance of urban areas and urban living conditions (Nagle 2013; Steele & Legacy 2017). Indeed, urban areas are the sites of the most dense expressions of infrastructure and urban dwellers almost completely depend on these networks (Graham 2010).

However, urban infrastructures are not always perceived as critical. The invisibility of infrastructure, together with urban habits and routines regarding waste, obfuscates the flow of waste through a city and conceals the relevance for urban dwellers of unhindered waste streams. This waste flow takes place in sewers and pipes, grinders and incinerators, opaque bags thrown in underground collectors and garbage trucks, processes modern urban dwellers tend not to notice – except when driving behind a truck in a small street or when waste stops flowing (Nagle 2013). Here, maintenance creates one of the moments through which municipal waste infrastructure becomes (temporally) visible. Breakdown and crisis, as instances of failed maintenance, also create moments in which infrastructure reappears. However, this appearance doesn't necessarily lead to insights concerning previously hidden features and embedded meanings (Young 2021b).

In this chapter, we investigate the maintenance of socio-technical systems that handle waste, especially but not exclusively in advanced urban settings. These systems include sewage systems, drains and canals all the way to recyclable and biodegradable waste, landfills replete with miscellaneous and electronic waste, waste-to-energy incinerators, the shipping and transportation of waste, nuclear waste and less obvious instances of waste, such as greenhouse gases. The focus of our investigation is on the ethical implications of maintaining socio-technical systems that handle municipal solid waste, to discuss how these systems have become problematic from a moral perspective and to explain why updating these systems to comply with new societal goals and visions is difficult, especially in HIC. We do this on the basis of the subsequent two-step process of analysis and normative argument.

The first step of the analysis consists of laying out the methodological framework for assessing existing infrastructure, based on Henke and Sim's reflexive repair of infrastructure (2020) as well as Mark Thomas Young's dynamic conception of maintenance (2021a, 2021b). From this basis, we explore how waste itself is being viewed and how that view is representative of current socio-political structures, assumptions, values and norms. This gives a descriptive outlook of what waste is and can be, and how it manifests itself in today's industrialised societies. Here we draw on anthropological, sociological and philosophical scholarship that has sought to understand the phenomenon of waste through different theoretical lenses (e.g., Thompson 1979; Hawkins 2010; Douglas 2002; Nagle 2013; Scanlan 2005). We then introduce the normative claim of how waste maintenance should be viewed. Here we seek to evaluate our interaction with waste, arguing how the engagement with and view of waste should and could change. Specifically, we claim that the maintenance of municipal waste infrastructures should be reimagined as an active and dynamic process, rather than caring for a stable, rigid artefact.

Since we believe the issues discussed here are particularly relevant for urban areas in HIC countries, because of high levels of stratified socio-material organisation, in which infrastructure appears as stable and permanent (Graham & Thrift 2007), we give examples mostly from these countries. We argue that first and foremost, socio-technical systems of municipal waste management in the Western Hemisphere are made to be invisible, and hidden from the public, and their operation is taken for granted, while the continuous effort to keep this system in operation is often neglected. This feature of invisibility has co-evolved with modern understandings of civilian culture and urban progress. Our claim is that the invisibility of waste infrastructure fixes the socio-ethical perception of waste as something to be avoided, something contaminated, thereby fixing its disvalue for society. This modern perception of waste, we claim, is problematic in terms of achieving a more sustainable relationship with waste in light of ecological concerns, health concerns and climate change. The modern perception, facilitated by the current design of municipal waste infrastructures in HIC, hampers the realisation of new visions of waste. A circular economy, for example, should be able to reconceptualise waste for its potential as a resource and something that has not lost its value. Despite steps taken in this direction, a shift in the collective understanding of waste in advanced urban areas has yet to occur. With 56% of the world's population already living in cities (World Bank Website)[2], the need to change this understanding is urgent. One step towards achieving this reconceptualisation is through the dynamic maintenance and reflexive repair of waste infrastructures (Young 2021a, 2021b; Henke & Sims 2020). In this study, we combine these insights with a focus on the values and meanings embedded in waste infrastructures. Our conclusions amount to a first step that we deem essential for new visions of waste to gain a foothold in advanced urban contexts.

The structure of the chapter outlining this argument is as follows. First, we introduce our methodological background considerations, outlining key concepts of maintenance and infrastructure. Then, we discuss waste as a dynamic and normative concept. In Sections 12.3 and 12.4, we explore how current waste management infrastructures tend to reproduce a specific waste imaginary, linked to a modern, urban ideal of tidiness. In these sections, we look at municipal waste mostly from a HIC perspective. Section 12.5 provides a further theoretical reflection by looking at waste infrastructure through the lens of socio-technical systems. This section highlights how values are embedded, afforded and resisted through the use of such systems. The subsequent Sections 12.6 to 12.9 introduce how value change affects these systems, subsequently presenting budding alternative visions of waste realised through dynamic maintenance and reflexive repair. The concluding section gives recommendations for further research.

12.2 Maintenance and Infrastructure

In recent years, maintenance has emerged as an important topic in the ethics and philosophy of technology, challenging the dominant focus on design and innovation. Maintenance, scholars stress, should be understood as more than simply preserving the function of artefacts (Graham & Thrift 2007; Jackson 2014; Young 2021a, b). Instead, maintenance often involves (re)design and innovation, creativity and adaptivity. Therefore, rather than as static objects in which designers' ideas are realised and carefully guarded, artefacts should be seen as processes of change in which maintenance interferes with, reinforces, guides or blocks these processes (Graham & Thrift 2007; Young 2021a).

Conversely, maintenance can also be understood as a challenge to the ideology of consumer societies, in which innovation, creative destruction, planned obsolescence and excessive growth (and thus lots of waste) are ultimate goals. Instead of a narrative of technological progress, maintenance offers an ideology of care for artefacts and, thus, for the people who depend on these artefacts (Denis & Pontille 2015). The maintenance of urban infrastructure is no exception. Rather than building new infrastructures that deliver new public goods, most of the time cities are places where existing infrastructure is updated, altered, repaired and maintained, often within tight budgets. This maintenance is an aspect of daily, urban life. However, its relevance is often neglected, even though maintenance carries the burden of keeping intact material and social orders in urban contexts (Graham & Thrift 2007).

Urban infrastructure, that is, the pipes, drains, cables, road networks, etc. plus the agents and institutions running and maintaining them, is often invisible to the everyday user. Despite this common association, the (in)visibility of infrastructure remains a contentious topic (Larkin 2013). Star's (1999) influential study on the ethnographic background of infrastructure has reinforced the idea that infrastructure goes unnoticed, as a seamless web (Hughes 1986), in order to serve its function. Likewise, Edwards (2003) sees infrastructure as the invisible background and the substrate of modernity. At the same time, however, infrastructures do often become noticeable and apparent when we interact with them. This might come in moments of breakdown, which open new perspectives on the technologies we use, and are often instructive of innovation trajectories (Jackson 2014). Furthermore, as Jackson (2014) implies, it is in maintenance and repair that we might encounter relations of value and social order that are often obscured under the 'smooth functioning of complex sociotechnical systems' (p. 231), although this encounter is contested (Young 2021b). But breakdown and maintenance are not the only ways in which infrastructures reappear. As repair scholars Henke and Sims write:

> Infrastructures are never truly invisible, but their salience to us, including their role in shaping power and privilege, shifts in and out of our consciousness based on their operation and to what extent we depend on or are oppressed by them.
>
> (Henke & Sims 2020, 143)

A second aspect of infrastructure that is relevant to our discussion, besides the issue of (in)visibility, is fragility. Nagle's (2013) analysis of urban sanitation workers in New York shows that garbage collection is a fragile assemblage of people, punching clocks, bureaucratic measures, weather forecasts, noses, luck, intuitions, truck maintenance schedules and grunting citizens. In her account, the 'smooth functioning' of this system is far from obvious. Rather, it seems to be in a constant mode of crisis, always about to come to a halt and disintegrate. It can only appear smooth from the outside. Indeed, maintaining a technological artefact or a socio-technical system can be understood as the continuous abatement of letting that object or system go to waste, despite seemingly solid and robust infrastructures. In the case of urban sanitation and many other urban structures, what needs to be maintained, furthermore, are not only machines but also work relations and routines, a company culture and a precarious relationship with other urban dwellers. What is maintained, then, is a socio-material world (Sormani, Bovet & Strebel 2019). Furthermore, this maintenance is not always successful. Rather, the system breaks down repeatedly, making waste suddenly a public concern (Nagle 2013). As a consequence, the politicisation of waste in urban contexts is often made possible by unsuccessful and neglected maintenance and repair of waste management systems. The fragility of the system is precisely what enables its politicisation.

Maintaining socio-technical systems means ensuring their continuity, alongside the inherent values and agents that underpin that system.[3] Evidently, infrastructure manifests human values and norms (Edwards et al. 2009). Furthermore, infrastructures, given both their centrality and invisibility, continuously reproduce certain values and norms, certain ways of social being, cooperation and existence. Their repair in the form of maintenance often leads to those values and norms being reified through adaptation and slight, which in turn makes the infrastructure ever more resilient to change. In a similar vein, Henke and Sims problematise the fact that repair as maintenance reproduces and embeds current '... material and discursive investments and assumptions embedded in those infrastructures ...' (2020, 121). Repair as maintenance could, in this sense, lead to the further entrenchment of the status quo.

This resistance to change is a third aspect important for our analysis, besides invisibility and despite fragility. Resistance is closely related to Henke and Sims' (2020) idea of resilience in infrastructure design: the protection against destabilising feedback loops in infrastructure use. According to Henke and Sims, growing resilience is a consequence of the two ways in which infrastructure is reflexive (2020, 124f). One, infrastructure is causally reflexive, that is, infrastructures are shaping the very environments within which they operate. For instance, the required additional infrastructure to maintain sewer pipes further entrenches the initial infrastructure on a physical and societal level. Two, the infrastructure's causal reflexivity requires the engineers, operators and policymakers involved in the design and operation of the infrastructures to be 'self-aware' (Henke and Sims 2020, 125). Accordingly, anticipating disruption paradoxically enables disruptive behaviour. For example, waste infrastructure that anticipates poor recycling behaviour (e.g., through the introduction of further tracking systems or additional sorting after the collection) and seeks to remedy those disruptions reflexively becomes more resistant to a necessary overhaul or rethinking of its central assumptions on how people view waste.

The disruptions are reflexively fended off, and life can continue business-as-usual without the need to reflect on the values that drive the behaviour and attitudes towards waste, as well as the need for the infrastructure in the first place. The degree to which disruptions succeed in discontinuing urban flows seems to determine, to a large extent, the degree to which citizens are able to engage with the politics, ideology and embedded values of waste and waste infrastructures (Graham 2010). A steady and uninterrupted urban flow eliminates the need for reflection on a socio-political level.

Concluding, urban infrastructures are essential to (modern) society and resistant to change. However, the age of the Anthropocene and environmental reckoning of societies built on ever-increasing consumption has shown the limitations and dangers of these infrastructures. Despite their continued function and delivery of energy, urban sanitation and stability, these systems are also 'broken'. Fossil-fuel-based energy systems, for example, show all of the above characteristics of an incredibly resilient piece of infrastructure that at the same time has become an existential threat to humanity. Similarly, albeit to a much lesser existential degree, current waste infrastructures are struggling to break free from now problematic and unsustainable perceptions of resource depletion – what we refer to as a 'modern' waste imaginary (see Section 12.6). The question then arises whether we can change these current infrastructures, whether we can reimagine waste and waste management to account for meaningful sustainability and social justice – all this without proposing a design from scratch or letting waste management systems break down and disintegrate completely. Maintenance seems to fit this role well. It allows the questioning of socio-technical regimes and socio-material routines and practices in urban environments without necessarily falling for the intrusive dogmas of innovation and design. Maintenance, breakdown and repair of infrastructure, furthermore, are able to engage citizens. They might successfully bring to the fore the inconspicuous culture and politics of waste.

12.3 Reflexive Repair and Dynamic Maintenance

Scholarship on maintenance and repair provides us with at least two possible answers to the question of how to change critical infrastructure, such as municipal waste management systems. In both these answers, the focus is on challenging the existing material and social order through the maintenance of infrastructures that always already reinforce these orders. One, through 'reflexive repair' (Henke & Sims 2020), and two, through a more dynamic and procedural conception of maintenance (Young 2021a, 2021b).

The guiding principle behind 'reflexive repair' is mainly discursive, that is, making the agents maintaining infrastructure projects aware of the underlying power dynamics as well as the values purported through these projects. Reflexive repair means '... asking questions that bring the sometimes obscure but always present properties of infrastructures to the surface, critically appraising the embedded dynamics of power, discourse, and materiality that are built into these sociotechnical structures' (Henke & Sims 2020, 143). Reflexive repair opens up a space for debate, for interaction among those stakeholders who otherwise would not communicate with one another. Such interactions might lead to the realisation that maintaining a given kind of infrastructure in a specific way through repair is not a value-neutral undertaking. Rather, it is the re-assertion of specific worldviews and specific ideas of urban living and cooperation. A re-assertion of ways of disclosing the city and disclosing urban life. Preceding that re-assertion by a reflexive and discursive engagement could highlight problematic power dynamics and injustices.

Henke and Sims (2020) summarise their views by proposing that reflexive repair will be able to 'repair infrastructural repair itself', providing a critical, political form of repair. Thus, reflexive repair proactively considers the limitations and (unintended) consequences of repair and maintenance, effectively negotiating maintenance amongst stakeholders with different aims and needs. Henke and Sims are aware of the difficulties of putting their approach into practice. For instance, we can question the manner in which a representative group of stakeholders is brought together in a globalised economy with complex supply chains. As an example, vast amounts of discarded 'fast fashion' from the Global North end up in Africa and increase economic inequality (Brooks 2019). Second, the unprotected disassembling of container ships in Bangladesh (Jackson 2014) and informal recycling of e-waste in developing countries (Ádám et al. 2021) show that discarded artefacts have become a global issue with a global supply chain, in which questions of justice are still disregarded. Can reflexive repair consider all these global consequences? And how would a hierarchy of repair preferences be justified? We do not attempt to answer these questions concerning 'waste justice' here, but realising its complexity is important.

The second approach to changing infrastructures is based on Young's reconceptualisation of maintenance as a dynamic procedure, rather than a static continuation of a given socio-technical system. The main thrust of this reconceptualisation of maintenance stems from a fundamental shift in how technology is viewed. Many current approaches in the philosophy of technology still frame technological artefacts as expressions of human (usually engineers' and designers') intentions. For instance, the value-sensitive design approach (Friedman, Kahn & Borning 2002) prioritises intended values in the design process of HCI technologies. The proponents of Constructive Technology Assessment have stressed the importance of broadening the design, development and implementation processes (Schot & Rip 1997). Verbeek's (2011) mediation approach focuses on the design of human–world relationship through technology. In all these approaches, the focus is on the design stage of artefacts: how an object is brought into the physical world through construction, following an ideal, preconceived form. The artefact is complete and ready to use at its creation, and any deviation from the designed ideal form is remedied through repairs or adjustments. Maintenance here plays an auxiliary role, meant to aid a finished design to cope with reality. This is what Young (2021a) dubs 'technology as form'.

Thus, maintenance, under this framing, is conservative of an initial design and function. In this understanding, artefacts are discarded when this initial design and function can no longer be restored. These artefacts are beyond repair – the change that time and use have created have pushed them over the brink of 'preservation' (Young 2021a, 360). With the intent to move the focus from the designers to the maintainers, Young proposes maintenance as the primary drive and functioning of technology. Rather than highlighting the design and innovation phase of an artefact, more emphasis should be placed on the continuous process of maintenance through repair, readjustment and upkeep. Technology should be seen as 'a process sustained by constant human activities of maintenance and repair', that is, 'technology as a process' (Young 2021a, 102).

Hence, we should think of the process of making, using and maintaining a technology as a process of growth, guided throughout different, fluctuating phases of interaction with the artefact. The following quote exemplifies this idea:

> Technologies are understood as fluid entities which constantly respond to the changing environments in which they exist, and which therefore require guidance through time in order to appear for us as stable entities.
>
> (Young 2021b; 364)

This fluidity and room for growth, coupled with reflexive repair is in focus when we argue that waste management systems can be reimagined to represent different visions of waste. In other words, we claim that a dynamic and reflexive conception of technology as a process gives theoretical room to re-conceptualise waste infrastructure/management systems and open up these systems for the negotiation of new, sustainable visions of waste. In the coming sections, we also show that this reconceptualisation encounters considerable sociomaterial resistance. However, we first need to establish what we mean by 'waste'.

12.4 A Dynamic Concept of Waste

Having established the concepts that we need to address and problematise municipal waste management, we now turn to waste itself. For the focus of this chapter, we chose municipal solid waste as a distinct category of waste, aside from fluid and gaseous municipal waste. Municipal solid waste has several characteristics that we will explore in relation to municipal waste management and waste infrastructures. Additionally, we show different perspectives on waste and then highlight how different valuations of waste have emerged over time. Finally, we show how the modern imaginary of waste hampers the materialisation of new visions of waste, as, for instance, proposed in the European Union.

Based on philosophical, anthropological and sociological scholarships, we identify four main characteristics of waste. Waste is 1) a socially determined product, 2) a normative category, consisting of ethical and aesthetic components, 3) an environmental agent and 4) inevitable. These four aspects give respective explanations of waste's ontological standing and ethical significance (1 & 2), its practical implications (3) and why waste merits debate in the first place (4).

First, waste is a socially determined product. Importantly, waste is not a predetermined, ontological category, something that can be objectively distinguished. Rather, waste is the product of a relational process of categorisation (Hawkins 2010; Douglas 2002). Structurally, we can argue that waste, like dirt, always is 'matter out of place' (Douglas 2002). However, that doesn't determine what counts as matter out of place in a certain (cultural) context. Thus, what is considered waste is culturally dependent (e.g., Edgerton 2007). The category of waste is itself a social product, something that has to be agreed upon and negotiated to be 'waste'. This means that waste itself is a dynamic concept that changes based on historical context, societal norms and values and the possibilities present in societies to reinterpret what constitutes waste. However, this dynamism of waste is curtailed by fixed processes, institutions and infrastructures for waste. Thus, while waste is nominally dynamic, real processes and structures fix its meaning through reproduction and static engagement.

This nominally dynamic character of waste becomes especially poignant when considering the ethical and aesthetic relevance of waste. Waste is an ethical and aesthetic category. From an aesthetic perspective, waste is something that is unpleasant, even disgusting and unclean. We want to avoid contact with waste. The idea of waste evokes sentiments of disgust, uncleanliness, potential danger through contamination, etc. These aesthetic judgments also have moral implications. Especially, if these judgments are mobilised to produce emotional reactions and relationships with waste, they fall clearly into the category of the ethical (Roeser & Todd 2014).

In those cases, these reactions have real-world effects that bear upon other people, close-by or far-off, affecting their lives and livelihoods.

Waste surrounds us, while always being something undesirable. Designating something as waste entails that it no longer has use, has lost its initial purpose or is spent. On the ethical significance of waste, we follow Hawkins when she writes that '[w]aste is now a field of activity structured by legislated and normative moralities, by disciplinary codes that order conduct in the interest of wider objectives: from reduction of landfill to global ecological survival' (2010, 22). The way we engage with, categorise and maintain waste has important moral implications, since waste is a normative category: it tells us how to interact with a given artefact or phenomenon. We focus on this criterion of waste in the following sections.

Thus, the ethical and normative characters of waste highlight the third and fourth practical and moral criteria of waste tied to its maintenance. From this third perspective, waste is relevant as an environmental phenomenon. The production and effective management of waste is seen as crucial in dealing with environmental degradation. Spilled and improperly processed waste can emerge as pollution and environmental hazard. This aspect of waste is further compounded by environmentalists' and climatologists' realisations that one of the biggest threats to humanity yet – the continuous emissions of CO_2 through combustion – is the ultimate kind of waste through its double invisibility. Not only is it invisible in the physical sense, as it cannot be perceived by the human eye and is odourless, that is, our senses cannot make sense of it. It is also invisible in its direct impact.

Finally, waste seems inevitable. There will always be matter out of place, as waste and dirt establish material and social orders and borders. The very human condition seems trapped in the necessity to produce waste and to create excess. The inevitability of waste production is particularly glaring in industrialised consumption-based societies, where waste rears its head undeniably through packaging, sewage and broken gadgets – societies in which, as Nagle (2013) puts it, 'San[itation] workers are key players in maintaining the most basic rhythms of capitalism' (46). This connects to Schumpeter's (2008 [1950]) more general idea of 'creative destruction', the destruction of old forms of wealth to create new ones, that is, the constant upheaval of economic structures in capitalist societies. The production of waste becomes necessary for economic progress.

12.5 Modern Waste Management: An Institutional Perspective

So far, we have argued that the dynamic nature of waste is most properly complemented by a dynamic concept of maintenance that allows infrastructure to be aligned with new visions of waste. However, as we have indicated, infrastructures provide resistance to change as well. The coming sections pay attention to some of the forms of resistance that urban infrastructures exhibit.

Given the sheer production and amount of waste, modern municipal waste management systems *need* to be invisible. In this aspect, they differ from other systems operating in our urban backgrounds. Insensibility is deliberately designed into municipal waste management systems (Hawkins 2007). Rather than retreat into the background of modern urban life, municipal waste management infrastructure always already has retreated. Recent garbage crises that affluent societies have faced mostly concerned the re-emergence of the sensibility of waste and the re-emergence of the question of how to best deal with this sensibility. More philosophical and ethical questions about waste often escape the political agenda. That is, infrastructural innovation is directed towards technological efficiency and efficacy regarding a set of predetermined goals taken as exogenous 'givens': health, tidiness and, therefore, insensibility. The existing rationale for waste management was, and to a large extent still is, 'expand and upgrade', based upon design principles committed to universal, reliable and affordable service (Moss & Marvin 2001, p. 5/6). Figure 12.1 shows a schematic overview of the four phases of solid waste management in urban areas in HIC, phases that are (in)formed by infrastructures, technologies, habits and institutions.

One way to understand the neglect of philosophical and ethical questions concerning waste is provided by the literature on socio-technical regimes. In this institutional perspective, the focus on efficiency, expansion and upgrading is seen as part of the socio-technical regime of urban solid waste management in HIC, constituting its paradigmatic centre. This regime is the result of the co-evolution of infrastructure and institutions.

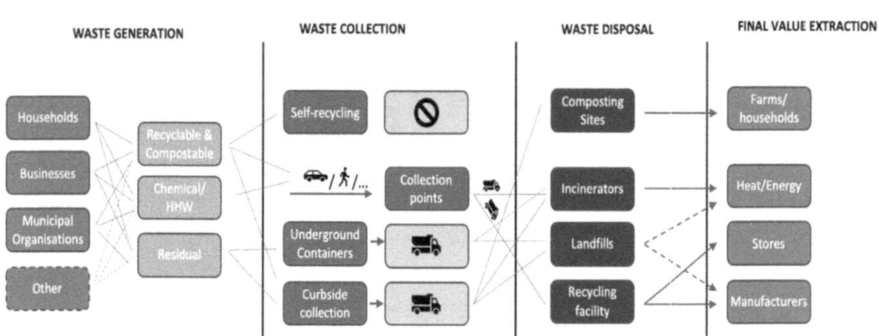

Figure 12.1 Schematised Municipal Waste Infrastructure.

It has its own 'logic' and 'grammar' (Fuenfschilling & Truffer 2014). Problems and their solutions come about as salient through this logic, which forms 'coherent arrangements of beliefs, norms, values and practices' (ibid. p. 773).

We have seen, furthermore, that waste management systems (Figure 12.1) come into the public eye when they break down or when their continued use and operation leads to morally problematic situations, such as disease, pollution, the involvement of organised crime and increased global inequality. In a socio-technical regime, these problems are often interpreted and dealt with in preconfigured ways, along already-established innovation pathways. The robustness of this institutional logic, its endurance, is determined by both internal and external events, such as the arrival of new actors and niche technologies, changes in the distribution of power and the social impact of crises and disasters (Geels 2010).

The institutional perspective adopted in the socio-technical regime's literature explains how institutions and technologies co-evolve in socio-technical systems, such as municipal waste management systems. From this perspective, the extent to which a complete reimagination and system transition is deemed necessary depends at least in part on the success of other (and earlier) attempts of the regime to cope with problematic situations caused or endured by the system. These attempts might have left the material design of the municipal waste management system largely intact. We can think here of failed attempts at the improvement of the institutional context (new rules, guidelines, laws, etc.) or failed efforts at educating city dwellers about good and bad waste habits (stickers on garbage bins, codes on packaging materials, waste separation folders, school trips to landfills, etc.). In these cases, new objectives and goals in waste management are pursued at the supply side of waste, changing the waste behaviour of citizens through education and regulation. In the case these new goals amount to substantial revisions of the ultimate objectives of waste management, we refer to these new objectives as 'values'. The pursuit of new values is often associated with socio-technical transitions, in which new practices, rules and technologies emerge (Geels 2004).

We depict the changes in these practices, rules and technologies in Figure 12.2. In this visualised trinity of (re-)education, institutional (re)design and material (re)design, the latter option, the material (re)design of parts of the socio-technical system of waste management, is often the most time-consuming and costly. We are dealing in these infrastructural systems with a materialisation of norms and values associated with waste, that have 'locked in' certain waste practices. The underground sewage pipes, the closed-off garbage trucks or covered trash cans, all represent physical manifestations of a certain understanding of waste that leaves its mark on waste practices. At the same time, physical infrastructures embedding

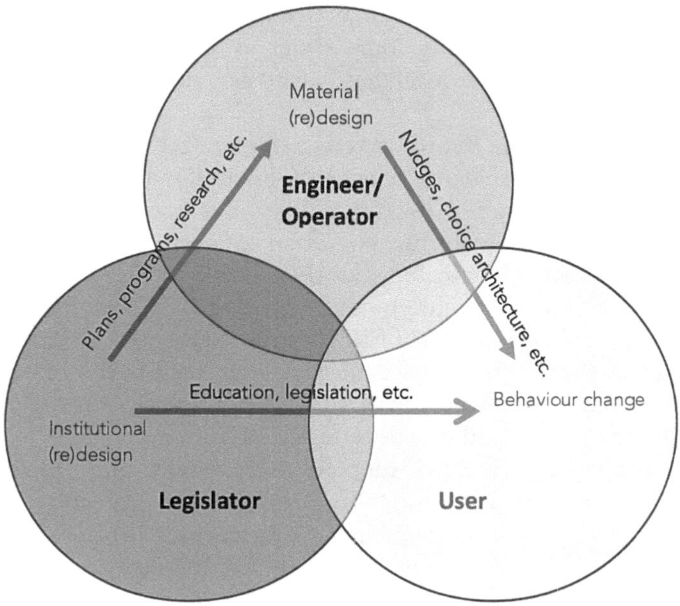

Figure 12.2 Visions of changing (waste) practices in socio-technical systems.

entrenched norms are very difficult to change, which is why acting on them is often the last option to be considered.

This leads to a paradoxical situation: *citizens are instructed to change their behaviour and practices while existing infrastructure keeps reinforcing that behaviour and those practices*. Institutionally, therefore, citizens, are motivated to change, while materially they are motivated to keep on doing what they did. A purely institutional perspective on change is therefore not enough. If cities want to successfully implement new waste visions, all three aspects of Figure 12.2 need to be put into operation. Institutional change, citizen education and motivation and material change need to go hand-in-hand. This means the inclusion of diverse stakeholder groups, such as citizens and maintainers, and the development of new ideas concerning the interaction of these groups. In our view, maintenance offers an underestimated opportunity for effectively bringing to the fore the material aspects of socio-technical transitions that are inspired by new visions of urban waste.

Finally, upholding a modern vision of municipal waste ('Burning, Burying, Be rid of it') through the static maintenance of waste infrastructures is costly. Most of these costs might be borne by future generations, the environment or subaltern social groups. Some of these costs might be invisible (air pollution and CO_2 emissions of waste transport) or might not be experienced collectively (the shipping of waste into other regions/countries). Hence, local invisibility is often associated with affluent parts of the city or affluent parts of the world, while waste becomes more visible in marginalised localities. Affluence, in this sense, determines the period during which municipal waste is visible and still affirmed and a confrontation and negotiation can take place.

This is a development long in the making. As a number of scholars have pointed out, the modern imaginary of urban living and waste focused around conceptions of tidiness that emerged throughout the nineteenth century in cities in Northern America and Europe (Chakrabarty 1992; Melosi 2004; Hawkins 2010; 2007). These conceptions have solidified in infrastructural design choices for modern urban waste management systems and continue to inform contemporary choices.

12.6 Modern Imaginary of Waste

The modern imaginary of urban living and waste emerging in the nineteenth century is the 'tidy city' (Chakrabarty 1992; Hawkins 2010; 2007). In this tidy city, the management of waste (and health) has become a public concern. This means that urban sanitation is no longer an individual responsibility, as it was prior to the nineteenth century, but a communal responsibility and a service that the city provides (Melosi 2004). Furthermore, responsible waste practices become an important aspect of civic life and citizen culture. The administrative treatment of waste and public health becomes a sign of urban progress. As the Dutch environmental ministry's homepage puts it: 'It is the role of governments to prevent the waste of raw materials and energy and to ensure that human health and the environment are not harmed by waste'(Ministerie van Infrastructuur en Waterstaat 2022). Modern waste management is a public concern.

A wide range of scholars of waste have shown how this nineteenth-century imaginary of waste is both persistent and resistant in the western hemisphere. It has set societies on socio-technical pathways of waste management systems that have led to both technological and institutional lock-in. The force of this imaginary of tidiness is dependent upon several of its features. First, its geopolitical power. As Newell (2015) and Chakrabarty (1992) argue, the assumed connection between cleanliness and civilisation has justified colonisation and subjugation of the 'dirty native' in the nineteenth and early twentieth centuries. Filth today is still associated with lower social classes (Berg 2015; Forty 1986), and with 'othering'.[4]

Second, overconsumption, and therefore waste, started to be taken as a fact of civilian life. Nineteenth-century affluent citizens began showing their wealth by buying things they wanted, but didn't need and could discard at will (Benjamin 1999).[5] Third, the medical sciences of virology and bacteriology emerged at the end of the century, as modern public health science established relations between (human) dirt, disease and death that are still prevalent today (Melosi 2004). Finally, the emergence of municipal sanitation had both ethical and aesthetic consequences. The 'tidy city' still is an attractive epithet. It attracts tourists, yuppies and supports increasing levels of gentrification (Hawkins 2007). Clean, tidy and orderly urban districts preserve the otherness of waste – as something that is both aesthetically and morally disavowed.

On this basis, we introduce the modern waste imaginary. In this imaginary, waste is something to be discarded; something that is both aesthetically unpleasant, morally bad and unwanted. It is out of place and should be put away, either geographically or excised and obliterated as a whole. Even apparently more reflective approaches, such as the Zero Waste plans of cities like San Francisco and Singapore uphold this image: 'Zero Waste' here does not refer to a completely circular (urban) economy or 100% biodegradable consumer goods. Rather, these Zero Waste plans rely heavily on waste-to-energy incinerators. These incinerators reduce waste mass up to 90% (Setoodeh Jahromy et al. 2019). Since the leftover material can be used for road construction and leftover metals can be separated and recycled, these Zero Waste plans almost realise Zero Waste.

However, as we detail in the latter sections of this chapter, these plans fall short of realising a new vision of waste on the supply side of waste, for example, with urban dwellers. In the next section, we introduce how socio-technical systems can entrench or hinder the expression and materialisation of new values. This provides theoretical tools to explore how current waste infrastructures either reproduce the modern waste imaginary or do not go far enough where they seek to go beyond it, given the inherently static underlying assumptions of current waste management systems.

12.7 Value Change

A growing amount of recent scholarship recognises that socio-technical systems, since they're bound to exist for multiple generations, will see some level of value change. In short, although these systems might have been designed while keeping certain social, ethical and public values in mind, the relative importance of these values, their conceptualisation or the (design) norms associated with their realisation, might change (van de Poel 2021; van de Poel & Taebi 2022; van de Poel & Kudina 2022). We can even imagine completely new values emerging in society that start to weigh down on the operation of certain systems, such as energy or waste systems (Taebi et al. 2020).

The literature on value change argues that morally problematic situations, whether caused by these socio-technical systems or not, create an urgent call to reassess the design of those systems, with this value change in mind. These might be values new to the context (e.g., new to the context of waste management systems, such as ecological sustainability) or completely new values as abstract goals (e.g., intergenerational justice). That is, the redesign of parts of the system of waste management, deemed necessary because of the problematic situations the current design has caused, is often guided by new requirements that stem from new values. As is evident from our discussion so far, the realisation of these new values cannot be achieved through institutional change only. Despite a recognition that municipal solid waste management needs to transition to more sustainable, just and ecological practices, urban waste management infrastructures continue to express an imaginary of tidiness that trumps the expression of these new values. Both education and institutional (re)design (Figure 12.2) have been unable to realise major shifts in collective beliefs and practices concerning waste in European urban areas. To date, the focus has been primarily on how to deal with waste, and visions are built around answers to this question. Yet, another question often goes unmentioned: to what extent should we change what we consider waste in the first place?

For instance, in Europe, the Smart City concept developed in the early 2000s, hardly pays attention to the reimagination of municipal solid waste, despite a focus on issues such as urban pollution, sustainability and sanitation. Smart waste collection does receive attention in urban contexts (e.g., Ali et al. 2020), but seems to focus more on Internet-of-Things solutions in which garbage cans communicate with garbage trucks (e.g., Bhor et al. 2015) and sensors determine the level of air pollution. The focus here is on efficiency and reliability, rather than new values emerging in relation to waste management.

Furthermore, the EU directive of 2015 established a hierarchy of goals in relation to waste management: prevention, reuse, recycling, recovery of energy and disposal. However, many member states have difficulties in achieving these goals and often focus on waste-to-energy processes instead of higher order goals (EU 2018). The failure to achieve a more radical transformation of municipal waste practices can be understood to stem in part from limitations in the political directives themselves. For example, one of the research foci of the Horizon 2020 programme was the circular economy [CE] and urban metabolism, which developed valuable insights into the challenges of CE in urban contexts, mostly from a governance perspective (e.g., Obersteg et al. 2019; Remøy et al. 2019). At the same time, however, the programme failed to address the affordance and resistance of waste management systems or the potential for a collective reimagining of waste and related practices.

Hence, waste is often assumed as a given fact of urban life. Contemporary debates concerning the ethics of waste highlight problematic attitudes to waste, delivering a moral appeal to change these attitudes (Hawkins 2010). Yet, the extent to which the current treatment of and attitude towards waste are seen as problematic is as much the result of the history of waste (see Section 12.6), as of the persuasive power that circulating new perspectives on waste are able to exercise. This persuasive power itself, we argue, depends both on the possibilities to render visible the problematics of the current operation of municipal waste management systems and on the historical meanings of waste still designed into these systems, in the form of embedded values. We call these two aspects *infrastructural affordance* and *infrastructural resistance*.

A collective reimagination of waste, deemed necessary for collective support of changes in socio-technical systems of waste management and the burdens these changes might impose on citizens, requires a collective confrontation with municipal waste in which new values can be brought forward as essential for the accumulation, collection and disposal of solid waste in urban contexts. For such a confrontation to take place, it will be essential to explore new ways of experiencing waste. However, despite a range of art projects involving waste, the conditions for such positive 'waste experiences' are still largely absent in affluent societies. One of the reasons for this may be found in the physical aspects of waste infrastructures, where the resistance presented in the infrastructurally embedded historical meanings of waste weakens the persuasive power of new visions.

Reconceptualising the maintenance of municipal solid waste management systems as reflexive repair and a dynamic process can play two roles here. First, it can undermine the resistance to value change designed into these systems by rendering them visible during the act of maintenance and allowing for new forms of political engagement with them, thus bringing to the surface the problems associated with their operation. Second, maintenance can be seen as a process of redesign of urban infrastructure instead of replacement/repair and short-term fixes, making maintenance an important strategy for bringing political and ethical questions to the fore and thus creating *infrastructural affordance*. However, the redesign of critical urban infrastructures is often avoided because of costs and risks to urban flows of utilities. While we do not claim to have a solution to this problem, we argue that the creation of public acceptance for new visions of waste can only be achieved through the 'visibilization' of waste infrastructures, a process of creating *infrastructural affordance* in which maintenance can play a crucial role.

12.8 Some Politics of Waste Infrastructures in the EU

As noted above, the affordance and resistance of contemporary municipal solid waste management systems determine the extent to which new meanings of waste can establish themselves in urban contexts. In a very basic sense, this would encompass the transition from a waste and dirt-rejecting system to a waste and dirt-affirming system of beliefs and practices (Douglas 2002) in the city, supported by a waste-affirming vision. Anthropologically, dirt affirmation refers to societies in which some forms of pollution and corruption are 'enshrined in sacred places and times' (Douglas 2002), whereas in dirt-rejecting societies dirt loses all identity and cannot acquire new functions before it is completely disintegrated and undifferentiated, when it has stopped being dirt.[6] Importantly, waste serves as a proxy for dirt in the modern waste imaginary. Most advanced urban areas are dirt rejecting in a non-religious sense, as places where 'order, efficiency, and perfection' (Scanlan 2013, p. 2) leave no room for the reintroduction of garbage as an object or even commodity before it has completely disintegrated in landfills or incinerators. Before it has stopped being garbage, that is. Nonetheless, we argue, that some level of dirt affirmation must occur to effectively negotiate waste and open up a critical space for a collective reimagination of what constitutes waste.

However, existing waste infrastructures are both robust and durable and infrastructure tends to become more resilient over time (Henke & Sims 2020, 121). Late twentieth-century problems with landfills and levels of toxicity and contamination of water tables have led many European countries to focus on waste-to-energy incinerators (McCauley 2009). The current ubiquity of incinerators is a consequence of the mixed nature of most solid municipal wastes, stalled efforts at sorting out different types of wastes, protests against shipping waste abroad and failed attempts at large-scale recyclable plastics. Nevertheless, many of these countries have also seen (local) forms of resistance against these incinerators that, in some cases, have led to further innovations (e.g., reduction of fly ash and better filters). The EU directive 2018/851 recognises that many member states have not yet developed the necessary waste management infrastructures to achieve a circular economy. It does not, however, provide reasons for this failure to adapt. Overall, the directive 2018/851 pays little attention to raising public awareness (and visibility) of waste. Action point (30) states:

> The promotion of sustainability in production and consumption can contribute significantly to waste prevention. Member States should take steps to make consumers aware of that contribution and encourage them to participate more actively in order to improve resource efficiency. As part of measures to reduce waste generation, Member States should include continuous communication and education initiatives to raise awareness on the issues surrounding waste prevention and littering and may include the use of deposit-refund schemes and the setting of quantitative targets, and provide, as appropriate, adequate economic incentives to producers.
>
> (EU 2018)

The EU directive 2018/851 focuses on education and institutional design to raise awareness (see Figure 12.2) about the negative effects of waste. Awareness, we have argued, is something that is hampered by the design of modern waste management systems and characterised by the processes of 'invisibilization' of municipal waste. Efforts at prevention, reusage and recycling (EU directive 2015) are all hindered by these processes.

Problems of citizen education and the collective reimagination of waste come together in the design of waste infrastructures in affluent countries. To gain collective support for the proposed and much-needed changes in the way these countries deal with waste, waste perceptions in advanced urban contexts must first be laid bare as both historically contingent and antiquated. This means that anthropological, sociological, psychological and urban geographical insights concerning waste and its management must be combined to address local, economic, environmental and social issues. Here we believe that both the dynamic reconceptualisation of maintenance as well as the reflexive repair approach can be of use.

What both of these approaches to maintenance highlight is the fact that infrastructure is more than a simple network of technologies, and that its upkeep is a value-laden process. This realisation, we believe, is paramount to making room for the re-imagination of waste infrastructure, and the way we experience waste itself. As Henke and Sims put it,

> [I]nfrastructural repair is not just about fixing things but also relationships and negotiation. An approach to reflexive repair that focuses only on technical fixes and eschews conversation misses opportunities to build common discourses and identities around a complex and urgent problem; talk also allows us to listen and learn when we disagree and misunderstand.
>
> (2020, 134)

Thus the remaining two sections of this chapter explore the potential avenues of reconceptualising our modern vision of waste through waste affirmation.

12.9 Waste Affirmation and Municipal Solid Waste Management

It is difficult to imagine what a waste-affirming modern society would look like. Indeed, such a society might be considered 'postmodern' in the sense that it would make previously established boundaries and strict dichotomies between concepts and categories permeable and porous. In this light, Douglas (2002) asks how it is possible that something as destructive as dirt can also have a creative force. Douglas (2002), citing William James (1902), claims that the most complete systems of thought need to find some way of affirming what has been rejected, whether it is matter out of place (Douglas 2002), objects that have lost their aura (Benjamin 2021) or the abject excrements of the body (Kristeva 1982). A more complete philosophy of the city would account for the presence of secular rituals in which the 'mixing up and composting of polluting things' (Douglas 2002) takes place without recourse to technological perspectives on waste. The practical question is how such rituals or, more generally, practices of waste affirmation can be firmly established in urban environments. Do we need to design crises of waste, or are other venues possible?

As we have shown in the previous sections, cities are not designed for zero waste visions, but rather for zero visibility waste visions. This zero visibility paradigm problematises the efforts to challenge the modern imaginary of tidiness, with its hygienic and aesthetic connotations, and problematises the efforts to remind citizens that waste is an integral aspect of urban living that, at the same time, has negative consequences. Waste infrastructures resist new interpretations and visions of waste. Before becoming waste affirming, we need to become waste admitting. Besides education and institutional (re)design (see Figure 12.2), waste infrastructures should also play a part in this admission. Several authors have explored what dirt affirmation would look like from a material, architectural perspective (e.g., Campkin 2013; Shonfield 2014). In these elaborations, dirt affirmation is related to Foucault's (1984) thoughts on heterotopias, places of otherness that function as mirrors to the city, as counter-sites. In these 'othering' spaces, we could find the celebration of dirt and waste, the denial of purity and tidiness, an inverted city, turned inside out, the 'garbage cities' presented in documentaries such as Waste Land (Walker 2010) or Plastic China (Wang 2016). However, despite the existence of heterotopias (Foucault names places of 'deviation' such as psychiatric hospitals, rest homes, cemeteries and zoological gardens), it is unclear how these sites, except as places of contestation and experimentation, could contribute to changing urban waste practices and beliefs.

The 'broken world' thinking of Jackson (2014) may offer a new paradigm here for thinking about urban problems and solutions. We refer to his oft-quoted idea that we should aim at:

> an appreciation of the real limits and fragility of the worlds we inhabit – natural, social, and technological – and a recognition that many of the stories and orders of modernity (or whatever else we choose to call the past two-hundred-odd years of Euro-centred human history) are in the process of coming apart, perhaps to be replaced by new and better stories and orders, but perhaps not.
>
> (221)

One such story that needs to be dismantled is the myth of urban infrastructures as stable, and permanent. This myth of fixed and stable infrastructure (Graham & Thrift 2007), while cities expand, and urban flows change, obscures the extraordinary everyday effort of keeping the city 'running'. The ongoing process of infrastructure, exempt from a narrative of progress and innovation, needs to be brought into focus. Post-modern cities, in this sense, are 'cities of repair' (Graham & Thrift 2007, 10.) Visibilisation of waste, we have argued, comes in two forms. On the one hand, maintenance can serve to lay bare waste infrastructures by enhancing the visibility of maintenance activities. However, this first form of visibilisation only temporally allows for critical engagement with waste. The second form is more permanent. It focuses on dynamic maintenance as an act of redesigning waste infrastructures to allow for engagement with waste. The resensibilisation of waste does not mean that we return to filthy streets filled with manure and all kinds of litter. It would mean that through this resensibilisation, waste becomes something relatable, positive and affirmative. What is repaired and maintained are positive meanings of waste that have become obfuscated by modern imaginaries of tidiness and unlimited progress. We can think here of maintenance and repair as a continuous laying bare of waste streams in urban contexts comparable to the way the Centre Pompidou in Paris lays bare its inner tubing – its flows of air, energy and waste.

12.10 Conclusion

Waste and maintenance are inadvertently connected. We hope to have shown that this connection can serve as a point of exploration to problematise current western urban societies' relationship and imagination of waste. Perceiving waste as something aesthetically and ethically problematic inhibits the transition to a more environmentally aware vision of consumption patterns. Our current vision of waste rejection is further entrenched into the very infrastructure we rely on to rid ourselves from our waste, which makes the re-imagination of waste all the more difficult. A dynamic conception of maintenance alongside a reflexive approach to infrastructure repair could aid us in reconfiguring our understanding of what waste is and how we ought to treat it. It would create infrastructural affordance for new practices related to new values. However, these approaches alone will not be sufficient. Instead, it will take radical new visions of waste that go beyond the modern understanding on institutional, societal and technological levels in order to re-imagine waste as a necessary reminder and remainder of our consumption, as something that we can gladly claim.

Notes

1 See Steinert (this volume).
2 See https://data.worldbank.org/indicator/SP.URB.TOTL.IN.ZS [last accessed January 2024]
3 See Steinert (this volume).
4 Indeed, modernity is often associated with these strict categorisations and dualisms, such as the distinctions between mind and body, nature and culture, public and private and agency and structure (e.g., Latour 2012; Douglas 2002; Bauman 2003; Jackson 2014; Giddens 2020).
5 For example, the word 'gadget' started being used late in the nineteenth century as a term for an object one cannot remember the name of, although the exact origins are debated (Merrin 2014).
6 We are aware that dirt and waste are not synonymous. In Douglas' work, the two concepts are not always easy to keep apart. On the one hand, dirt seems to be an interpretation and evaluation of superfluous things, materials, words and bodily aspects, as excessive 'things' that confuse and endanger accepted social classifications. On the other hand, as we have noted in the introduction, waste is always in immediate danger of becoming dirt. Douglas herself gives examples of human matter in food such as hairs in soups. For the sake of our argument, we do not think that a strict conceptual boundary is necessary.

Bibliography

Ádám, B., Göen, T., Scheepers, P.T., Adliene, D., Batinic, B., Budnik, L.T., Duca, R.C., Ghosh, M., Giurgiu, D.I., Godderis, L. and Goksel, O., 2021. From inequitable to sustainable e-waste processing for reduction of impact on human health and the environment. *Environmental Research*, *194*, p. 110728.

Ali, T., Irfan, M., Alwadie, A.S. and Glowacz, A., 2020. IoT-based smart waste bin monitoring and municipal solid waste management system for smart cities. *Arabian Journal for Science and Engineering*, *45*(12), pp. 10185–10198.

Bauman, Z., 2003. A sociological theory of postmodernity. In *Intimations of Postmodernity* (pp. 215–232). Routledge.

Benjamin, W., 1999. *The arcades project*. Harvard University Press.

Benjamin, W., 2021. *One-way Street: And other writings*. Verso Books.

Berg, A., 2015. Waste streams and garbage publics in Los Angeles and Detroit. In *Global Garbage* (pp. 92–110). Routledge.

Bhor, V., Morajkar, P., Gurav, M., Pandya, D. and Deshpande, A., 2015. Smart garbage management system. *International Journal of Engineering Research & Technology (IJERT)*, *4*(03).

Brooks, A., 2019. *Clothing poverty: The hidden world of fast fashion and second-hand clothes*. Bloomsbury Publishing.

Campkin, B., 2013. Placing "matter out of place": Purity and danger as evidence for architecture and urbanism. *Architectural Theory Review*, *18*(1), pp. 46–61.

Chakrabarty, D., 1992. Of garbage, modernity and the citizen's gaze. *Economic and Political Weekly*, pp. 541–547.

Dant, T. and Bowles, D., 2003. Dealing with dirt: servicing and repairing cars. *Sociological Research Online*, *8*(2), pp. 1–17.

Denis, J. and Pontille, D., 2015. Material ordering and the care of things. *Science, Technology, & Human Values*, *40*(3), pp. 338–367.

DIRECTIVE (EU) 2018/851 OF THE EUROPEAN PARLIAMENT AND OF THE COUNCIL of 30 May 2018 amending Directive 2008/98/EC on waste (Text with EEA relevance) https://eur-lex.europa.eu/legal-content/EN/TXT/HTML/?uri=CELEX:32018L0851&from=EN

Douglas, Mary. 2002. *Purity and danger: An analysis of concepts of pollution and taboo*. Routledge. https://doi.org/10.4324/9780203361832

EC, 2015. COM (2015) 614 final. COMMUNICATION FROM THE COMMISSION TO THE EUROPEAN PARLIAMENT, THE COUNCIL, THE EUROPEAN ECONOMIC AND SOCIAL COMMITTEE AND THE COMMITTEE OF THE REGIONS Closing the loop - An EU action plan for the Circular Economy. Brussels 2.12.2015. Available at https://eur-lex.europa.eu/resource.html?uri=cellar:8a8ef5e8-99a0-11e5-b3b7-01aa75ed71a1.0012.02/DOC_1&format=PDFFoucault, M. and Miskowiec, J., 1986. Of other spaces. *Diacritics*, *16*(1), pp. 22–27.

Edgerton, David. 2007. Creole technologies and global histories: rethinking how things travel in space and time. *Journal of History of Science and Technology*, *1*(1), 75–112.

Edwards, P.N., 2003. Infrastructure and modernity: Force, time, and social organization in the history of sociotechnical systems. *Modernity and Technology*, *1*, pp. 185–226.

Edwards, P.N., Bowker, G.C., Jackson, S.J. and Williams, R., 2009. Introduction: an agenda for infrastructure studies. *Journal of the Association for Information Systems*, *10*(5), p. 6.

EU 2018. Directive (EU) 2018/851 of the European Parliament and of the Council of 30 May 2018 amending Directive 2008/98/EC on waste (Text with EEA relevance) https://eur-lex.europa.eu/legal-content/EN/TXT/?uri=celex:32018L0851

Forty, A., 1986. *Objects of Desire: a history of commodity design*. Pantheon Books.

Friedman, B., Kahn, P. and Borning, A., 2002. Value sensitive design: Theory and methods. *University of Washington technical report*, 2, p. 12.

Fuenfschilling, L. and Truffer, B., 2014. The structuration of socio-technical regimes—Conceptual foundations from institutional theory. *Research Policy*, *43*(4), pp. 772–791.

Geels, F.W., 2004. From sectoral systems of innovation to socio-technical systems: Insights about dynamics and change from sociology and institutional theory. *Research Policy*, *33*(6–7), pp. 897–920.

Geels, F.W., 2010. Ontologies, socio-technical transitions (to sustainability), and the multi-level perspective. *Research Policy*, *39*(4), pp. 495–510.

Giddens, A., 2020. Modernity and self-identity: Self and society in the late modern age. In *The new social theory reader* (pp. 354–361). Routledge.

Graham, S., 2010. When infrastructures fail. In *Disrupted cities* (pp. 13–38). Routledge.

Graham, S. and Thrift, N., 2007. Out of order: Understanding repair and maintenance. *Theory, Culture & Society*, *24*(3), pp. 1–25.

Guy, S., Marvin, S. and Moss, T., 2001. *Urban infrastructure in transition: Networks, buildings, plans*. Earthscan.

Harmer, L., Cooper, T., Fisher, T., Salvia, G. and Barr, C., 2019. Design, Dirt and Disposal: Influences on the maintenance of vacuum cleaners. *Journal of Cleaner Production*, *228*, pp. 1176–1186.

Hawkins, G., 2007. Waste in Sydney: Unwelcome Returns. *PMLA*, *122*(1), pp. 348–351.

Hawkins, G. 2010. *The ethics of waste: How we relate to rubbish*. Lanham: Rowman & Littlefield Publishers.

Hawkins, G. and Muecke, S. eds., 2002. *Culture and waste: The creation and destruction of value*. Rowman & Littlefield Publishers.

Henke, C.R. and Sims, B., 2020. *Repairing infrastructures: the maintenance of materiality and power*. Mit Press.

Hird, M.J., 2022. A public sociology of waste. In *A public sociology of waste* (pp. 83–98). Bristol University Press.

Hughes, T.P., 1986. The seamless web: technology, science, etcetera, etcetera. *Social Studies of Science*, *16*(2), pp. 281–292.

James, W., 1902. *The varieties of religious experience*, New York: The Modern Library.

Jackson, Steven J. 2014. Rethinking repair. In *Media technologies: Essays On Communication, Materiality, and Society*, 221–239. MIT Press.

Kan, S., 2015. *Symbolic immortality: The Tlingit potlatch of the nineteenth century*. University of Washington Press.

Kristeva, J., 1982. *Powers of horror: An essay on abjection*, New York: Columbia University Press.

Larkin, Brian. The politics and poetics of infrastructure. *Annual Review of Anthropology* 42 (2013): 327–343.

Latour, B., 2012. *We have never been modern*. Harvard University Press.

Lejeune, C., 2019. Interruptions, lunch talks, and support circles: An ethnography of collective repair in steam locomotive restoration. *Repair Work Ethnographies: Revisiting Breakdown, Relocating Materiality*, pp. 221–251.

McCauley, D., 2009. Wasting energy? Campaigns against waste-to-energy sites in France. *Environmental Politics*, *18*(6), pp. 917–938.

Melosi, M. V., 2004. *Garbage in the Cities: Refuse reform and the Environment*. University of Pittsburgh Press.

Merrin, W., 2014. The rise of the gadget and hyperludic me-dia. *Public Culture*, *10*(1), pp. 1–20. https://doi.org/10.1215/17432197-2397209

Ministerie van Infrastructuur en Waterstaat. 2022. *Rijksoverheid*. 02 09. Accessed 09 02, 2022. https://www.rijksoverheid.nl/ministeries/ministerie-van-infrastructuur-en-waterstaat

Moss, Timothy, and Simon Marvin. 2001. *Urban infrastructure in transition: Networks, buildings and plans*. Routledge.

Nagle, R., 2013. *Picking up: on the streets and behind the trucks with the sanitation workers of New York City*. Macmillan.

Newell, S., 2015. Dirty familiars: colonial encounters in African cities. In Lindner, C., Meissner, M. (eds.) *Global Garbage* (pp. 47–63). Routledge.

Obersteg, A., Arlati, A., Acke, A., Berruti, G., Czapiewski, K., Dąbrowski, M., Heurkens, E., Mezei, C., Palestino, M.F., Varjú, V. and Wójcik, M., 2019. Urban regions shifting to circular economy: Understanding challenges for new ways of governance. *Urban Planning*, 4(3), pp. 19–31.

Raggi, Virginia, interview by Mark Lowen. 2021. Can Rome be rescued from the rubbish? - BBC News (3 October).

Remøy, H., Wandl, A., Ceric, D. and Timmeren, A.V., 2019. Facilitating circular economy in urban planning. *Urban Planning*, 4(3), pp. 1–4.

Roeser, Sabine, and Cain Todd. 2014. *Emotion and value*. OUP Oxford.

Scanlan, J., 2005. *On Garbage*. London: Reaktion.

Scanlan, John. 2013. Introduction to *Aesthetic Fatigue: Modernity and the Language of Waste*, 2–24, edited by Clark, John F. M., and Scanlan, John. Cambridge Scholars Publishing.

Schot, J. and Rip, A., 1997. The past and future of constructive technology assessment. *Technological Forecasting and Social Change*, *54*(2–3), pp. 251–268.

Schot, Johan, and Arie Rip. 1997. The past and future of constructive technology assessment. *Technological Forecasting and Social Change*, *54*(2–3), 251–268. https://doi.org/10.1016/S0040-1625(96)00180-1

Schumpeter, J. A., 1976. *II. Capitalism, Socialism, and Democracy, 1942*.

Schumpeter, Joseph A. *Capitalism, Socialism, and Democracy*: Third Edition. Harper Perennial Modern Thought, 2008 [1950].

Seale, K., 2015. The paradox of waste: Rio de Janeiro's Praça XV flea market. In *Global Garbage* (pp. 80–91). Routledge.

Schumpeter, Joseph A. 2008. *Capitalism, socialism, and democracy: Third edition*. Harper Perennial Modern Thought.

Setoodeh Jahromy, S., Jordan, C., Azam, M., Werner, A., Harasek, M. and Winter, F., 2019. Fly ash from municipal solid waste incineration as a potential thermochemical energy storage material. *Energy & Fuels*, *33*(7), pp. 5810–5819.

Shonfield, K., 2014. Two architectural projects about purity. In *Architecture* (pp. 29–43). Routledge.

Sormani, P., Bovet, A. and Strebel, I., 2019. Introduction: When things break down. In *Repair work ethnographies: Revisiting breakdown, relocating materiality* (pp. 1–27).

Star, S.L., 1999. The ethnography of infrastructure. *American Behavioral Scientist*, *43*(3), pp. 377–391.

Steele, W. and Legacy, C., 2017. Critical urban infrastructure. *Urban Policy and Research*, *35*(1), pp. 1–6.

Strasser, S., 2000. *Waste and want: A social history of trash*. Macmillan.

Taebi, B., Kwakkel, H.J., and Kermisch, C., 2020. Governing Climate Risks in the Face of Normative Uncertainties. *WIREs Climate Change* 11 (5): e666. https://doi.org/10.1002/wcc.666

Taylor, A. (Director), 2008. Examined Life [Documentary].

Taylor, C. 2004. *Modern social imaginaries*. Duke University Press.

The World Bank. 2022. *Urban population (% of total population)*. Accessed 01-03-2024. https://data.worldbank.org/indicator/SP.URB.TOTL.IN.ZS

Thompson, M. 1979. *Rubbish theory: The creation and destruction of value*. Oxford University Press, 1979.

Un-Habitat. *Solid waste management in the world's cities*. Un-Habitat, 2010.
Van de Poel, I. 2021. Design for value change. *Ethics and Information Technology 23*, no. 1: 27–31.
Van de Poel, I., and Kudina, O., 2022. Understanding Technology-Induced Value Change: a Pragmatist Proposal. *Philosophy & Technology 35*, no. 2: 1–24.
Van de Poel, I. and Taebi, B., 2022. Value change in energy systems. *Science, Technology, & Human Values*, 47(3), pp. 371–379.
Verbeek, P. P., 2011. *Moralizing technology: Understanding and designing the morality of things*. Chicago/London: University of Chicago Press.
Walker, Lucy, dir. 2010. *Waste Land* [Documentary]. Midas Filmes.
Wang, J. 2016. Plastic China [Documentary].
Weinberg, Alvin M. 1966. Can Technology Replace Social Engineering? *Bulletin of the Atomic Scientists 22* (10): 4–8. https://doi.org/10.1080/00963402.1966.11454993
Young, M. T. 2021a. Maintenance. *In The Routledge handbook of the philosophy of engineering*, edited by Michelfelder, Diane P., and Neelke Doorn. (pp. 356–368). Routledge Handbooks in Philosophy. New York London: Routledge, Taylor & Francis Group.
Young, M. T. 2021b. Now You See It (Now You Don't): Users, Maintainers and the Invisibility of Infrastructure. In *Technology and the City*, edited by Michael Nagenborg, Taylor Stone, Margoth González Woge, and Pieter E. Vermaas (vol. 36, pp. 101–119). Philosophy of Engineering and Technology. Cham: Springer International Publishing. https://doi.org/10.1007/978-3-030-52313-8_6

13 Repairing AI

Taylor Stone and Aimee van Wynsberghe

13.1 Introduction

This chapter is a call to see our thinking about artificial intelligence (AI)[1] as broken and in need of repair. Not literally broken in the sense of no longer being usable or functioning. Quite to the contrary, we are seeing an increasingly rapid uptake of innovative applications for AI, especially with advances in machine learning (ML). Rather, we will argue that when AI is positioned as an increasingly integral component of digital infrastructures – and a critical infrastructure itself – there is something that urgently requires fixing. This fix, however, is not one of efficiency, usability, transparency or other concerns often debated in AI ethics. Nor is it only about literal fixes to hardware or software (although how we approach such acts of maintenance and repair are implicated here). Rather, it is repairing how we *think about* AI: as a material infrastructure that is shaping, and shaped by, its physical environment. AI carries an ecological footprint and its increasing proliferation, even in the name of laudable targets such as achieving the Sustainable Development Goals, will have material impacts that will affect its own 'operating system', namely the built and natural environments in which it is embedded. To understand and address these material impacts, we need a re-orientation of AI ethics. This chapter builds directly on calls for a new conceptualisation of the ethical and political facets of AI (e.g., Crawford, 2021). This re-orientation requires that we look beyond calculable or checklist-friendly issues such as explainability, privacy and accountability, and instead, appreciate that '... AI applications must not be conceived in isolation but within a larger network of social and ecological dependencies and relationships' (Hagendorff, 2022, p. 852).

Repair as a conceptual lens can have far-reaching implications regarding the re-orientation of AI ethics.[2] Here we focus on how the philosophy of maintenance and repair can help to frame the burgeoning idea of Sustainable AI and the intertwined facets of utilising AI *for* sustainable

DOI: 10.4324/9781003316213-15

ends and appreciating AI as an infrastructure that itself needs to be sustainably designed, maintained and decommissioned. Towards this goal, the chapter will build on the recent distinction made by van Wynsberghe (2021) between *AI for sustainability* and the *sustainability of AI* (Figure 13.1). The former utilises or develops AI in the service of sustainable goals; the latter concerns the sustainability of AI systems themselves. Building on the two-part definition of Sustainable AI, this distinction will be expanded to explore *Sustainable AI as acts of repair* and *repairing AI*. The former concerns the practical use of AI for the monitoring, upkeep and adaptation of critical infrastructures, whereas the latter requires conceptualising AI as an increasingly critical urban infrastructure itself requiring care and repair. Through this lens, AI can be considered 'broken' in two ways. First, as being an increasingly unsustainable and unjust infrastructure that is itself a cause of the environmental degradation which it is attempting to mitigate. Second, as being conceptualised as an externality, or worse a 'cloud', that is not itself embedded within and reliant upon the material world. We will therefore argue that you cannot have AI *for* sustainability without first addressing the sustainability *of* AI.

As AI is increasingly used for achieving the Sustainable Development Goals (i.e., AI *for* sustainability), the upkeep of those systems becomes increasingly critical (i.e., the ongoing repair and maintenance *of* AI). This becomes especially clear as AI-enabled systems and artefacts are deployed as solutions to the pivotal challenges of the twenty-first century, including combating climate change, preserving biodiversity and managing unprecedented urbanisation (Rolnick et al., 2022). While this can include efforts focused on efficiency and optimisation (e.g., smart energy grids), we are also seeing innovations utilising AI for the management and monitoring of natural areas and/or 'natural infrastructures' (McDonald and Beatley, 2021). In these instances, AI becomes integral to the management and monitoring of wilderness, wildlife or (urban) ecosystems and a core driver of restoration

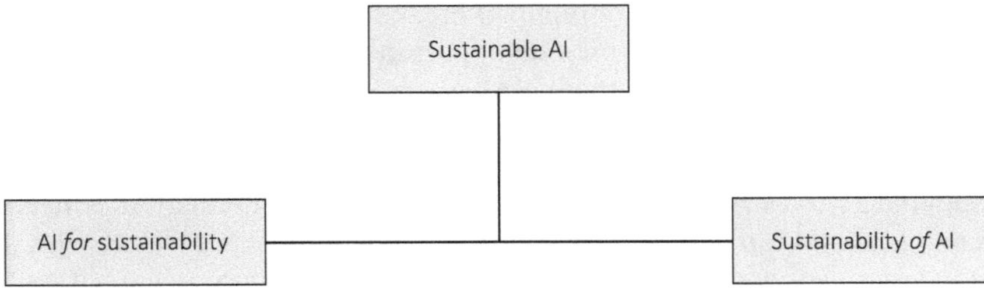

Figure 13.1 Sustainable AI as AI *for* sustainability and the sustainability *of* AI.
Adapted from van Wynsberghe (2021).

and rewilding strategies (Cantrell *et al.*, 2017). We return throughout to the case study of utilising AI for ecological restoration efforts – as an example of Sustainable AI as an act of repair, broadly construed.

The focus of this chapter is the interrelated notions of *maintenance* and *repair* and their relevance to AI ethics. Maintenance is typically framed as an activity through which we provide stability to something over time, although more recent scholarship also emphasises a process-oriented view that highlights the changes to artefacts and infrastructures such practices facilitate (Young, 2021a). This leads towards questions of repair, which can be defined as 'the subtle acts of care by which order and meaning in complex sociotechnical systems are maintained and transformed, human value is preserved and extended, and the complicated work of fitting to the varied circumstances of organizations, systems, and lives is accomplished' (Jackson, 2014, p. 222). What will be emphasised throughout is the layered meaning of repair, namely both its technical and social functions – something critical when considering the convergence of AI and sustainability. We thus follow Henke and Sims (2020, p. 3), who situate the process of repair – and in particular repairing infrastructures – as one of 'restoring both social and material order'. In this sense, repairing an artefact or system is not only an act of restoring functionality but also often serves to repair social relations and values. For our purposes, we build on this conceptualisation to delineate two levels of repair: as the literal upkeep of the hardware and software of AI and as a conceptual lens for thinking about the trajectory of AI as a material infrastructure. We will show that this second, conceptual level is critical and must accompany any literal interventions in AI (development, design, maintenance or decommissioning), if we are to make significant progress towards achieving Sustainable AI.

A focus on repair expands existing scholarship within the ethics of AI, where the design stage is largely prioritised. This comes at the cost of understanding how technologies change and evolve through time *and* the processes and practices through which this change occurs. As we will discuss, the continual process of (almost) breakdown and repair can serve to reinforce the status quo or provide opportunities to re-consider the environmental impacts of AI. Situating AI as a material infrastructure offers a different ontological perspective, emphasising the dynamic, complex and evolving nature of this system. Combined with the literal upkeep and developments to hardware and software, this reminds us that AI is never 'complete' and can never reach a static end goal of 'sustainable'. Rather, it is intertwined with the systems, structures and environments it will impact – necessitating a reflexive approach to the relationship between sustainability and AI (see Henke and Sims 2020).

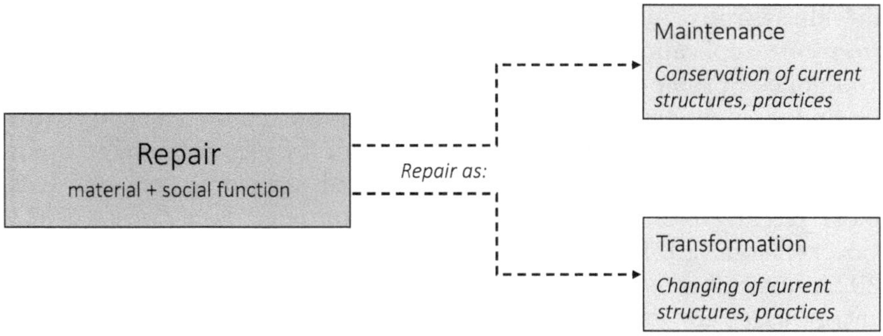

Figure 13.2 Two potential targets for acts of repair.
Adapted from *Repairing Infrastructures* (Henke & Sims, 2020).

Following Jackson's notion of 'broken world thinking' (2014), we should remember that such a perspective is not just a pessimistic outlook, but is also hopeful in appreciating the potential for change carried within these acts of maintenance and repair. It is these very processes and practices that can offer a reorientation of AI. For this, the contrasting notions of *repair as maintenance* and *repair as transformation* as proposed by Henke and Sims (2020) are applied and extended to the topic of AI. Rather than striving to maintain current structures and relations, repair as transformation instigates processes that go beyond the upkeep and conservation of current practices and structures, instead spurring deliberate change (Figure 13.2). Within transformative repair exists the possibility to question and re-think the physical structures that infrastructures support. Taking a reflexive and interactional approach to the relation between AI *for* sustainability and the sustainability *of* AI, it must be appreciated that the innovative uses of AI for goals such as ecological restoration must occur in tandem with a transformation of the digital infrastructures, power dynamics and sociopolitical systems that shape this technology.

13.2 Sustainable AI

With the growing number of AI/ML models in use across the globe, the ethics of AI has emerged as a transdisciplinary area of research at the intersection of philosophy, policy, innovation and practice to address the potential for negative outcomes associated with the pervasiveness of AI/ML applications. The field of AI ethics has observed trends in analysis from dystopian scenarios in which AI is thought to be the end of mankind to AI and robots taking all our jobs (see Coeckelbergh, 2020, 2021; van Wynsberghe, 2021). More recent trends have focused on how to design AI in 'value sensitive' ways that mitigate specific sociopolitical concerns raised from one application to another – for example, developing frameworks and policies promoting positive goals such as fairness and accountability or aiming to mitigate privacy infringements or the reinforcement of harmful biases in data sets. While these sociopolitical issues continue to dominate discourse in/on AI ethics, systemic and structural concerns have more recently been given increasing attention (e.g., Bolte, *et al.*, 2022; Coeckelbergh, 2021; Crawford, 2021; Hagendorff, 2022; Kaack, *et al.*, 2022; Strubell *et al.*, 2019). Practical issues have been raised regarding energy usage for data processing, data storage and the energy required (with resulting carbon emissions) for training AI, particularly for large ML algorithms. Further, often-hidden issues such as mining for rare minerals raise concerns over resource depletion and the life cycle of electronic waste. This can lead to political contestations over environmental justice, as well as individual autonomy and freedom. As a result, scholars have urged a shift in focus within AI ethics towards environmental justice issues, which is referred to broadly as Sustainable AI (van Wynsberghe, 2021).

To be sure, impacts such as racial discrimination resulting from the unquestioned use of historical data sets in training models are very real and very dangerous risks; however, they are the risks for those exposed to AI as a product. They are the risks for a society that believes they have, or should have, power over the way in which technologies are made and used. They are not the risks raised by those who are further up in the supply chain of AI. The concern above is not (yet) a concern for those working in the mines to procure the minerals necessary for the physical infrastructure of AI nor is it a concern for those who have electronic waste mountains in their backyards. The field of Sustainable AI is meant to target the latter set of concerns and ask: 'What is it that we, as a global community, *want* to sustain?'. This is a descriptive question in so far as we must know the actual environmental damages caused by the infrastructure of AI and normative in so far as we must be willing to suggest what are (in)appropriate consequences.

Sustainable AI has been defined by van Wynsberghe (2021) as a movement towards greater ecological integrity and social justice throughout the design process of AI. In focusing on the interrelations and interconnections between the use of AI for sustainable ends and the sustainability of developing and using AI in and of itself, Sustainable AI presents a kind of duality to the discussion. It is not possible, nor desirable, to look at only the positive or negative aspects of AI; rather, they are two sides of the same coin. For example, in a country where mining is a main source of income for millions of families, the social structure of entire villages revolves around the practice of mining. Moreover, the social conditions of life (clean drinking water, contamination of land) are impacted by these practices.

Thus, reframing AI ethics to focus on issues such as ecological impacts, energy usage and environmental justice forces one to dive deeper into the hidden costs and hidden demographics.

As the literature on Sustainable AI grows, we see a refining of the concept. Recently, it has been argued that the very notion of sustainability – when applied to something as complex as AI – must be understood as a dynamic property of systems (Bolte *et al.*, 2022). Crucially, this requires conceiving of AI as a key node of our digital infrastructures *and* novel infrastructure itself, that is, AI will use existing infrastructures to function and will itself also become an infrastructure upon which businesses, public services and individuals will rely (see Robbins and van Wynsberghe, 2022). Recognising AI as an infrastructure calls upon us to investigate the maintenance of said infrastructures. What needs to be maintained, is it possible to maintain all aspects, and critically what needs to change? In what follows, we expand on the ideas proposed in the abovementioned work, namely that Sustainable AI, as a conceptual framework, can provide tools that AI ethics alone cannot or has not. In particular, we will show how an examination of Sustainable AI through the lens of repair allows for a further elucidation of AI as a dynamic and embedded infrastructure and highlights a set of challenges and opportunities for the future of AI.

13.3 Maintenance, Repair and Sustainable AI

Ethical analyses of technology are often focused on the design phase, overlooking how artefacts and systems change, adapt and evolve over time. For example, value sensitive design has developed a series of methods for identifying and incorporating moral values into the development and design of (novel) technologies (Friedman and Hendry, 2019). While a useful advancement for the ethics of technology, there is an implicit assumption that, if successful, moral values will become embedded in the artefact in question. A specific ontological view of technology is presumed, in which artefacts have a set of stable characteristics (Young, 2021b). However, there are many critiques to this assumption, including the 'multistability' of objects (Albrechtslund, 2007), concerning the disjunction between the design and use phases of an artefact, as well as the notion of value change and its effect on the evaluation and use of technologies (van de Poel, 2021). As van de Poel (2020) points out, we should avoid overly determinist *and* constructivist perspectives but rather appreciate that technologies such as AI will be shaped by the co-evolution of technology and society. This requires an abandonment of completeness as a substantive goal for value-sensitive analyses of technology, instead appreciating that any interventions will change the relative meaning, importance and priority of the values in question (Stone, 2021).

Yet, with a complex and dynamic technology such as AI, we should not presume that this interaction ends with the design phase. Where van de Poel discusses the possibility of unanticipated consequences and changing evaluative judgements as drivers of change, he does not elaborate on the practices and processes *through which* this coevolution occurs. Any innovation must necessarily be sustained *through* repair, and likewise, repair and maintenance *become* sites of creativity and innovation (Jackson, 2014). Further, as Graham and Thrift (2007) point out, what begins as 'repair' may quickly evolve into 'improvement' or 'innovation'. To look beyond the design phase is to appreciate repair as a key mechanism for society–technology interaction and co-evolution, and as a moment where the social and environmental impacts of AI can either be challenged or reified.

Before moving on to discuss repair as a driver of the co-evolution of AI and society, it is first useful to offer a brief clarification and general definition. By way of clarification, it is important to appreciate that repair as discussed here is not only about moments of complete breakdown. There is a tendency within the study of infrastructures to over-emphasise moments of breakdown, as this makes their use and function temporarily visible (Jackson, 2014; Star, 1999). For example, our infrastructure of artificial lighting, while formative to – and ubiquitous throughout – our world after dark, readily fades into the background of lived experience. It is largely unnoticed until moments of breakdown occur in the form of blackouts, and we are temporarily plunged into what feels like an 'artificial' darkness (Nye, 2010). Through these experiences, the invisibility of infrastructure is temporarily disrupted, and they are momentarily brought into the spotlight. Yet, a focus on such (often sensational or dramatic) moments overlooks the constant upkeep, monitoring and incremental changes that are critical to the longevity and functionality of infrastructures (Young, 2021b). Hence why Jackson (2014) situates broken world thinking at the generative and continual interface between breaking down and restoring.

How, then, to approach repair as a constant and critical – if taken for granted – practice? As a first step, it is useful to understand the relation between repair and *maintenance*, conceptualised as 'an activity through which we provide stability to something over time' (Young, 2021a, p. 356). However, what exactly this stability entails, as Young points out, is an open question requiring philosophical analysis – and, one particularly relevant to discourse surrounding sustainability. A common view of maintenance is as an act of conservation (Young, 2021a), in the sense of acting *against* the passage of time; working to combat the inevitable processes of breakdown, erosion, corrosion, decay, malfunction, obsolescence, etc. This aligns with the conservative notion of *repair as maintenance*

(Henke and Sims, 2020), aiming to maintain existing infrastructures and relations and to upkeep systems and artefacts in largely the same way as they currently operate.

While often focused on the restoration and preservation of existing systems and relations, maintenance and repair can also instantiate processes of change and evolution through adjustment, tweaking, adaptation or more radical changes to systems and artefacts. This highlights the temporal and dynamic nature of technology, as processes extended and guided through time. Put otherwise, it asks us to appreciate that no technology or system is completely static, but necessarily evolves and changes. The replacement of brakes and tyres on your bicycle, the rust growing on tools left outside and the need for software updates on computers all remind us that artefacts are not static. Such a perspective blurs the boundary between the 'designing/making' and 'using' phases of a technology. Instead, we see the continuous processes of adaptation and change via external influences and deliberate choices as fundamental to the interaction of artefacts with their world (Young, 2021a, 2021b). This is a useful reminder that all artefacts are dynamic and necessarily interact with the world in some ways, and the results of these interactions often necessitate maintenance and repair. Deliberate acts of repair, however, need not only be concerned with maintaining the status quo. Henke and Sims (2020) contrast *repair as maintenance* with the more ambitious and idealistic notion of *repair as transformation*. While the former aims to maintain current structures and relations (both social and material), repair as transformation instigates a process of repair that goes beyond the conservation of current structures, instead striving for change. Within transformative repair exists the possibility to question and re-think the physical and social structures that infrastructures support, including those related to environmental justice and power relations.

What then does a focus on repair offer to the nascent concept of Sustainable AI? And in particular, what does a process-oriented, transformative perspective on the *potential* of repair surface regarding the ethical framing of (Sustainable) AI? To analyse AI through this lens, we propose to extend and expand the overarching framework proposed by van Wynsberghe (2021). We will argue that repair is where AI *for* sustainability and the sustainability *of* AI interact (Figure 13.3) – at times literally as implied above, but critically also in how we position the relationship between AI and sustainability. Practically, we cannot reap the full benefits of utilising AI for sustainable goals without transforming the infrastructure of AI. Conceptually, this requires reflexivity as a central consideration for the ethical impacts of AI, particularly in how we choose to maintain those systems which are themselves utilised for sustainable applications.

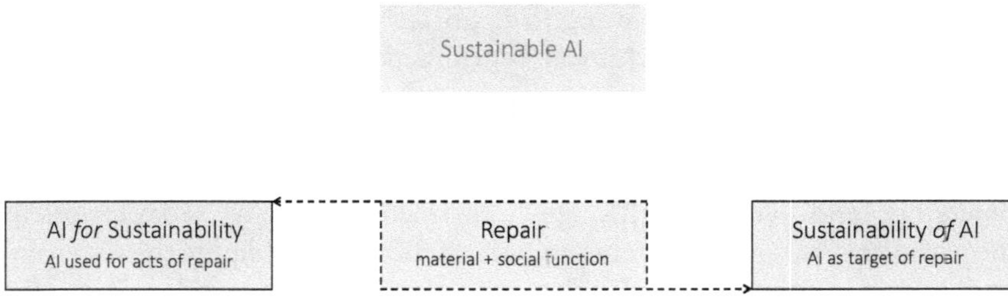

Figure 13.3 Repair as the interface between AI *for* sustainability and the sustainability *of* AI.

13.3.1 AI *for* Sustainability as Acts of Repair

Considered through the lens of repair, AI *for* sustainability concerns the practical use of AI-enabled innovations for the monitoring, upkeep and optimisation of critical infrastructures. This can include the use of AI-enabled services and autonomous robots to better organise mundane but necessary maintenance, such as road repair (Mukherjee et al., 2021) and garbage collection (Zhang et al., 2020). ML techniques are also being developed for preventative actions, for example, the World Bank's project to identify vulnerable housing in the Global South (Wahba, n.d.). ML research has also focused on predictive maintenance of digital infrastructures via autonomous systems (Khan et al., 2020). Here, AI is a tool to improve the practices of maintenance – either via providing stability through continual and efficient upkeep or as a driver of the process of change via monitoring and intervention or by serving to adapt artefacts over their lifespan.

AI can be utilised for a variety of tasks falling under the broad categorisation of maintenance, but here we are particularly interested in how the application of AI *for* sustainability instantiates moments of repair. We note that, arguably, many acts of maintenance and repair aimed at optimising a system or gaining efficiencies (in the sense of reducing energy demands or materials used, extending lifetimes of a product, etc.) can fall under an instrumental notion of sustainability. However, here we will look more specifically at how Sustainable AI can be seen as acts of *repair as transformation.* Henke and Sims (2020), in discussing the IPCC reports, describe them as effectively being 'massive planetary repair' manuals (p. 140). Building on this notion, we can see that many of the applications and initiatives that utilise AI towards environmental ends are essentially attempts to enact *planetary repair* on some scale.[3] This arguably also underlies increasing calls for interdisciplinary approaches to climate change mitigation and adaptation with the aid of ML (e.g., Rolnick *et al.*, 2022), including the initiatives of organisations and movements such as AI for Good (aiforgood.itu.int), Climate Change AI (www.climatechange.ai), Microsoft's AI for Earth (www.microsoft.com/en-us/ai/ai-for-earth) and Global Forest Watch (globalforestwatch.org).

Within and beyond the work of these organisations, an increasing number of projects and initiatives take aim at more specific problems, such as biodiversity preservation and ecological restoration. While this may expand the notion of *repair* as typically used in infrastructure studies, it does align with Henke and Sims' (2020, p. 3) definition of repair as acts of 'restoring both social and material order'. In this sense, repairing an artefact or system is not only an act of restoring functionality but also often serves to repair social relations and values. Utilising AI for ecological restoration and preservation then further extends repair to also include environmental relations and values. As Dauvergne discusses in *AI in the Wild* (2020), there are already many such initiatives underway. This includes the use of ML for the monitoring of threatened or endangered areas, such as tracking illegal fishing boats via remote sensing to see where overfishing is taking place, monitoring the depletion of rainforests or tracking the movement of wildlife – a tool also increasingly employed by groups such as the WWF (World Wildlife Fund, 2020). It also includes various interventions, such as re-seeding areas using drones and culling invasive species along the Great Barrier Reef using semi-autonomous submarines.

Monitoring and interventions have also been proposed for urban areas and the assistance of urban rewilding efforts. This may open up new uses for ML, with ramifications for urban planning and design – for example, Owens and Wolch (2019) discuss how urban wildlife can be tracked and monitored via camera traps with deep learning pattern recognition, towards better understanding their movement and the creation of wildlife corridors. More speculatively, they discuss how autonomous vehicles could serve to reduce roadkill deaths. This could be coupled with other design directions for autonomous vehicles, such as a reduction in road lighting (Stone *et al.*, 2020) towards the preservation of nocturnal habitats. As another example, there are proposals to create an *Internet of Nature* through which embedded sensors, urban robots, remote sensing and AI algorithms work together to monitor and improve the health of urban forests (Galle *et al.*, 2019; Galle and Nitoslawski, 2020).

The main takeaway from these varied examples is that, at its core, the use of AI *for* sustainability is in many ways acts of repair; of aiming AI innovation towards the target of planetary repair. This necessitates a conscientious reflection on the ethics and politics of said initiatives. First, are these applications truly instantiating instances of *repair as transformation* or serving to reinforce existing power structures and exploitative practices (Dauvergne, 2020)? For example, the proposed *Internet of Nature* mentioned above, as a type of programmed green infrastructure, has been critiqued for perpetuating existing power imbalances along with socially and environmentally unsustainable practices (Gabrys, 2022). A second question to ask is how exactly sustainability is framed and bounded in these contexts. Improving the efficiency of current systems is undoubtedly desirable and worthwhile in many contexts, and necessary given that we have an urgent need to reduce greenhouse gas emissions. However, to simply optimise the systems, infrastructures and power relations that helped lead us into our current crises can hardly be considered sustainable in any meaningful sense. Granted, an *Internet of Nature* could lead to profoundly positive changes for urban ecosystems and wildlife, as well as the well-being of humans living in those cities. Yet, the full extent of the social and ecological impacts of the infrastructure that would power this system must also be considered. Put otherwise, there is a need to ask how to repair the very systems we are utilising to enact planetary repair.

13.3.2 Repairing AI

The above discussion highlights the opportunities and challenges created by the utilisation of AI for sustainable goals. However, this is only one side of the coin. The sustainability *of* AI requires a conceptualisation of AI *as* a critical infrastructure itself (Crawford, 2021; Robbins and van Wynsberghe, 2022). The process-oriented perspective on technologies (Young, 2021a) reminds us that AI artefacts are *not* stable but will necessarily adapt, evolve and ultimately be decommissioned and demolished. This draws attention to the adaptability, vulnerabilities and ongoing upkeep of both the digital *and* physical infrastructures underlying AI systems. Equal consideration must be given to critical material components such as IoT devices, data centres, transcontinental cables and satellites alongside codes and algorithms. Yet these considerations cannot be viewed in isolation from their uses and applications, especially in the name of sustainability. We must appreciate that AI does not actually operate within a disembodied 'cloud' but rather within material artefacts and infrastructures that themselves require monitoring, upkeep, replacement, etc. In our ethical analyses of AI, we too readily forget that AI includes physical components and human labour. This complicates the analysis of AI as a discrete category rather than one closely intertwined with larger digital infrastructures and reliant upon seemingly external critical infrastructures (e.g., energy systems). This requires an ethics of (Sustainable) AI that is *material* – in the sense of interrogating physical artefacts and effects. This fact has been emphasised in infrastructure studies but seemingly overlooked in AI ethics:

> ... it is by concentrating on the material architectures of electricity generation and supply that, in many ways, we can begin to 'see' 'cyberspace' for what it is – not an ethereal domain of 'virtual' bits and bytes, but a gigantic, materialized and electrically powered system requiring massive amounts of continuous and concerted maintenance and repair.
>
> (Graham and Thrift, 2007, p. 13)

This integration of digital infrastructures brings with it vulnerabilities and cascading effects that need to be considered, and which fall under the umbrella of repair.

As an example of the material presence of digital infrastructure, consider data centres. Due to issues of security and competition, companies actively strive to keep data centres – the material home of 'the cloud' – largely invisible. This also makes them difficult to research. The companies that have released images of their data centres often present staged pictures in which massive buildings are seemingly in harmony with their surrounding environment (Holt and Vonderau, 2015). Yet, this overlooks both the political and environmental impacts of these material homes of our digital world. A recent report from the International Energy Agency (IEA, 2022) estimates that data centres are responsible for 1–1.5% of global electricity use and approximately 0.6% of global greenhouse gas emissions. Likely, only a fraction of this is directly attributable to AI; however, the exact amount is not known due to lack of data and challenges in identifying the boundaries of AI uses (Kaack *et al.*, 2022). Data centre energy demands are projected to increase rapidly in the coming years, as part of more worrying energy forecasts for ICT generally. Some models predict ICT-related emissions could jump from about 2% to over 20% within a generation (Jones, 2018). To put into perspective, in 2010 it was estimated that if 'the cloud' were a country, it would have the fifth largest energy consumption in the world (Greenpeace 2012; Holt and Vonderau, 2015). While energy required for computing power is a major issue, again the material upkeep of data centres requires broader consideration: air conditioning can account for up to 40% of data centre energy usage. This has geographical and political implications for data centres as they are often situated at sites with cheap electricity and where climatic conditions offer 'free' cooling (Holt and Vonderau, 2015).

The above numbers are indicative of the sheer scale of the materiality required for our current digital infrastructure. Again, we must begin to see critical nodes such as data centres *not* as static artefacts, but as complex arrangements of servers and cables housed within specialised buildings – which will necessarily undergo moments of breakdown and repair. Such a perspective necessitates a shift in how we approach the ethics and politics of AI. It is imperative that the ongoing upkeep and required maintenance of data centres, including the servers themselves and the surrounding built environment (e.g., cooling and other features of the building), are included in a holistic conception of Sustainable AI specifically, and AI ethics generally. These are nodes within a complex and dynamic system. And if we indeed see the system supporting AI as in a constant state of (almost) breakdown and repair rather than stable, we can begin to appreciate that the maintenance of AI's material components is critical to any conception of Sustainable AI. While the increasing material impact of digital infrastructure can be concerning, this is not meant to be a condemnation of *all* uses and applications of AI. Indeed, many – including some examples of ecological restoration mentioned above – may very well be desirable despite their energy or material demands. However, the more we rely upon AI *for* sustainability, the more critical the sustainability *of* AI becomes.

Yet, there is also a more profound implication underlying the increasing concerns regarding the energy demands of AI and associated digital infrastructures. It is headline-grabbing to point towards these quantifiable and growing numbers, which can indeed serve as useful in calls for greater awareness, new policies focused on reductions in energy usage, etc. However, this focuses solely on the literal notions of maintenance and repair as defined above. As Kaack *et al.* (2022) emphasise, it is the system-level impacts that are both harder to quantify and predict and that will likely have larger impacts on critical challenges such as greenhouse gas emissions. Kaack *et al.* mention important considerations such as rebound effects and the risks of 'lock-in', which point towards the need for proactive and anticipatory actions. Because of this, the processes and practices that sustain our AI infrastructure cannot be seen in isolation from the downstream uses and applications, for this reinforces dualistic thinking and overlooks the material embeddedness of AI. This serves as a useful reminder that the environmental problems we currently face are not 'out there' to be addressed and solved, but a product of our own choices and actions.

> It is not a question of our encountering the crisis and resolving it through technology. The crisis is not simply something we can examine and resolve. We *are* the environmental crisis. The crisis is a visible manifestation of our very being... The environmental crisis is inherent in everything we believe and do; it is inherent in the context of our lives.
>
> (Evernden, 1993, p. 128)

Accepting this means we must then see AI and digital infrastructure as inseparable from the problems they are purporting to solve.

This is not just a practical necessity, but one with profound potential to re-shape our digital infrastructure. Here, we again invoke the idea from Henke and Sims (2020) that repair has a material *and* social function. Consider again the example of the *Internet of Nature* briefly mentioned above (Galle *et al.*, 2019). This can be seen as an innovative (and in many ways promising) use case of urban AI for ecological restoration, utilising data analytics, predictive and real-time monitoring and urban robotics to improve urban ecosystems, biodiversity, the health and well-being of citizens, etc. But it is also introducing a new layer of infrastructure to the already complex urban fabric – it will necessarily respond to, and in turn impact, the urban environments in which it is deployed. Further, its successful integration into a city will make it hard to define exactly where the boundaries of the *Internet of Nature* lie. This makes quantifying the exact impacts of such interventions difficult (e.g. in attempting any sort of cost–benefit analysis that compares energy and material demands with local benefits). It will also engender a social change, affecting the relationship between 'natural' and 'artificial' components of a city. These changes could indeed offer a desirable urban future; the point is that the ethical questions that arise extend well beyond issues such as explainability or transparency, or even quantifying energy usage.

Through such an example, we are forced to confront the reality that we must continually reflect upon the systems that are themselves increasingly being used for planetary repair. This highlights a paradoxical feature of repairing infrastructure during the Anthropocene, that 'we cannot develop data and theories about "the world" without the very systems that are creating wide-scale change' (Henke and Sims, 2020, p. 122). To confront our current situation, we need to think about what Henke and Sims call 'repairing repair', or '... a reorientation of repair itself, toward an ethic that considers the consequences of infrastructural globalism through the complex interactions of power, scale, and time that are built into the sociotechnical structures of modernity' (p. 137). Here, repair can be considered as a reflexive act on two levels: in a causal sense (i.e., that the operation of infrastructures cause impacts that (re-)shape their own operating environment) and in the sense of self-aware thinking (i.e., of those designing and managing the systems to appreciate potentially negative feedback loops). Put otherwise, to adopt a *repair as transformation* approach to Sustainable AI, we need a reflexive and iterative approach to how the sustainability *of* AI affects the efficacy of AI *for* sustainability, and vice versa (Figure 13.4).

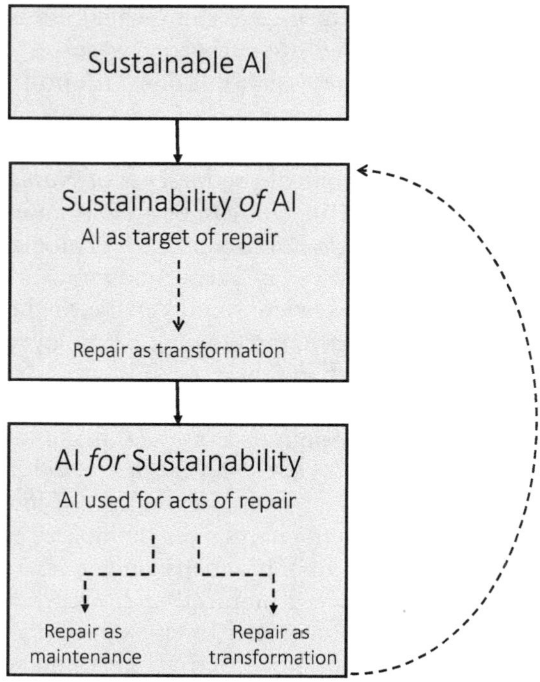

Figure 13.4 Repair as reorienting the relationship between AI *for* sustainability and the sustainability *of* AI. While the use of AI for sustainability can target maintenance or transformative repair, it relies on a transformative approach to the repair of AI and associated digital infrastructures. Once a transformative approach is initiated, AI-enabled innovations can themselves be used to support and improve the sustainability of AI itself.

13.4 Conclusion

This chapter outlines a research agenda that takes AI ethics out of the cloud and into the world of material things – as a tool to repair our material world, and likewise as a material infrastructure itself.

For this, it investigated the notion of repair and its usefulness for thinking through the conceptualisation of Sustainable AI. It asks us to understand technologies not as stable entities, but rather as evolving and dynamic processes. Because of this, the constant upkeep of technologies – and the resulting change and adaptation this instigates – are central considerations for the ethics of AI. Repair should be understood to have both a material and social function, with the potential to either maintain and reinforce the status quo or to become a means of transforming AI in more sustainable and socially just directions. This requires understanding the notion of repairing AI as having two interrelated meanings: the literal upkeep of AI infrastructure and a conceptualisation of AI as embedded within its own operating system. The latter opens a necessary shift in thinking about the ethics of AI that focuses on how it shapes – and in turn is shaped by – the material environments in which it operates. This layered notion of repairing AI can advance the nascent notion of Sustainable AI specifically and support more general calls for a re-orientation of AI ethics towards systemic and structural issues. Building on the definition of Sustainable AI as having two branches, namely the application of AI *for* sustainability and the sustainability *of* AI itself (van Wynsberghe, 2021), we have shown that repair is a moment where the *of* and *for* come together. You cannot have the *for* without the *of*; the benefits of utilising AI *for* sustainability cannot be fully realised without also repairing the digital and material infrastructure of AI itself. This serves as a stark reminder that the tools we increasingly use for planetary repair are themselves embedded within their own operating systems. Overlooking this important reality risks reinforcing a dualistic mode of thinking that has arguably contributed to the environmental crises we now face.

Notes

1 According to the European Commission's High-Level Expert Group on Artificial Intelligence (2019, p. 1): 'Artificial intelligence (AI) refers to systems that display intelligent behaviour by analysing their environment and taking actions – with some degree of autonomy – to achieve specific goals'. Following from this, there are a range of methods in which this so-called intelligence can be achieved from symbolic reasoning to machine learning (ML) and neural networks, to name a few. Here, we will not go further than this broad definition when referring to AI or ML, as this chapter is not directly addressing the development or use of specific models but rather our framing of this rapidly proliferating technology – particularly how we situate its material processes, structures and impacts.

2 Other sociopolitical aspects of AI could also be assessed through the conceptual lens of repair, such as the exploitation of workers, threats to democracy or the treatment of animals (e.g., Crawford, 2021; Coeckelbergh, 2021; Hagendorff, 2022). We acknowledge these social dimensions and see usefulness in the framing of AI developed here for these impacts and ethical problems, but will focus explicitly on AI's environmental and material impact – using the recurrent case study of AI for ecological restoration – to provide an initial sketch and framing of how 'repairing AI' can advance and extend Sustainable AI.

3 The concept of planetary repair has previously been used to describe '… a conceptual provocation for an era of global environmental destabilization and as a conflictual turn in mainstream conservation practice – particularly as technofuturist visions of ecological design and geoengineering gain new adherents' (Knuth, 2019, p. 488). In this context, however, we draw this term from the discussion of reflexive repair in the Anthropocene by Henke and Sims (2020, chapter 5). Practically, it reflects contemporary thinking around ecological restoration as a means of repairing the planet, required alongside sustainable transitions of socio-technical systems (Cooke *et al.*, 2023) – an idea arguably

reflected in the AI-focused initiatives and organisations referenced above. We acknowledge that this term can be suggestive of viewing Earth as an artefact. While an in-depth discussion of this issue is outside the scope of this chapter, the terminology of planetary repair is meant to evoke critiques of nature–artefact dualistic thinking generally (Vogel, 2015), and specifically the need to better appreciate the material impacts of digitalisation on the Anthropocene (Creutzig *et al.*, 2022). Further, it can be seen as a response to recent calls to question our ontological categories of 'nature' and 'technology' as they become increasingly entangled in the Anthropocene, and the need for philosophers of technology to adopt a 'planetary horizon' given that '... technology is not just embedded in ecology anymore but is increasingly becoming its very foundation, such that the future habitability of the Earth is becoming the technological question par excellence' (Lemmens *et al.*, 2017, p. 124).

References

Albrechtslund, A. (2007) 'Ethics and technology design', *Ethics and Information Technology*, 9(1), pp. 63–72. Available at: https://doi.org/10.1007/s10676-006-9129-8

Bolte, L., Vandemeulebroucke, T. and van Wynsberghe, A. (2022) 'From an ethics of carefulness to an ethics of desirability: going beyond current ethics approaches to sustainable AI', *Sustainability*, 14(8), p. 4472. Available at: https://doi.org/10.3390/su14084472

Cantrell, B., Martin, L.J. and Ellis, E.C. (2017) 'Designing autonomy: opportunities for new wildness in the Anthropocene', *Trends in Ecology & Evolution*, 32(3), pp. 156–166. Available at: https://doi.org/10.1016/j.tree.2016.12.004

Coeckelbergh, M. (2020) *AI Ethics*. MIT Press.

Coeckelbergh, M. (2021) *Green Leviathan or the Poetics of Political Liberty: Navigating Freedom in the Age of Climate Change and Artificial Intelligence*. Routledge. Available at: https://www.routledge.com/Green-Leviathan-or-the-Poetics-of-Political-Liberty-Navigating-Freedom/Coeckelbergh/p/book/9780367747794 (Accessed: 12 July 2022).

Cooke, S.J. et al. (2023) 'Ecological Restoration in Support of Sustainability Transitions: Repairing the Planet in the Anthropocene', in M.U. Hensel et al. (eds) *Introduction to Designing Environments: Paradigms & Approaches*. Cham: Springer International Publishing (Designing Environments), pp. 93–112. Available at: https://doi.org/10.1007/978-3-031-34378-0_6

Crawford, K. (2021) *Atlas of AI*. New Haven: Yale University Press.

Creutzig, F. et al. (2022) 'Digitalization and the Anthropocene', *Annual Review of Environment and Resources*, 47(1), pp. 479–509. Available at: https://doi.org/10.1146/annurev-environ-120920-100056

Dauvergne, P. (2020) *AI in the Wild: Sustainability in the Age of Artificial Intelligence*. Cambridge: The MIT Press.

European Commission's High-Level Expert Group on Artificial Intelligence (June 2019) 'A definition of AI: Main capabilities and scientific disciplines', *European Commission*. Available at: https://digital-strategy.ec.europa.eu/en/library/definition-artificial-intelligence-main-capabilities-and-scientific-disciplines (Accessed 5 March 2022).

Evernden, L.L.N. (1993) *The Natural Alien: Humankind and Environment.* University of Toronto Press.

Friedman, B. and Hendry, D. (2019) *Value Sensitive Design: Shaping Technology with Moral Imagination.* Cambridge: The MIT Press.

Gabrys, J. (2022) 'Programming nature as infrastructure in the smart forest city', *Journal of Urban Technology*, 29(1), pp. 13–19. Available at: https://doi.org/10.1080/10630732.2021.2004067

Galle, N.J. and Nitoslawski, S.A. (2020) 'Biophilic Cities for the 22nd Century', *Biophilic Cities Journal*, 3(2), pp. 12–21.

Galle, N.J., Nitoslawski, S.A. and Pilla, F. (2019) 'The Internet of Nature: How taking nature online can shape urban ecosystems', *The Anthropocene Review*, 6(3), pp. 279–287. Available at: https://doi.org/10.1177/2053019619877103

Graham, S. and Thrift, N. (2007) 'Out of order: Understanding repair and maintenance', *Theory, Culture & Society*, 24(3), pp. 1–25. Available at: https://doi.org/10.1177/0263276407075954

Greenpeace (April 2012) 'How Clean is Your cloud?', *Greenpeace International.* Available at: https://www.greenpeace.org/static/planet4-international-stateless/2012/04/e7c8ff21-howcleanisyourcloud.pdf (Access 5 March 2022).

Hagendorff, T. (2022) 'Blind spots in AI ethics', *AI and Ethics*, 2(4), pp. 851–867. Available at: https://doi.org/10.1007/s43681-021-00122-8

Henke, C.R. and Sims, B. (2020) *Repairing Infrastructures: The Maintenance of Materiality and Power.* Available at: https://doi.org/10.7551/mitpress/11771.001.0001

Holt, J. and Vonderau, P. (2015) '"Where the Internet Lives": Data centers as cloud infrastructure', in L. Parks and N. Starosielski (eds) *Signal Traffic: Critical Studies of Media Infrastructures.* University of Illinois Press, pp. 71–93.

IEA (2022) 'Data centres and data transmission networks', *International Energy Agency.* Available at: https://www.iea.org/reports/data-centres-and-data-transmission-networks (Access 10 May 2022).

Jackson, S.J. (2014) 'Rethinking Repair', in T. Gillespie, P.J. Boczkowski, and K.A. Foot (eds) *Media Technologies: Essays on Communication, Materiality and Society.* The MIT Press, pp. 221–239.

Jones, N. (2018) 'The Information Factories', *Nature*, 561, pp. 163–166.

Kaack, L.H. et al. (2022) 'Aligning artificial intelligence with climate change mitigation', *Nature Climate Change*, 12(6), pp. 518–527. Available at: https://doi.org/10.1038/s41558-022-01377-7

Khan, S. et al. (2020) 'On the requirements of digital twin-driven autonomous maintenance', *Annual Reviews in Control*, 50, pp. 13–28. Available at: https://doi.org/10.1016/j.arcontrol.2020.08.003

Knuth, S. (2019) 'Cities and planetary repair: The problem with climate retrofitting', *Environment and Planning A: Economy and Space*, 51(2), pp. 487–504. Available at: https://doi.org/10.1177/0308518X18793973

Lemmens, P. et al. (2017) 'Toward a Terrestrial Turn in Philosophy of Technology', *Techné: Research in Philosophy and Technology*, 21(2), pp. 114–126. Available at: https://doi.org/10.5840/techne2017212/363

McDonald, R. and Beatley, T. (2021) *Biophilic Cities for an Urban Century.* Cham, Switzerland: Palgrave MacMillan.

Mukherjee, R. et al. (2021) 'AI driven road maintenance inspection'. arXiv. Available at: https://doi.org/10.48550/arXiv.2106.02567

Nye, D.E. (2010) *When the Lights Went Out: A History of Blackouts in America*. Cambridge, MA: MIT Press.

Owens, M. and Wolch, J. (2019) 'Rewilding cities', in N. Pettorelli, S.M. Durant, and J.T. du Toit (eds) *Rewilding*. 1st edn. Cambridge University Press, pp. 280–302. Available at: https://doi.org/10.1017/9781108560962.014

van de Poel, I. (2020) 'Three philosophical perspectives on the relation between technology and society, and how they affect the current debate about artificial intelligence', *Human Affairs*, 30(4), pp. 499–511. Available at: https://doi.org/10.1515/humaff-2020-0042

van de Poel, I. (2021) 'Design for value change', *Ethics and Information Technology*, 23(1), pp. 27–31. Available at: https://doi.org/10.1007/s10676-018-9461-9

Robbins, S. and van Wynsberghe, A. (2022) 'Our New Artificial Intelligence Infrastructure: Becoming Locked into an Unsustainable Future', *Sustainability*, 14(8), p. 4829. Available at: https://doi.org/10.3390/su14084829

Rolnick, D. et al. (2022) 'Tackling Climate Change with Machine Learning', *ACM Computing Surveys*, 55(2), p. 42:1–42:96. Available at: https://doi.org/10.1145/3485128

Star, S.L. (1999) 'The Ethnography of Infrastructure', *American Behavioral Scientist*, 43(3), pp. 377–391. Available at: https://doi.org/10.1177/00027649921955326

Stone, T. (2021) 'Design for values and the city', *Journal of Responsible Innovation*, 8(3), pp. 364–381. Available at: https://doi.org/10.1080/23299460.2021.1909813

Stone, T., Santoni de Sio, F. and Vermaas, P.E. (2020) 'Driving in the Dark: Designing Autonomous Vehicles for Reducing Light Pollution', *Science and Engineering Ethics*, 26(1), pp. 387–403. Available at: https://doi.org/10.1007/s11948-019-00101-7

Strubell, E., Ganesh, A. and McCallum, A. (2019) 'Energy and Policy Considerations for Deep Learning in NLP'. arXiv. Available at: https://doi.org/10.48550/arXiv.1906.02243

Vogel, S. (2015) *Thinking like a Mall: Philosophy after the End of Nature*. Cambridge: The MIT Press.

Wahba, S.N. (n.d.) 'Housing Policies That Save (and Improve) Lives, Protect Assets, and Shield Economies'. Available at: https://thedocs.worldbank.org/en/doc/185371540492697571-0200022018/original/WorldBankGlobalProgramforResilientHousing.pdf

World Wildlife Fund (2020) 'Employing AI to evaluate wildlife populations on a global scale', *World Wildlife Fund*. Available at: https://www.worldwildlife.org/magazine/issues/winter-2020/articles/employing-ai-to-evaluate-wildlife-populations-on-a-global-scale (Accessed: 12 July 2022).

van Wynsberghe, A. (2021) 'Sustainable AI: AI for sustainability and the sustainability of AI', *AI and Ethics*, 1(3), pp. 213–218. Available at: https://doi.org/10.1007/s43681-021-00043-6

Young, M.T. (2021a) 'Maintenance', in D. Michelfelder and N. Doorn (eds) *The Routledge Handbook of the Philosophy of Engineering*. New York: Routledge, pp. 356–368.
Young, M.T. (2021b) 'Now You See It (Now You Don't): Users, Maintainers and the Invisibility of Infrastructure', in M. Nagenborg et al. (eds) *Technology and the City: Towards a Philosophy of Urban Technologies*. Cham: Springer International Publishing (Philosophy of Engineering and Technology), pp. 101–119. Available at: https://doi.org/10.1007/978-3-030-52313-8_6
Zhang, S. et al. (2020) *Regaining Amsterdam Canals for Waste Collection*. MIT Senseable City Lab.

Index

Pages in *italics* refer to figures and pages followed by "n" refer to notes.

FSC
www.fsc.org
MIX
Papier | Fördert gute Waldnutzung
FSC® C083411

Kolibri 360 GmbH
Ferdinand-Jühlke-Straße 7
99095 Erfurt, Deutschland
produktsicherheit@kolibri360.de